Dynamical and Chemical Coupling
Between the Neutral and Ionized Atmosphere

NATO ADVANCED STUDY INSTITUTES SERIES

Proceedings of the Advanced Study Institute Programme, which aims at the dissemination of advanced knowledge and the formation of contacts among scientists from different countries

The series is published by an international board of publishers in conjunction with NATO Scientific Affairs Division

A	Life Sciences	Plenum Publishing Corporation
B	Physics	London and New York
C	Mathematical and Physical Sciences	D. Reidel Publishing Company Dordrecht and Boston
D	Behavioral and Social Sciences	Sijthoff International Publishing Company Leiden
E	Applied Sciences	Noordhoff International Publishing Leiden

Series C – Mathematical and Physical Sciences

Volume 35 – Dynamical and Chemical Coupling Between the Neutral and Ionized Atmosphere

Dynamical and Chemical Coupling Between the Neutral and Ionized Atmosphere

Proceedings of the NATO Advanced Study Institute held at Spåtind, Norway, April 12–22, 1977

edited by

B. GRANDAL

Norwegian Defence Research Establishment, Kjeller, Norway

and

J. A. HOLTET

University of Oslo, Oslo, Norway

D. Reidel Publishing Company

Dordrecht-Holland / Boston-U.S.A.

Published in cooperation with NATO Scientific Affairs Division

Library of Congress Cataloging in Publication Data

NATO Advanced Study Institute on the Dynamical and Chemical Coupling of the Neutral and Ionized Atmosphere, Spatind, Norway, 1977.
Dynamical and chemical coupling between the neutral and ionized atmosphere.

(NATO advanced study institutes series : Series C, Mathematical and physical sciences ; v. 35)
Includes index.
1. Atmosphere–Congresses. 2. Ionosphere–Congresses. I. Grandal, B., 1948-
II. Holtet, Jan A. III. Title. IV. Series.
QC851.N34 1977 551.5'1 77-14235
ISBN-13: 978-94-010-1264-5 e-ISBN-13: 978-94-010-1262-1
DOI:10.1007/978-94-010-1262-1

Published by D. Reidel Publishing Company
P.O. Box 17, Dordrecht, Holland

Sold and distributed in the U.S.A., Canada, and Mexico
by D. Reidel Publishing Company, Inc.
Lincoln Building, 160 Old Derby Street, Hingham, Mass. 02043, U.S.A.

CONTENTS

PART III: COUPLING BETWEEN NEUTRAL AND IONIZED ATMOSPHERE

PART IV: AURORAL ZONE PHENOMENA

PREFACE

These proceedings are based upon the review lectures, the research talks and the accompanying discussion from the NATO Advanced Study Institute on "The Dynamical and Chemical Coupling of the Neutral and Ionized Atmosphere" held at Spåtind, Norway April 12-22, 1977.

In recent years, in spite of the many subdivisions of atmospheric physics it has become clear that a complete understanding of the subject matter requires an interdisciplinary effort of the scientists working on a variety of atmospheric phenomena.

This Advanced Study Institute was organized in order to meet this demand. A number of eminent scientists were invited to review the major subfields in atmospheric physics and expose the areas where the coupling between the neutral and ionized atmosphere is evident. These review papers were supported by current research results.

However, the material covered here is by no means complete considering the range of subject matter covered by the title. The responsibility for this inadequacy lies with the editors and the program committee, rather than with the individual contributors.

Oslo, June 1977.

B. Grandal J.A. Holtet.

PARTICIPANTS

Arnold, F.	Max-Planck-Institute for Nuclear Physics P.O.Box 1248, D-0069 Heidelberg F.R.G.
Bjørn, L.	Uppsala Ionospheric Observatory S-755 90 Uppsala Sweden
Brekke, A.	The Institute for Mathematical and Physical Sciences University of Tromsø P.O.Box 953, N-9001 Tromsø Norway
Christophe, J.	Institut d'Astrophysique 98 bis Boulevard Arago, Paris 14 France
Cole, K.D.	Division of Theoretical and Space Physics La Trobe University Bundoora, Victoria Australia 3083
During, G.	Defence Research Establishment Linnégt. 89, Fach, S-104 05 Stockholm Sweden
Egeland, A.	The Norwegian Institute of Cosmic Physics P.O.Box 1038 Blindern, Oslo 3 Norway
Folkestad, K.	Norwegian Defence Research Establishment P.O.Box 25, N-2007 Kjeller Norway
Friedrich, M.	Institut für Nachrichtentechnik und Wellen-ausbreitung Technische Hochschule Graz, A-8010 Graz Inffeldgasse 12 Austria
Grandal, B.	Norwegian Defence Research Establishment P.O.Box 25, N-2007 Kjeller Norway
Greenwald, R.A.	Max-Planck-Institut für Aeronomie Institut für Ionosphärenphysik, D-3411 Katlenburg - Lindau 3, P.O.Box 20 F.R.G.

Harang, O.	The Institute of Mathematical and Physical Sciences University of Tromsø P.O.Box 953, N-9001 Tromsø Norway
Hargreaves, J.K.	Environmental Sciences Department University of Lancaster Bailrigg, Lancaster LA1 4IQ United Kingdom
Herwig, T.	Physikalisches Institut der Universität Bonn Nussallee 12, D-5300 Bonn 1 F.R.G.
Hesstvedt, E.	Institute of Geophysics University of Oslo P.O.Box 1022 Blindern, Oslo 3 Norway
Hickman, D.R.	The Aerospace Corp. A6/2417 P.O.Box 92957, Los Angeles, Calif. 90009 U.S.A.
Holtet, J.A.	The Norwegian Institute of Cosmic Physics P.O.Box 1038 Blindern, Oslo 3 Norway
Inan, U.S.	Radioscience Laboratory Durand Bldg. 323, Stanford University Stanford, California 94305 U. S. A.
Jones, T.B.	Department of Physics University of Leicester, Leicester LE1 7RH United Kingdom
Jørgensen, T.S.	Danish Meteorological Institute Lyngbyvej 100, DK-2100 København Ø Denmark
King, J.W.	Appleton Laboratory Ditton Park, Slough SK 397 X United Kingdom
Labitzke, K.	Freie Universität Berlin Institut für Meteorologie Podielski Allee 62, 1 Berlin 33 F.R.G.
Larsen, T.R.	Norwegian Defence Research Establishment P.O.Box 25, N-2007 Kjeller Norway
Liszka, L.	National Board of Occupational Safety and Health, S-901 85 Umeå Sweden

Litvak, H.E.	Cooperative Institute for Research in Environmental Science University of Colorado, Research Laboratory Boulder, Colorado 80309 U.S.A.
Lundblad, J.Å.	Institute of Physics University of Bergen, Alleegt. 55, N-5014 Bergen U Norway
Marøy, S.	Institute of Physics University of Bergen Alleegt. 55, N-5014 Bergen U Norway
McCormick, K.S.	Communications Research Centre, P.O.Box 11490 Station H Ottawa, Ontario K2H 8S2 Canada
Mentzoni, M.	Institute of Physics, University of Oslo P.O.Box 1048 Blindern, Oslo 3 Norway
Moe, T.E.	Institute of Physics, University of Bergen Alleegt. 55, N-5014 Bergen U Norway
Moncrieff, M.W.	Atmospheric Physics Group Imperial College, London SW 7 United Kingdom
Offermann, D.	Physikalisches Institut der Universität Bonn Nussallé 12, D-5300 Bonn 1 F.R.G.
Parthasarathy, R.	Geophysical Institute University of Alaska Fairbanks, Alaska 99701 U.S.A.
Paulson, K.V.	Department of Physics University of Saskatchewan Saskatoon S7N OWO Canada
Piggott, W.R.	British Antarctic Survey Madingley Road, Cambridge CB3 OET United Kingdom
Potemra, T.A.	The John Hopkins University Applied Physics Laboratory Laurel, Maryland 20810 U.S.A.
Ranta, H.	Geophysical Observatory SF-99600 Sodankylä Finland

Rastogi, P.K.	The Institute of Mathematical and Physical Sciences University of Tromsø P.O.Box 953, N-9001 Tromsø Norway
Rawer, K.	Institut für physikalische Weltraumforschung D-7800 Freiburg im Breisgau Heidenhofstrasse 8 F.R.G.
Reagan, J.B.	Lockheed Palo Alto Research Laboratory 3251 Hanover Street, Palo Alto, California 94304 U.S.A.
Reid, G.C.	Aeronomy Laboratory, NOAA Boulder, Colorado 80302 U.S.A.
Roble, R.G.	National Center for Atmospheric Research P.O.Box 3000, Boulder, Colorado 80307 U.S.A.
Rodgers, C.D.	Clarendon Laboratory, University of Oxford, Oxford OX1 3PO United Kingdom
Rose, J.	Meteorological Institute Stockholm University Fack, S-104 05 Stockholm Sweden
Rusch, D.W.	Space Physics Research Laboratory University of Michigan Ann Arbor, Michigan 48109 U.S.A.
Schlegel, K.	Max-Planck-Institut für Aeronomie D-3411 Katlenburg, Lindau-3 F.R.G.
Schmerling, E.	NASA Headquarter 400 Maryland Avenue, S.W. Washington D.C. 20546 U.S.A.
Shepherd, G.G.	Centre for Research in Experimental Space Science York University, 4700 Keele Street Downsview, Ontario M3J 1P3 Canada
Stadsnes, J.	Institute of Physics University of Bergen Alleegt. 55, N-5014 Bergen U Norway
Steen Mikkelsen, I.	Danish Meteorological Institute Lyngbyvej 100, DK-2100 København Ø Denmark

Thorne, R.M.	Department of Atmospheric Sciences University of California Los Angeles, California 90024 U.S.A.
Thrane, E.V.	Norwegian Defence Research Establishment P.O.Box 25, N-2007 Kjeller Norway
Trøim, J.	Norwegian Defence Research Establishment P.O.Box 25, N-2007 Kjeller Norway
Walterscheid, R.L.	Department of Atmospheric Sciences University of California Los Angeles, California 90024 U.S.A.
Williams, R.G.	Physics Department, The University Southampton SO9 5NH United Kingdom
Witt, G.	Meteorological Institute Stockholm University Fack, S-104 05 Stockholm Sweden
Zimmerman, S.P.	Air Force Geophysics Laboratory Hanscom Air Force Base Bedford, Massachusetts 01731 U.S.A.

OPENING ADDRESS - NATO ADVANCED STUDY INSTITUTE ON DYNAMICAL AND CHEMICAL COUPLING OF NEUTRAL AND IONIZED ATMOSPHERE

Eivind V Thrane

Norwegian Defence Research Establishment
N-2007 Kjeller, Norway

Dear colleagues,

It is indeed with great pleasure that I, on behalf of the organizing committee, welcome you to this NATO Advanced Study Institute. You have all spent time and effort to get here, some of you have indeed travelled very far, we sincerely hope to make it worth your while. Throughout the years a series of these institutes has been arranged in Norway, all devoted to specific, limited topics related to the ionosphere, the magnetosphere and radio wave propagation. The present institute is, in one sense, an experiment. Its title, Dynamical and Chemical coupling of neutral and ionized atmosphere, indicates that, for the first time in our ASI's we have ventured to take an interdisiplinary approach to the problems of the upper atmosphere. The reason for this venture is, of course, that such an approach is necessary to solve these problems.

The broadening of scope has its dangers in that it is much more difficult to set up a coherent scientific programme that will interest all participants. However, we hope that bringing together people from different fields to learn from each other and exchange ideas, will prove fruitful and more than compensate for this drawback.

We have among us meteorologists, ionospheric physicists and plasma physicists, some of us are theoreticians, others experimentalists. We do have one thing in common, our interest in the atmospheric system. Von Karmann has reputedly said: "When a theoretician presents a new theory, nobody but the theoretician believes it, when an experimentalist presents a new set of data, everyone but the experimentalist believes in it". There may be some truth in this statement, but we are here, through lectures and discussion, to reach a balanced assessment of current theories and experimental data.

A NATO Advanced Study Institute is not a symposium, it is designed as a high level teaching activity. In contrast to a symposium, lecturers will have time to present their speciality in depth for colleagues and students, and there will be time for discussion. Roles as lecturer and atudent may change from sessions to session. You may feel that the meeting is a long one. It is one of the characteristics of a NATO-ASI and one of the conditions for holding it, that it lasts long enough for people to get to grips with the theme of the meeting.

In setting up the programme for an Advanced Study Institute the programme committee seldom feels that it has achieved what it set out to do. However, we hope that you will find some logical structure in the sessions. Although we have chosen a broad theme, we have tried to impose certain limits, concentrating in the mid- and high latitude atmosphere in the height range 30 - 160 km. These limits should be regarded as guidelines only and should certainly not stop us from pursuing a particularly interesting problem outside these bounds. We have divided the programme into four main sessions.

- I The neutral atmosphere
- II The ionosphere
- III The coupling between neutral and ionized atmosphere
- IV Atmospheric emissions and auroral phenomena

In each session a few key lecturers will introduce the topic and guide the discussions, for which there should be ample time. The schedule is not rigid, and we are open for suggestions for topics for special working groups.

Finally a word about the setting of this meeting. To arrange NATO ASI's in isolated spots has become tradition for us. Some of you have been with us before and know what is involved. The recipe is simple: You spirit a group of scientists away from everything except each other, give them a scientific programme as a guide for hard work, add some comfort, food, a dash of skiing, stir gently and let the whole thing brew for about 10 days. Fermentation will soon start and you obtain hopefully new knowledge, ideas and friendships.

PART I: THE NEUTRAL ATMOSPHERE

MORPHOLOGY OF UPPER ATMOSPHERE TEMPERATURES

C. D. Rodgers

Department of Atmospheric Physics,
Oxford University, Oxford, England.

INTRODUCTION

The temperature structure of the atmosphere changes on many scales of time and space. The most obvious scale is the annual cycle, but there are also changes on longer time scales such as the quasibiennial oscillation and interannual variability, and on smaller scales such as planetary waves (both forced and free), gravity waves and tides. Until recently the only way of studying the variation of temperature of the upper stratosphere and mesosphere on a large scale has been from rocket soundings. The distribution in time and space of these soundings leaves a lot to be desired, so that only the simplest climatology (e.g. CIRA 1972) has been available. With the launch of the Nimbus 6 satellite in 1975, a wealth of new information became available from the Pressure Modulator Radiometer (PMR) and Limb Radiance Inversion Radiometer (LRIR) (Sissala 1975). The description of the temperature structure of the upper stratosphere and mesosphere to be given here will be based mainly on results from the PMR. This data is still in a preliminary form at the time of writing and should be treated with caution. However, the general features will be correct even if there are some minor corrections to be made. The instrument is mounted on a sun synchronous satellite in a near polar orbit, making about 13 orbits per day, so that almost the whole globe is seen at fairly

B. Grandal and J. A. Holtet (eds.), Dynamical and Chemical Coupling, 3-16.

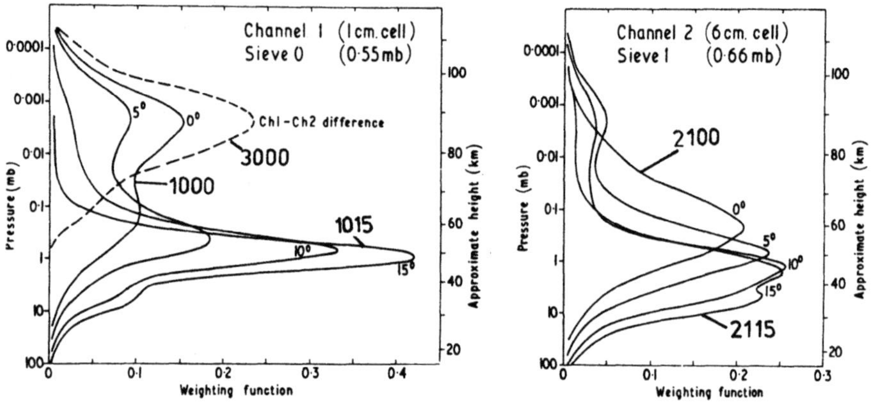

Fig. 1. Pressure Modulator Radiometer weighting functions.

coarse resolution in longitude twice per day. Measurements are made of emitted thermal radiation about every 500 km along the orbit. The measured radiance can be shown to be a weighted mean of the profile of Planck function (or source function) which depends on temperature in a known way. Typical weighting functions are shown in fig. 1, of which those marked 2115, 2100, 1015 and 3000 will be referred to in this paper. It is possible to derive a vertical temperature profile for the measurements, provided the limitations on vertical resolution are understood (Rodgers 1976).

ZONAL MEAN TEMPERATURE CROSS SECTIONS

The simplest aspect of the annual cycle of temperature distribution is the variation with time of the zonal mean temperature. Figs. 2-7 show this quantity for six dates at intervals of approximately one month. These may be compared with the CIRA 1972 cross sections, which

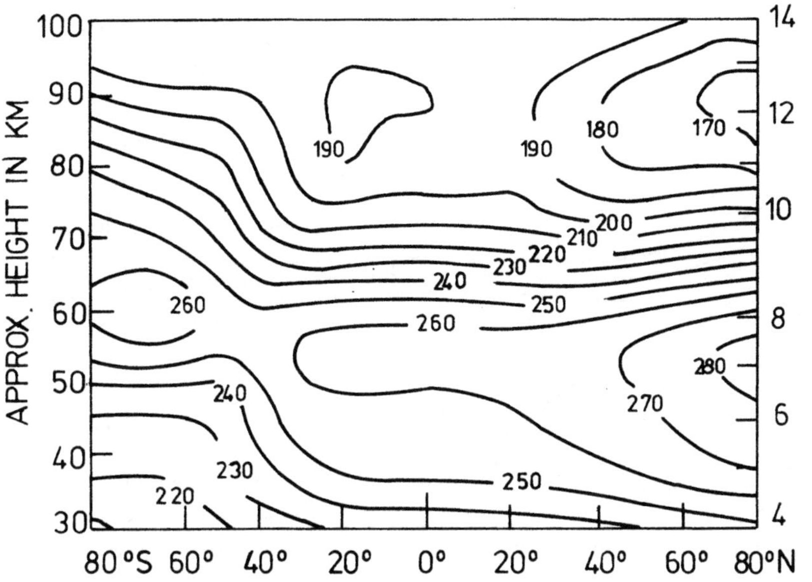

Fig. 2. Zonal mean temperature cross section 21 June 1975

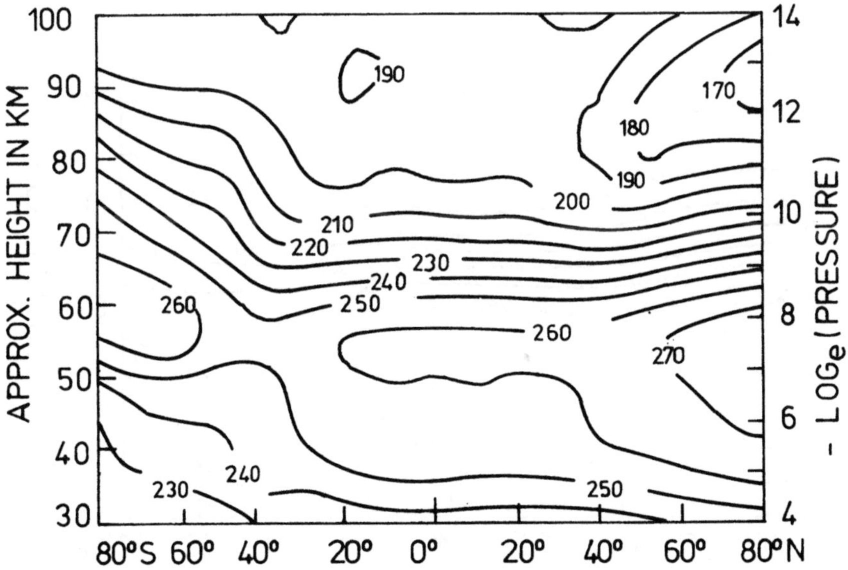

Fig. 3. Zonal mean temperature cross section 5 Aug.1976

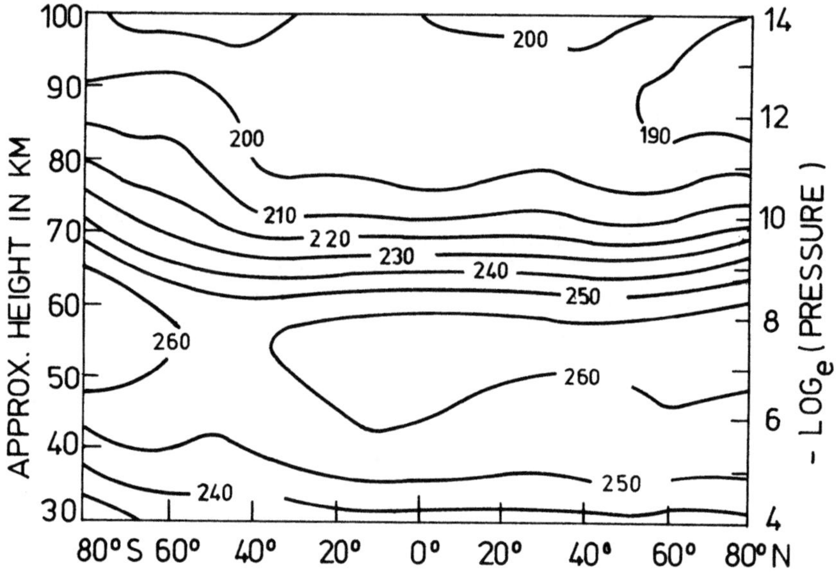

Fig. 4. Zonal mean temperature cross section 5 Sept.1976.

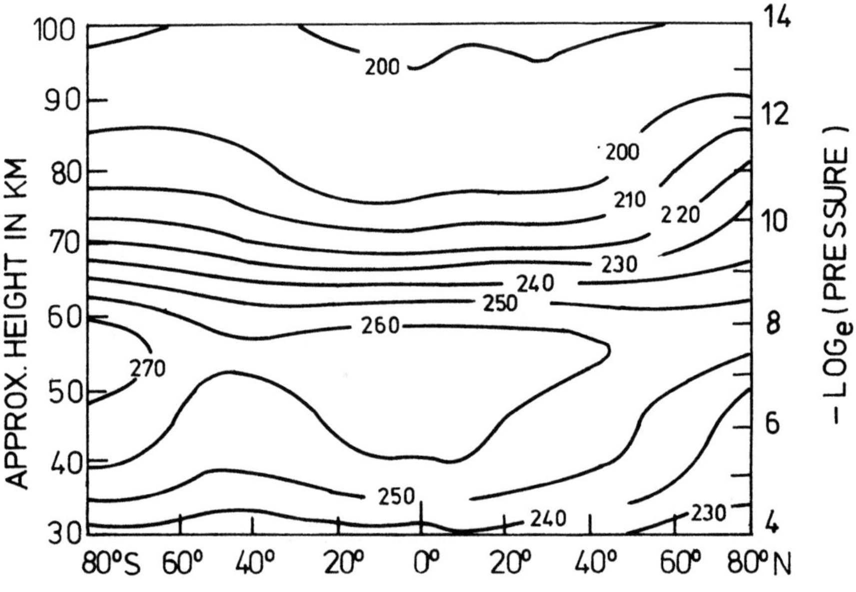

Fig. 5. Zonal mean temperature cross section 5 Oct.1976.

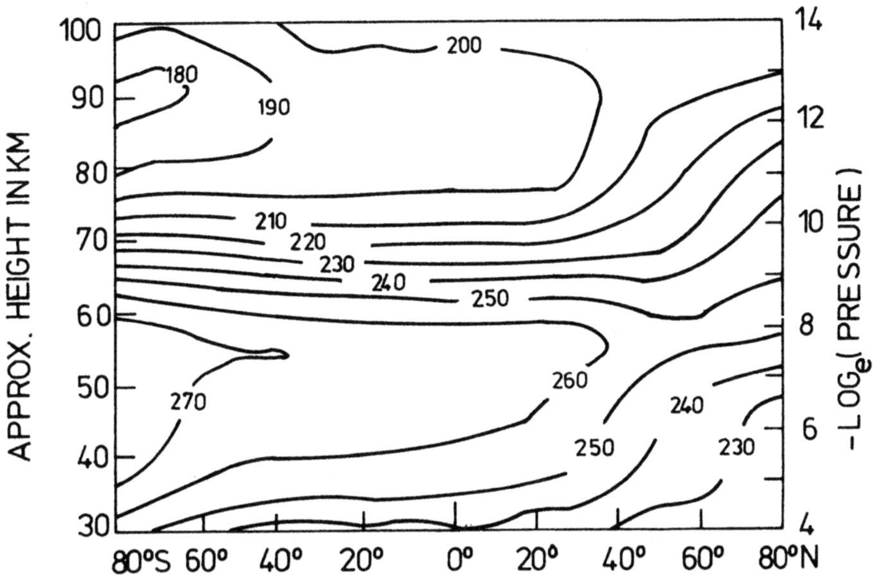

Fig. 6. Zonal mean temperature cross section 5 Nov.1976

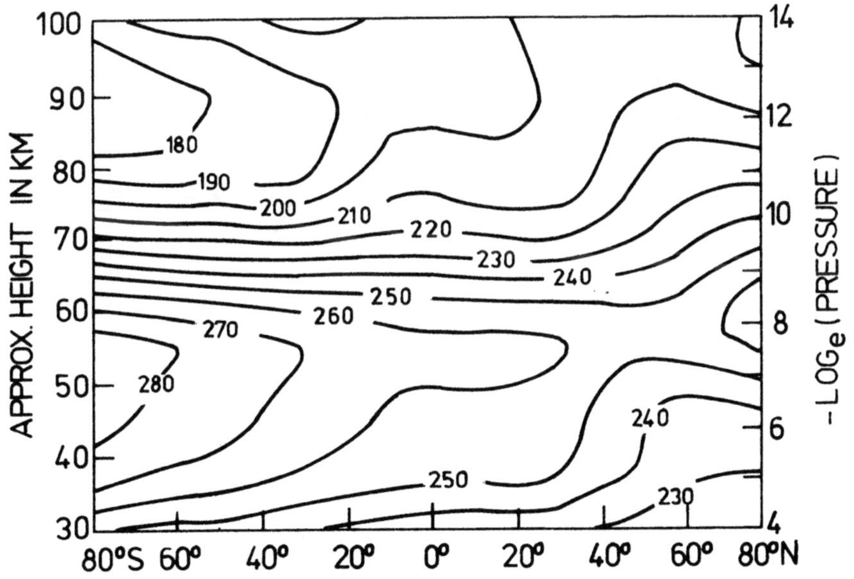

Fig. 7. Zonal mean temperature cross section 5 Dec.1976

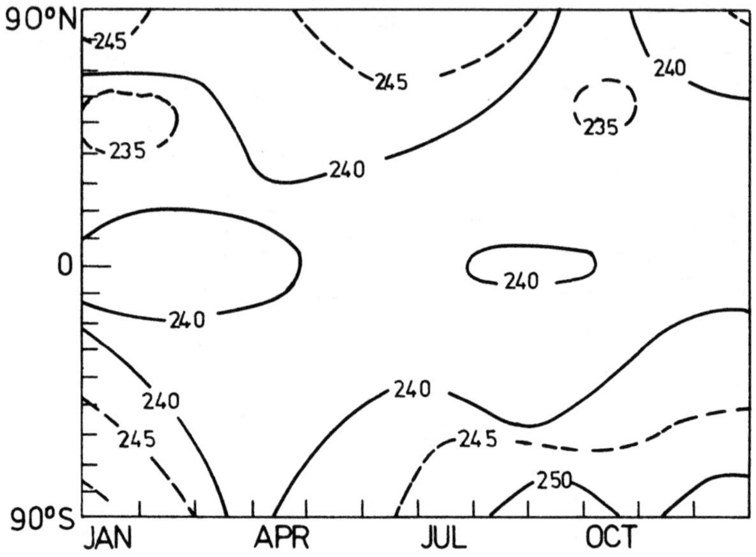

Fig. 8. Time/latitude cross section channel 2100 (60 km)

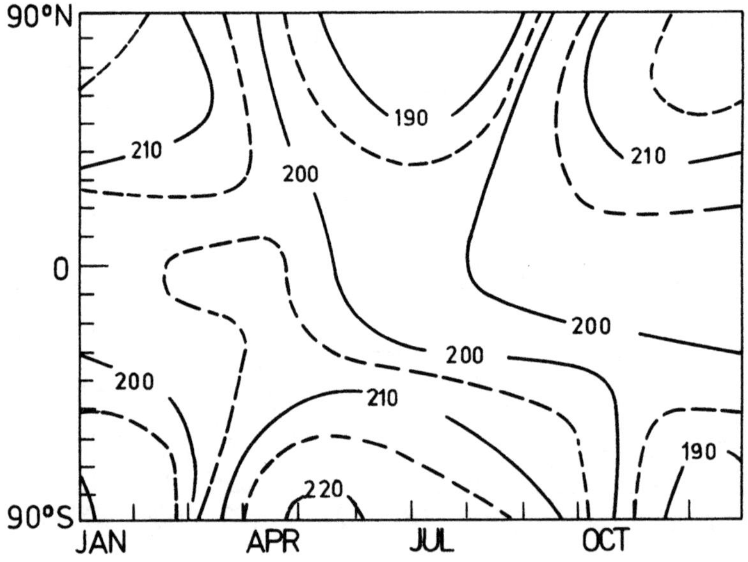

Fig. 9. Time/latitude cross section channel 3000 (85 km).

are based on northern hemisphere data only. The most obvious feature is that the temperature difference between the stratopause and mesopause is considerably larger in summer than in winter. The difference between the hemispheres at stratopause levels is due to the difference in solar heating of ozone. The difference at the mesopause is not completely understood, but is a consequence of the global circulation, as radiative heating would work in the opposite direction. One feature which is obvious here, but not so apparent in the CIRA 72 cross sections is the horizontal temperature gradient in the winter hemisphere between latitudes 40° and 60°. This is associated (through the thermal wind equation) with the polar winter vortex which has a peak westerly (eastward) flow at about 50 km or 60 km altitude. An obvious consequence of this flow is that the stratopause and mesopause are at higher altitudes over the winter pole than elsewhere. After a sudden warming when the westerly vortex has been destroyed, the zonal mean stratopause and mesopause are found to have descended.

The southern hemisphere temperatures are found to be somewhat different from northern hemisphere temperatures of the same season. This is due partially to the ellipticity of the earth's orbit producing a 6% modulation of the solar input, and partially to the different distribution of land and sea which excites much larger amplitudes of planetary waves in the northern hemisphere. Unfortunately it was not possible (for technical reasons) to construct twelve monthly zonal cross sections for this paper, but some of the effect can be seen in figs. 8 and 9 which give latitude/time cross sections of brightness temperatures for channels 3000 and 2100 whose weighting functions peak at about 85 km and 60 km respectively.

MONTHLY MEAN RADIANCE MAPS

Figures 10, 11 and 12 show monthly means of radiance measured in channels 2115, 2100 and 3000, whose weighting functions peak at about 40 km, 60 km and 85 km respectively. They are plotted on a northern hemisphere polar stereographic projection with the Greenwich meridian at the bottom. Four monthly maps are shown for each channel as being representative of the four seasons, and the

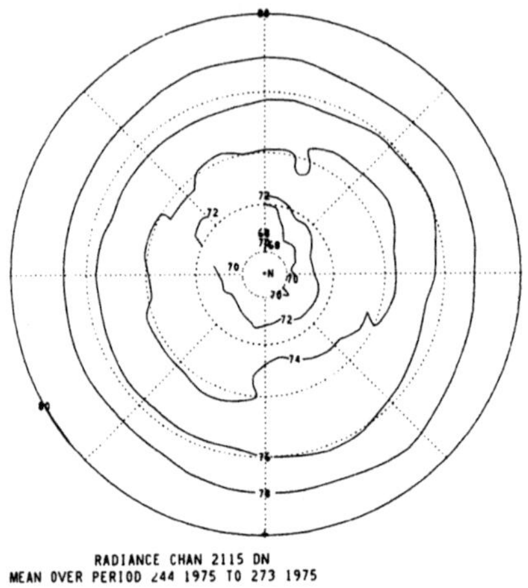

Fig.10a Sept.1975

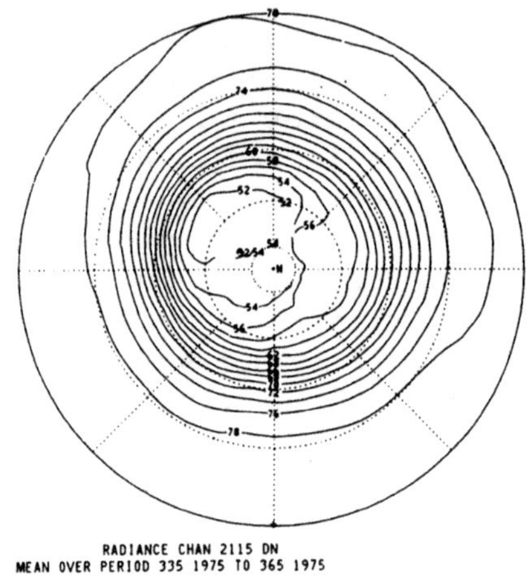

Fig.10b Dec. 1975

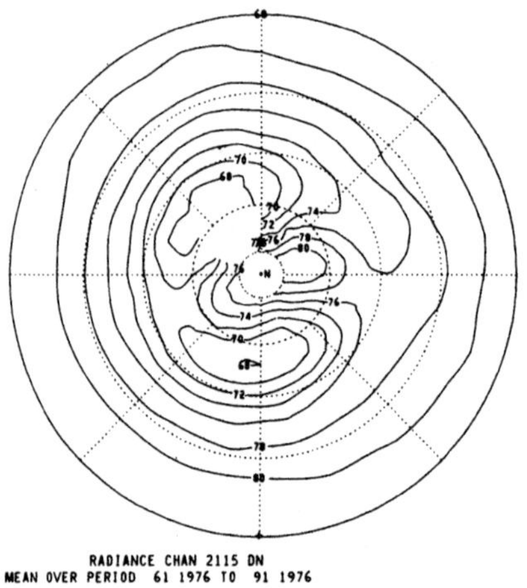

Fig.10c March 1976

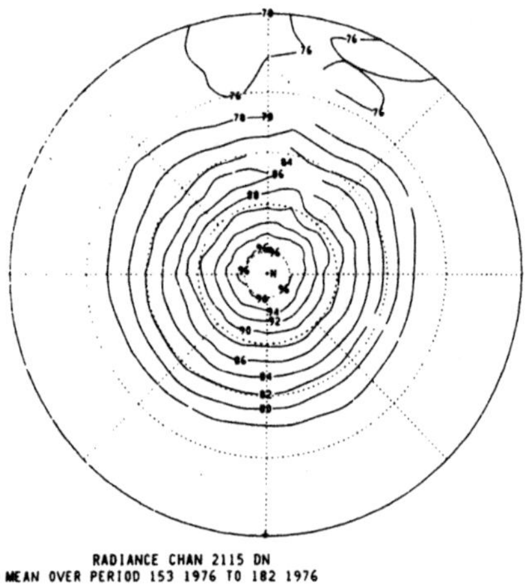

Fig.10d June 1976

Fig.10 Monthly mean radiance maps channel 2115 (40 km) Polar stereographic projection

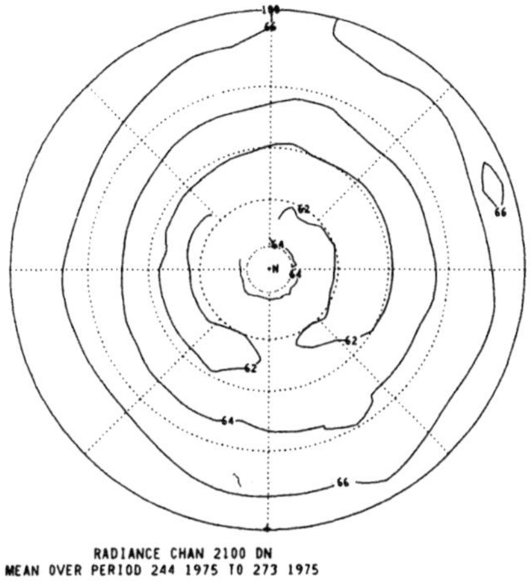

Fig.11a Sept.1975

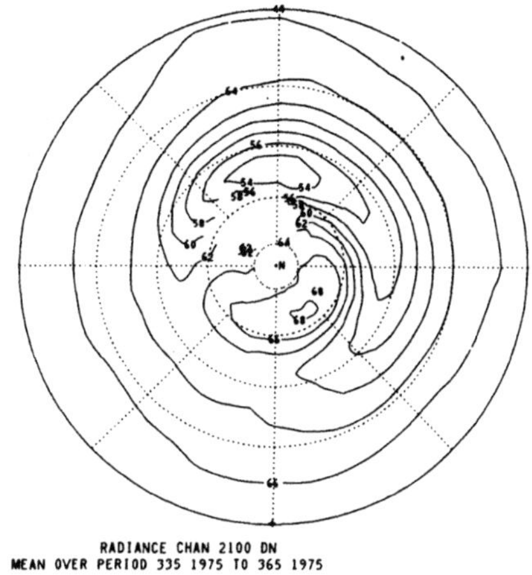

Fig.11b Dec. 1975

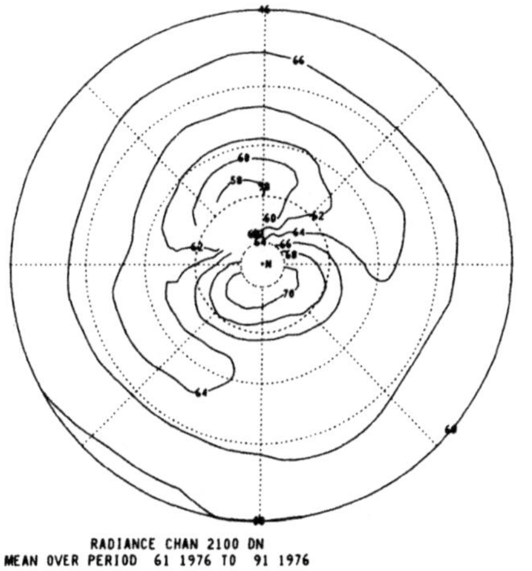

Fig.11c March 1976

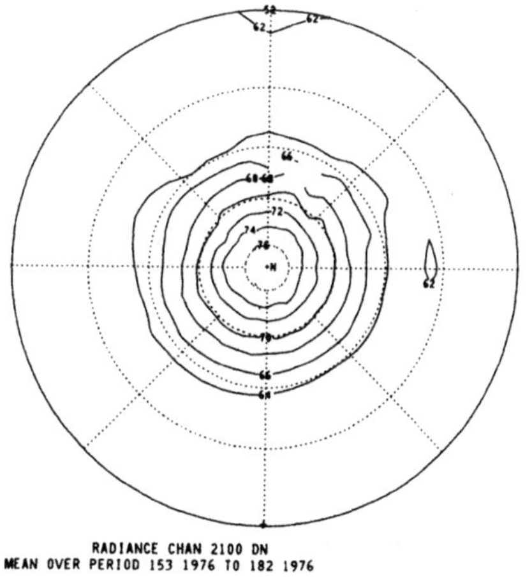

Fig.11d June 1976

Fig.11 Monthly mean radiance maps channel 2100 (60 km)
Polar stereographic projection

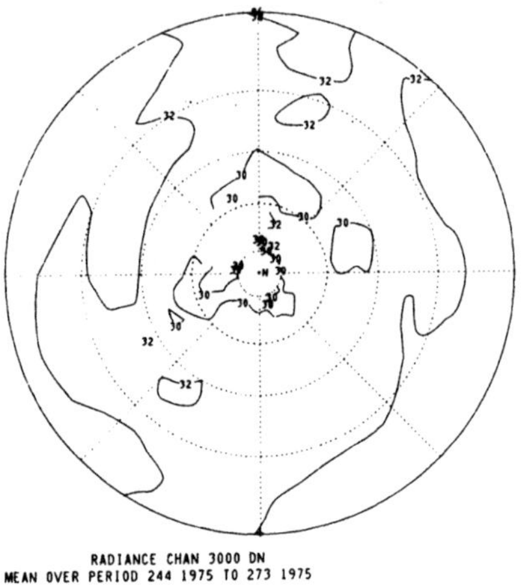

Fig.12a Sept.1975

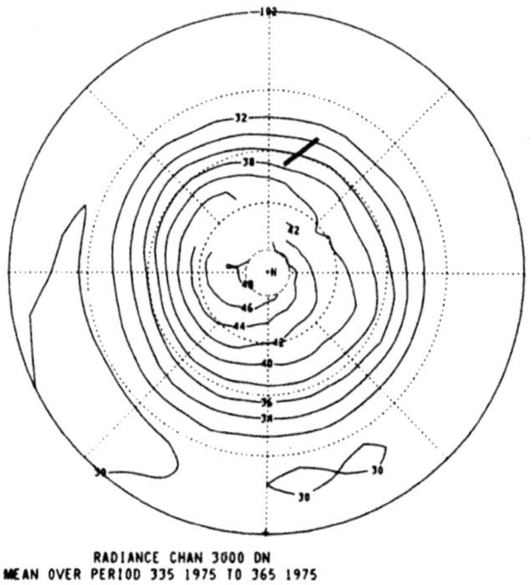

Fig.12b Dec. 1975

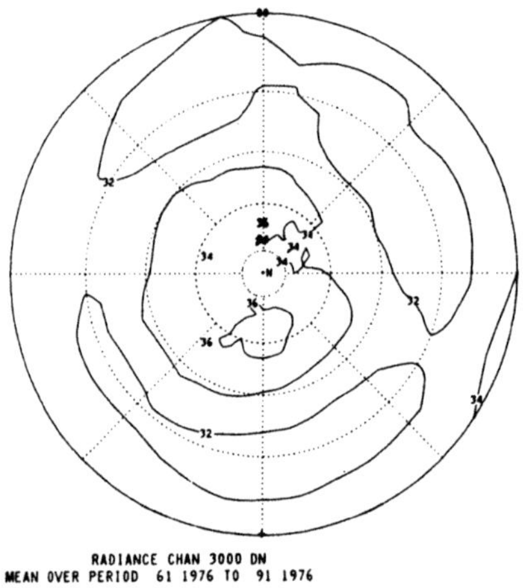

Fig.12c March 1976

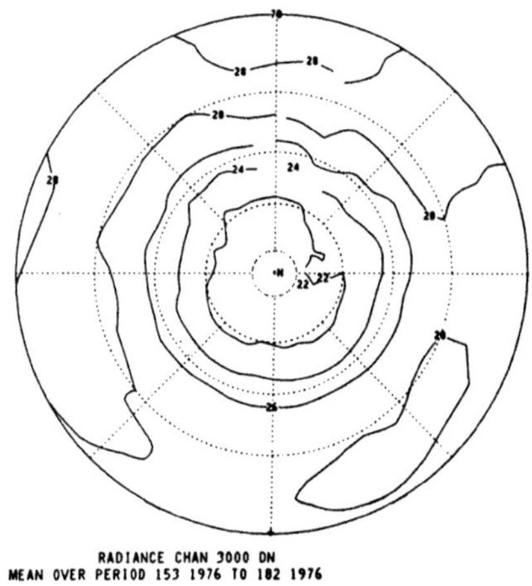

Fig.12d June 1976

Fig.12 Monthly mean radiance maps channel 3000 (85 km)
Polar stereographic projection

Table 1.

Conversion from Planck function at 668 cm^{-1}

Planck Function mW $m^{-2}ster(cm^{-1})^{-1}$	Temperature K
10	163.6
20	185.4
30	201.0
40	213.7
50	224.7
60	234.6
70	243.6
80	251.9

southern hemisphere is omitted in order to keep the number of illustrations within reasonable limits. The quantity plotted is Planck function in units of mW $m^{-2}ster^{-1}(cm^{-1})^{-1}$. This may be converted to brightness temperature with the aid of table 1; the resulting temperature may then be regarded as approximately the mean temperature over the appropriate weighting function from fig. 1.

The situation is at its simplest in the summer (e.g. June 1976) when winds are easterly (westward) and almost zonally symmetric. The radiance contours are almost symmetric with the high temperature at the pole at 40 and 60 km, and a low temperature at the pole at 85 km. The thermal wind equation states that the vertical gradient of wind is proportional to the horizontal temperature gradient. Therefore the peak of the summer easterly flow lies between 60 and 85 km.

By autumn (e.g. September 1975) solar heating is reduced, the winter circulation is developing and the 40 km channel is now cold over the pole. The 60 km channel has its lowest temperature in mid latitudes at about 50°N, and is still relatively warm at the pole whilst at 85 km the temperature is almost uniform over the whole hemisphere. In December the early winter circulation is firmly established with a strong temperature gradient towards a minimum at the pole at 40 km, and a somewhat smaller gradient towards a maximum at the pole at 85 km. The 60 km channel is rather more complicated, showing a zonal wavenumber one pattern with a warm area and a cold area. The implication here is that the altitude of the peak of the westerlies is above 60 km at low latitudes and below 60 km in the general region of the warm area. The peak is at 60 km at the temperature minimum in this channel.

At the end of winter there are disturbances at all levels as the final stratospheric warming takes place, making way for the summer circulation. It is difficult to give an intuitive argument to describe the wind field from these maps, but in general this is the period during which the summer easterlies are becoming established. The details of the winter patterns - such as locations and timings and intensity of stratospheric disturbances - differ between individual winters, but the general features remain similar.

A STRATOSPHERIC DISTURBANCE

To illustrate the effect of winter stratospheric disturbances on the temperature field, a cross section has been prepared through the major event that occurred at the end of December 1976. Fig. 13 is a longitude/height cross section of temperature around a latitude circle at 60°N, whilst fig. 14 is a cross section of departures from the zonal mean temperature profile.

The most obvious feature of the temperature plot is variation with longitude of the height of the stratopause, ranging between 40 and 75 km. Reference to fig. 6 shows that the normal for this time of year at 60°N is about 60 km. Correspondingly the mesopause height varies between 75 and over 100 km. In fact it seems as if stratopause and mesopause are no longer the clearly defined entities that they are at other times and places.

The departure of temperature from the zonal mean show the nature of the disturbance more clearly. There is a component of zonal wave number one whose phase is tilting westwards with height, as wouldbe expected from a planetary wave which is propagating energy upwards. There is also a strong hint of a node at about 65 km with a phase reversal above, which may correspond to the critical level of Charney and Drazin (1961) and Matsuno (1970).

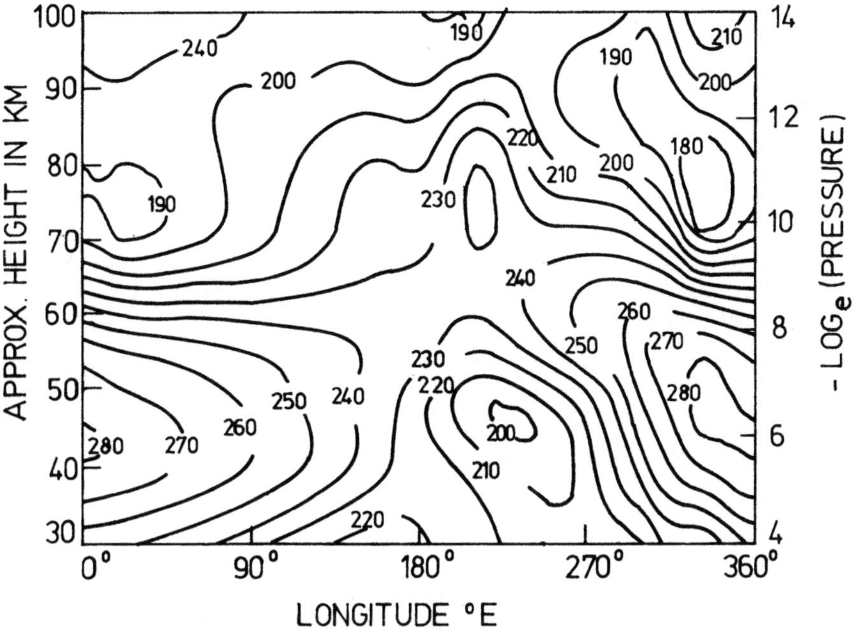

Fig.13. Temperature cross section at 60°N on 29 Dec.1976

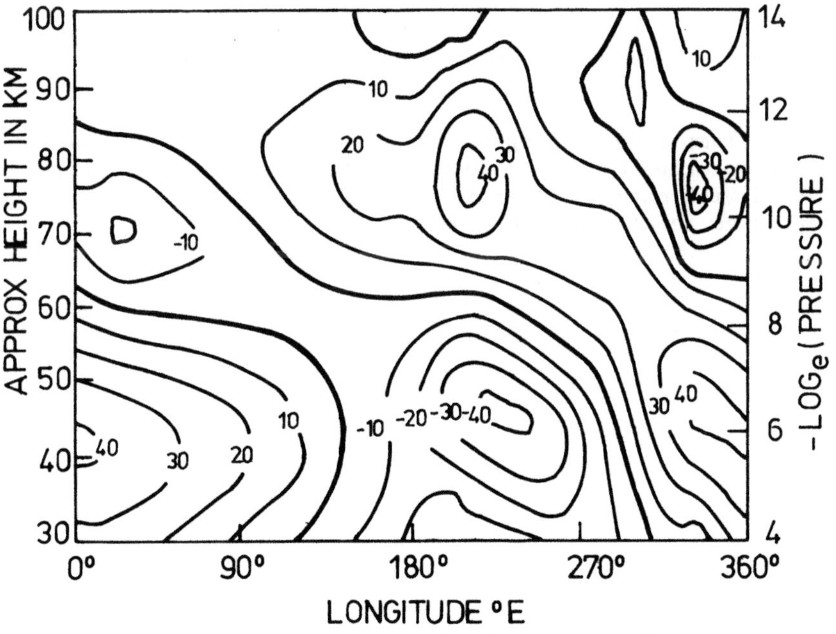

Fig.14. Departures from zonal mean at 60°N on 29 Dec.1976

ACKNOWLEDGEMENTS

I would like to thank Dr. J.J. Barnett for figures 10, 11 and 12, and Mr. R.J.D. Kirkman for all the retrievals of temperature profiles from measured radiances.

REFERENCES

Charney, J.G. and Drazin, P.G. 1961, J. Geophys. Res. 66, 83.
CIRA 1972. Cospar International Reference Atmosphere 1972, Akademie-Verlag, Berlin.
Matsuno, T. 1971. J. Atmos. Sci. 28, 1479.
Rodgers, C.D. 1976. Reviews of Geophysics and Space Physics, 14, 609.
Sissala, J.F., Ed., 1975. The Nimbus 6 Users Guide, Goddard Space Flight Center, Greenbelt, Maryland.

DISCUSSION:

Reagan: Large geomagnetic substorm and storm activity and the subsequent energetic electron precipitation should result in temperature changes in the altitude range 80-100 km at high latitudes. Are such temperature changes observed in your satellite data ? Correlations should be performed between satellite data on precipitating particles and temperature profiles to establish the cause and effect relationship between the former and the latter.

Rodgers: Temperature changes on a small space and time scale do occur in our top channel (~ 80 km) data, which are not obviously connected with changes lower in the atmosphere. We have not yet attempted to relate them to geomagnetic phenomena.

STRATOSPHERIC-MESOSPHERIC MIDWINTER WARMINGS

K. Labitzke

Institut für Meteorologie,
Freie Universität Berlin,
Berlin, F.R.G.

The most interesting synoptic-scale events of the circulation in the strato- and mesospere are the so-called midwinter "warmings". These are forced from the troposphere but a complete explanation has not yet been given.

DEFINITION OF MIDWINTER WARMINGS

For the description of the synoptics of a midwinter disturbance it is useful and common to distinguish between "major" and "minor" warmings. Major warmings are those events (Quiroz et al., 1975) in which poleward movement of planetary-scale thermal systems results in a reversal of the polar circulation ("breakdown") at an altitude at least as low as the 10-mb level (about 30 km), in association with a reversal of the meridional temperature gradient in the middle and lower stratosphere. The breakdown period is followed by a period of cooling and reintensification of the polar vortex in the stratosphere and lower mesosphere (Labitzke, 1972a; Labitzke and Barnett, 1973). This cooling must be of dynamic origin, as the temperatures reached in February/March are lower than in early winter. Minor warmings may reach comparable intensities in the upper stratosphere but do not lead to the breakdown of the circulation as defined above.

B. Grandal and J. A. Holtet (eds.), Dynamical and Chemical Coupling, 17-34.

VERTICAL TEMPERATURE CHANGES

The typical features of a major stratospheric midwinter warming (of the N.H.) have been summarized by Labitzke (1972a), given here in Figs. 1 and 2. Referring to these two illustrations, the following statements can be made:

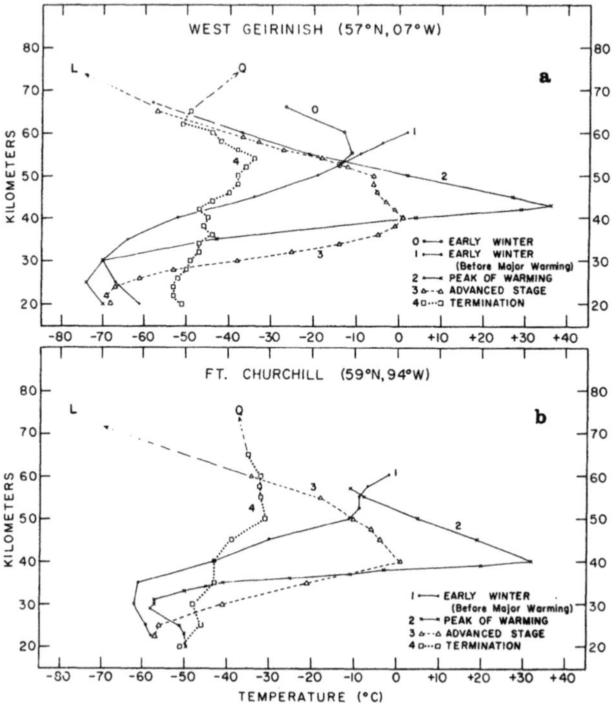

Figure 1 Typical temperature profiles (compiled from meteorological rocket sonde data from several winters) of different stages of the midwinter disturbances; a) for West Geirinish, U.K., b) for Fort Churchill, Canada.

1) The changes which happen during a stratospheric midwinter warming are not limited to the stratosphere but extend at least into the upper mesosphere. If the whole layer is considered, the word "warming" is a misnomer. Rather, temperature changes of opposite sign take place at the same time at different levels, presumably linked with vertical motions.

2) The events begin outside the regime of the Aleutian high, with rising temperatures around a very high stratopause at ~ 60 km (profile 1, Fig.1).

3) The stratopuse descends 20 km within several days while the temperature around it rises rapidly to a peak above +30°C. At the same time the upper mesophere, and often also the lower stratosphere, cool (profile 2 in Figs. 2 and 2, 5-20 December).

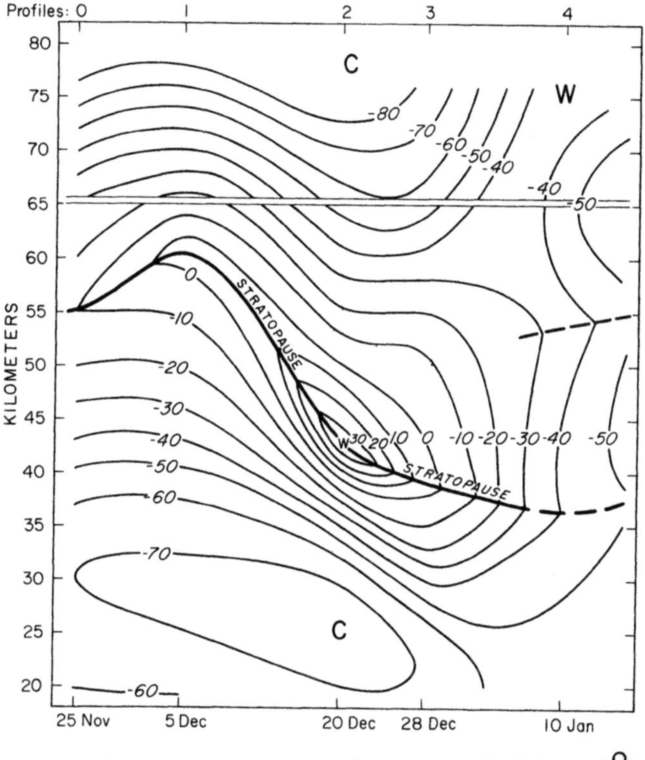

Figure 2 Schematic temperature variation (°C) with height and time during the course of a a major midwinter warming at about 60°N, outside the Aleutian anticyclone.

4) After the peak, the stratopause descends further, and the upper stratosphere, now at 30-40 km, warms slowly whereas the upper mesosphere stays cold and the lower mesosphere cools (profile 3 in Figs. 1 and 2, 20-28 December).

5) If the polar vortex breaks down (a major midwinter warming), the warming reaches the lower stratosphere (about 5 weeks after stage 1), the warm stratopause region is destroyed through cooling of the layer between 30 - 60 km, (late winter cooling period), and the upper mesosphere warms (profile 4 in Figs. 1 and 2, after 28 December).

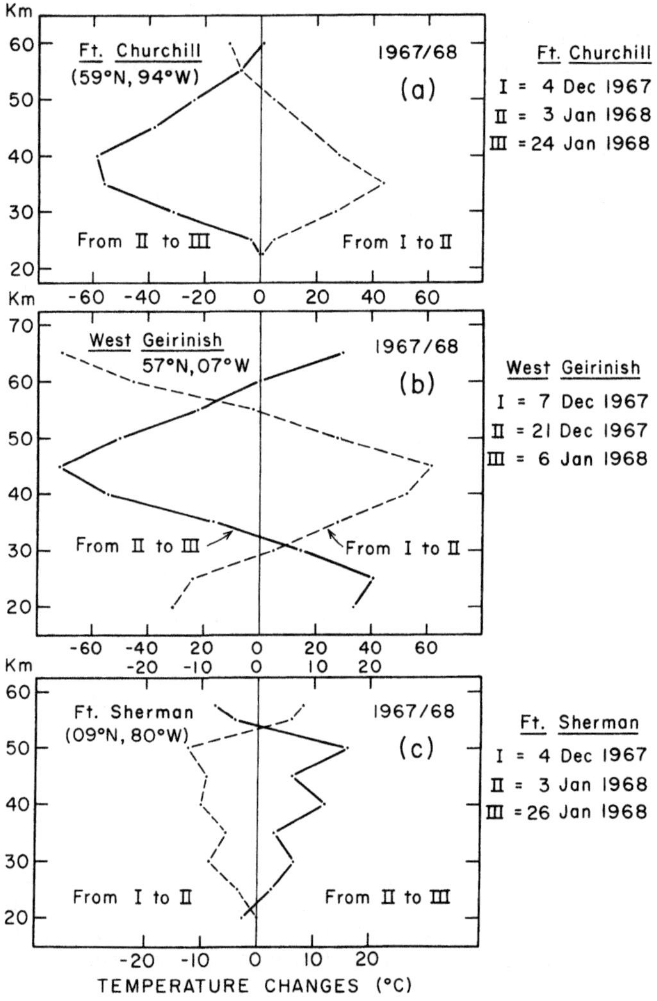

Figure 3 Vertical temperature changes derived from rocket sonde temperature measurements; the dates I, II, and III refer to early winter, peak of warming, and late winter cooling period, respectively. Note also the opposite temperature changes between high and low latitudes.

6) The mean profiles in Fig. 1 support Quiroz' idea (1969) and Leovy's computations (1964) that the upper mesosphere is cold at high latitudes in early winter and again in late winter and that the warm upper mesosphere observed in late January-early February is an interval connected with the breakdown of the stratospheric polar vortex, i.e., stage 4.

The described temperature changes of opposite sign between stratos- and mesosphere are summarized in Fig. 3 (Labitzke, 1972b). This behavior is an expected characteristic of stratospheric circulation changes driven by planetary waves, as explained by Dickinson (1975). As was also shown by Labitzke, a temperature distribution like in profile 2 of Fig. 1 (peak of warming) is not a sufficient condition for a <u>major</u> warming.

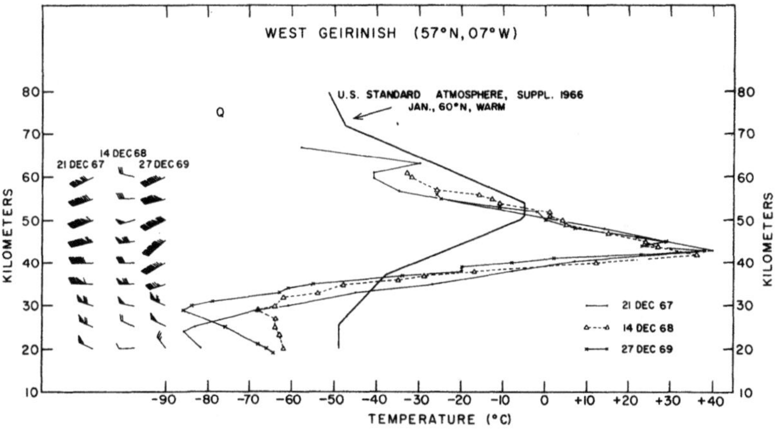

Figure 4 Three profiles of rocketsonde data at West Gerinish and U.S. Standard for January at 60 N (warm).

Fig. 4 shows three profiles of rocketsonde data at West Geirinish and the U.S. Standard for January at 60°N (warm). <u>Major</u> warmings developed during the winter 1967/68 and 1969/70, whereas the warming did not penetrate into the lower stratosphere in 1968/69. The only apparent difference between the two soundings associated with a complete breakdown and that which did not develop into a major warming is noticeable in the wind. The two former show considerably higher wind speeds above 30 km (between 50 and 55 km about six times higher) than the latter. - The midwinter warmings develop, however, south of the stratospheric polar jet stream (Matsuno, 1971).

Over the S.H., major midwinter warmings do not take place. Warmings of the upper stratosphere, however, have been observed regularly since satellite data have become available. But already early rocket sonde observations made at McMurdo Station during 1962 and 1963 caught midwinter warmings with intensities comparable to those over the N.H. In Fig. 5, a few selected profiles are shown and compared with the "early winter profile (1)" and with the "advanced warming profile (3)" of West Geirinish (Labitzke and van Loon, 1972).

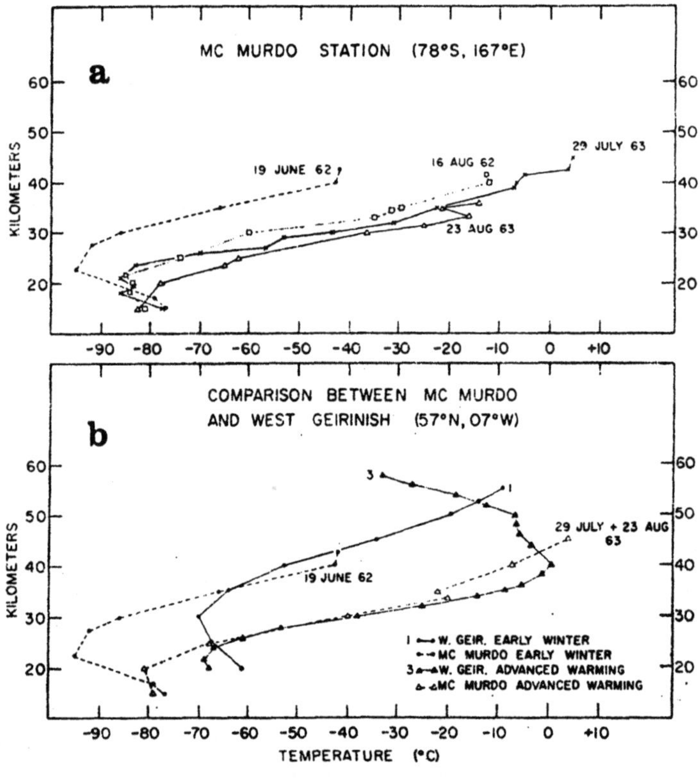

Figure 5 Selected rocketsonde temperature profiles at McMurdo Sound for the winters of 1962 and 1963, a; and comparisons between rocketsonde profiles at McMurdo Sound and West Geirinish, b.

HORIZONTAL CHANGES OF TEMPERATURE AND LEVELS OF CONSTANT PRESSURE

Ever since their discovery by Scherhag (1952) many have studied the midwinter warmings and a detailed documentation of the height and temperature distribution up to the 10-mb level is available for the N.H. for more than ten years, together with case studies. The investigation of the warmings has been greatly accelerated in recent years by the use of satellite infrared measurements from such instruments as the Satellite Infrared Spectrometer (SIRS), Vertical Temperature Profile Radiometer (VTPR), and Selective Chopper Radiometer (SCR). Together with conventional rocket- and radiosonde data, these measurements have given a comprehensive record of warmings in both hemispheres (Harwood, 1975; Labitzke, 1972a; Labitzke and Barnett, 1973; Quiroz, 1971, 1974a, 1975).

In a recent paper, Quiroz et al. (1975) distinguished two principal types of developments of major stratospheric warmings: Type (1) is the interaction of an eastwards travelling thermal wave with the standing wave which in the middle stratosphere is positioned over East Siberia. This standing wave accounts for the wave 1 perturbation of the pressure field known as the Aleutian anticyclone. In type (2), the travelling wave has a strong meridional trajectory in or near the area of the North Atlantic Ocean. In 1963, a typical example for type (2), the North Atlantic and East Siberian thermal systems arrived seemingly independently in the polar region, forming a single thermal system at 10 mb after a pronounced wave number 2 pattern was observed in the pressure field. In minor warmings which are fairly common in both hemispheres, the trajectory of warm cells tends to retain a basically zonal (eastward) direction, and local warmings may be observed in connection with the passage of a single warm-air system or as a consequence of the superposition of two warm cells. Infrequently, the trajectory may exhibit a strong poleward component, but polar warming in the middle and lower stratosphere is insufficient to reverse the circulation except possibly at very high stratospheric or mesospheric altitudes. Significant distortion or elongation of the polar vortex may occur, however, at lower stratospheric altitudes.

In the following, some synoptic features of the major warming of the winter 1970/71 will be discussed. This warming is typical for type (1) of Quiroz et al.

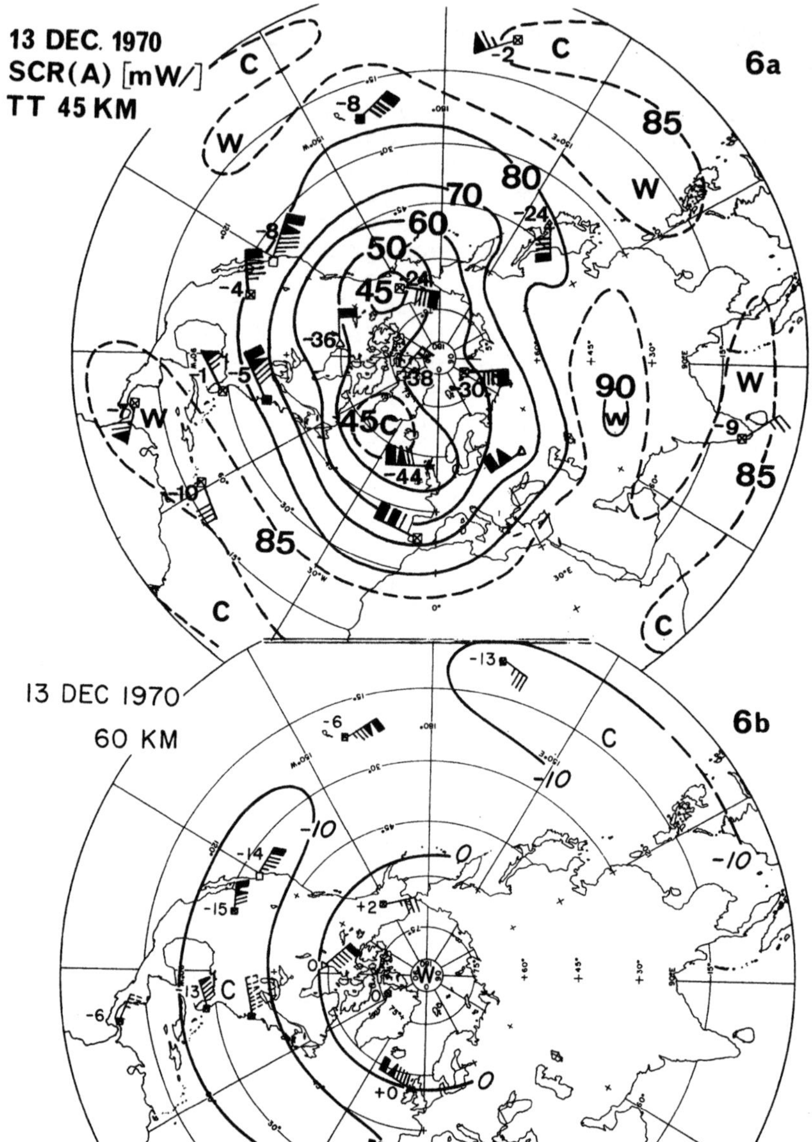

Figure 6a Isolines of radiances [mW $m^{-2}sr^{-1}(cm^{-1})^{-1}$] of channel A of the SCR-Experiment, and rocket sonde data (wind and temperature) at 45 km, for 13 December 1970.

Figure 6b Temperatures at 60 km and available rocketsonde data, for 13 December 1970.

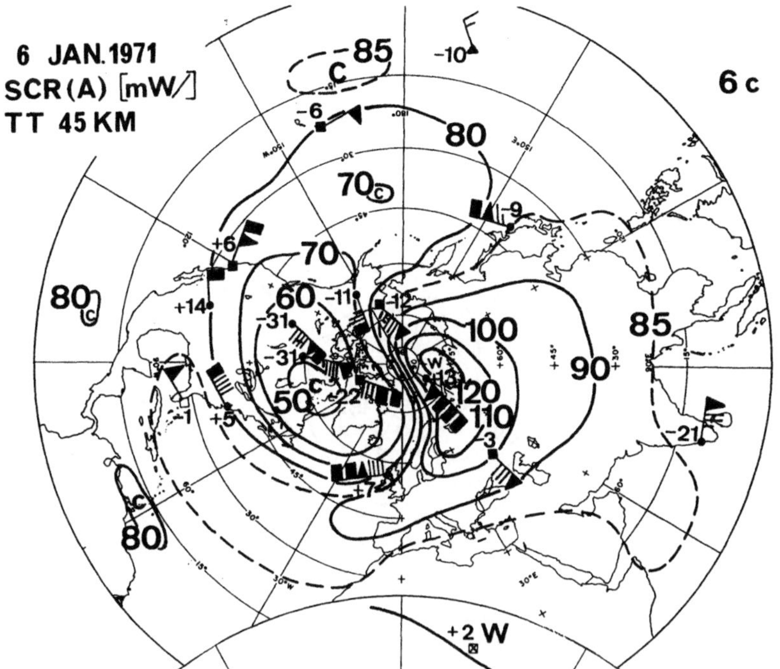

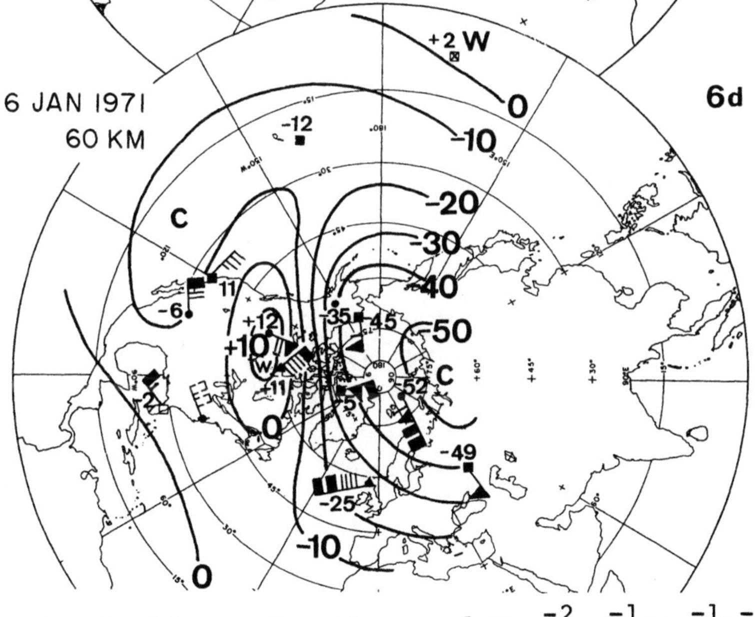

Figure 6c Isolines of radiances [mW $m^{-2}sr^{-1}(cm^{-1})^{-1}$] of channel A of the SCR-Experiment, and rocket sonde data (wind and temperature) at 45 km, for 6 January 1971.

Figure 6d Temperatures at 60 km and available rocketsonde data, for 6 January 1971.

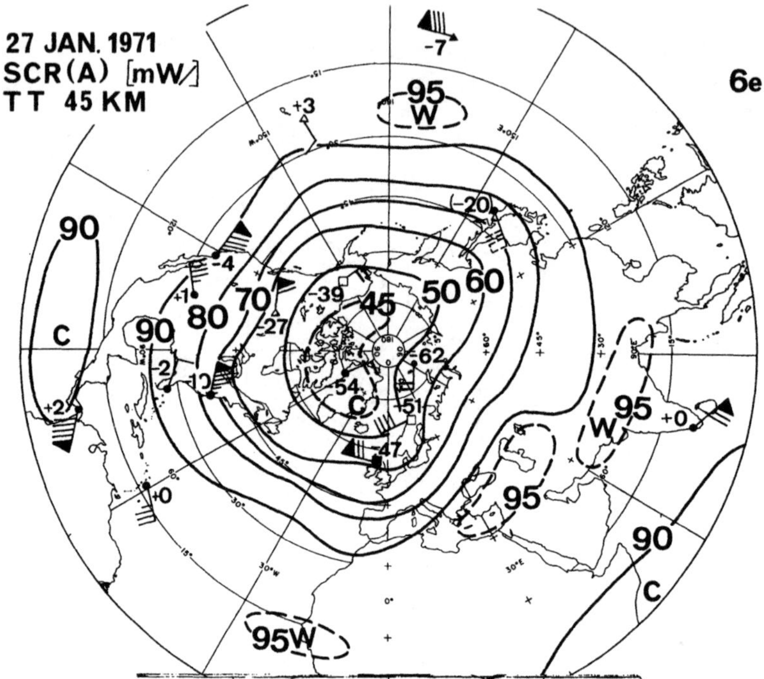

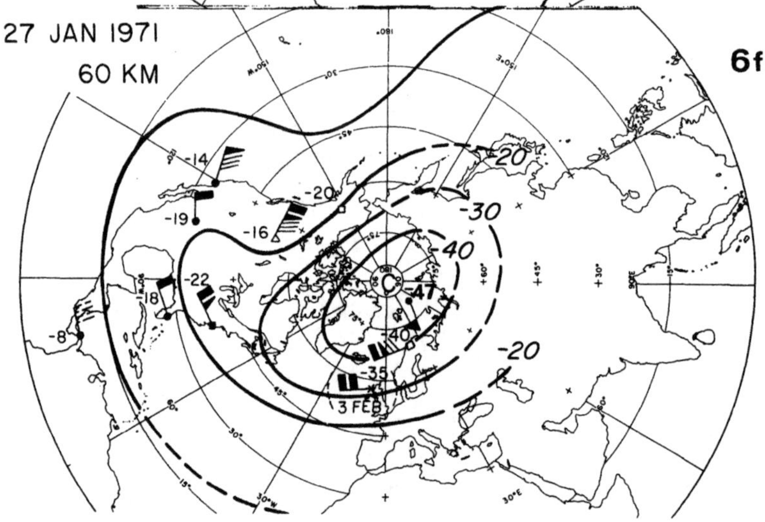

Figure 6e Isolines of radiances [mW $m^{-2} sr^{-1} (cm^{-1})^{-1}$] of channel A of the SCR-Experiment, and rocket sonde data (wind and temperature) at 45 km, for 27 January 1971

Figure 6f Temperatures at 60 km and available rocket-sonde data, for 27 January 1971.

For the description of the temperature pattern at the 45-km level (Fig. 6 a,c,e) radiance data [mW $m^{-2} sr^{-1} (cm^{-1})^{-1}$] of channel A of the SCR have been used. These data are highly correlated with the temperature of the 2-mb level (Quiroz, 1974b). Rocketsonde data are entered to give an impression about the circulation pattern.

For the 60-km level only few rocketsonde data were available, but they indicate the large scale temperature changes.

Figs. 6a and b depict the synoptic situation at the beginning of the warming, i.e., 13 December 1970. Fig. 6a shows the early winter pattern at 45 km with a well developed standing wave 1 and a cold polar vortex near Iceland. Fig. 6b shows the temperature pattern at 60 km which is completely reversed compared against Fig. 6a, with warm air over the polar region. During this stage the stratopause is very high over polar latitudes, close to 60 km (cf. profile 1 in Fig. 1). The circulation is still governed by a strong polar cyclone. Figs, 6c and d (6 January 1971) show the development into a major warming, after an eastwards travelling thermal wave (Barnett et al., 1971) had merged with the standing wave 1 over Siberia. Over Asia, particularly Siberia, the stratopause has descended now towards the 45-km level with strong warming around the stratopause region. In the center of the warming (high radiances) temperatures reached values of about +40°C (Petzoldt, 1973) and Heiss Island, situated somewhat off the center, reported +13°C. This agrees with stage 2 and profile 2 in Figs. 1 and 2, respectively. -At the same time the mesosphere (e.g., 60 km) cools above the warming, Fig. 6d, while the axis of warming, i.e., the stratopause, slopes from Siberia across the polar region towards Canada where extremely high temperatures for the 60-km level have been measured, above the cold center of the 45-level. - The circulation is disturbed accordingly.

Three weeks later, Fig. 6e and f (for 27 January 1971), the late winter cooling period is leading to the reintensification of the polar vortex, with low temperatures over polar and middle latitudes throughout the strato- and lower mesosphere, stage 4 and profile 4 in Figs. 1 and 2.

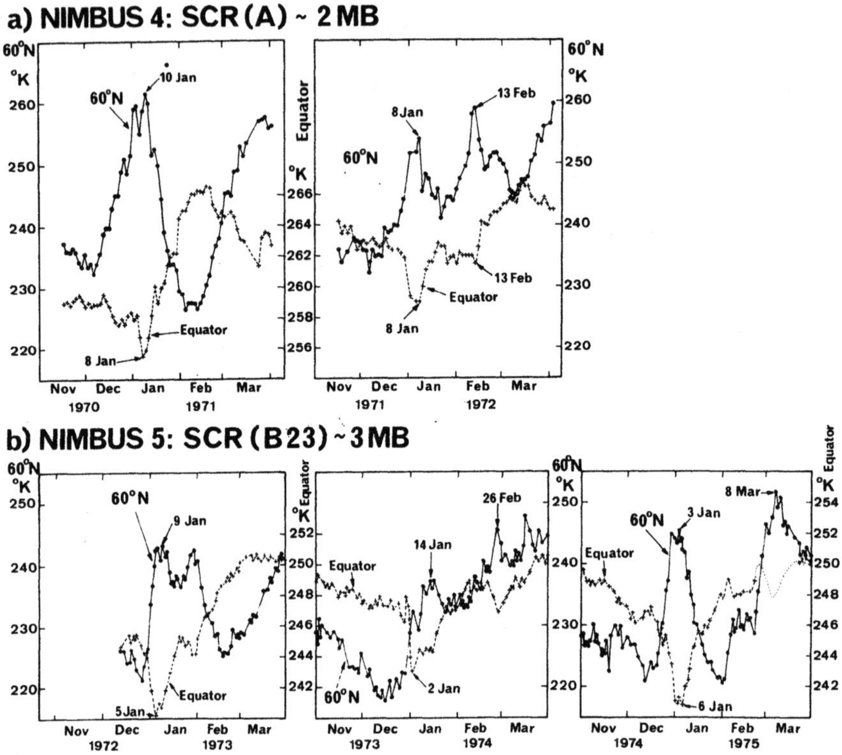

Figure 7 Daily zonal means of radiances (equivalent black body temperatures, °K) at 60°N and at the equator, for five winters.

At the same time, the temperature changes over the tropics and subtropics were out of phase with those over middle and high latitudes. This behaviour has been noted first by Fritz and Soules (1970) who used SIRS radiances, and is shown in Fig. 7 for five winters, using radiances received with the upper stratospheric channels of the SCR, flown on Nimbus 4 and 5. As Dickinson (1975) points out, "planetary wave induced temperature changes involve a horizontal redistribution of thermal energy with no change in the global mean values. Consequently, temperature variations in middle and high latitudes should be accompanied by opposite changes in the tropics".

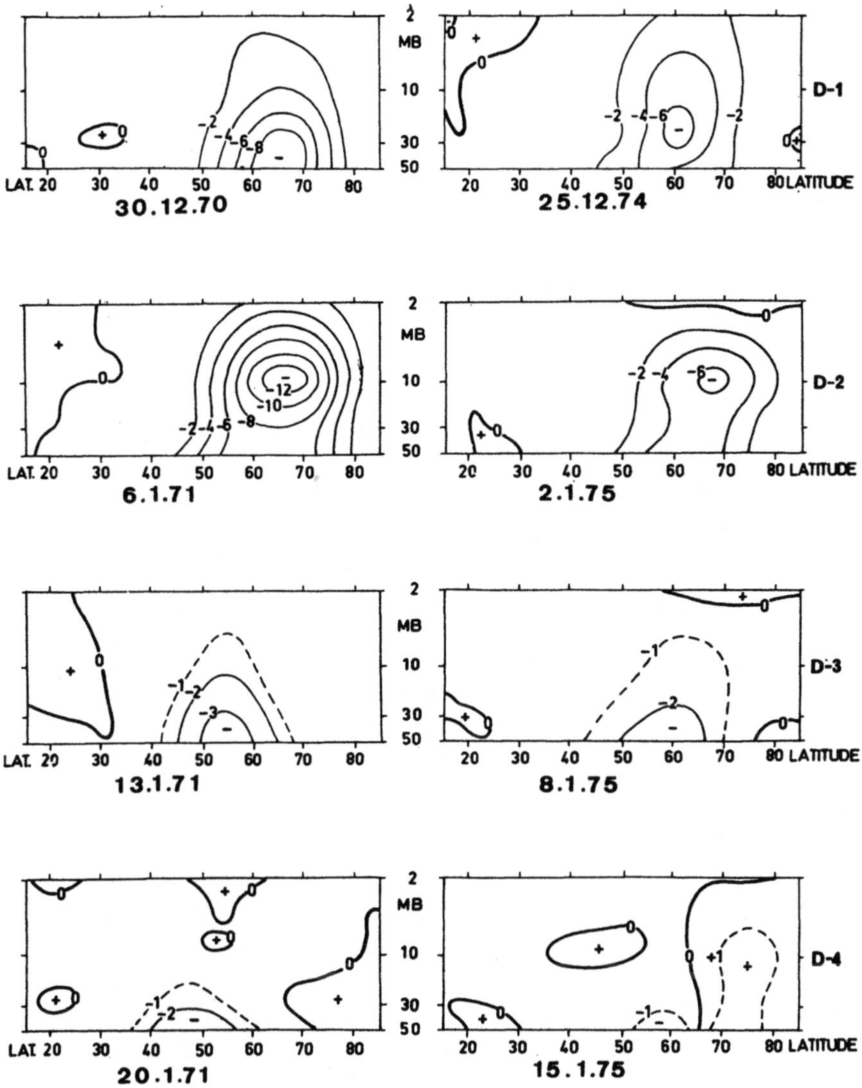

Figure 8a Geopotential flux
$(10^2$ erg cm^{-2} $sec^{-1})$ $[Z^*\omega^*]_{appr.}\cos\phi$.

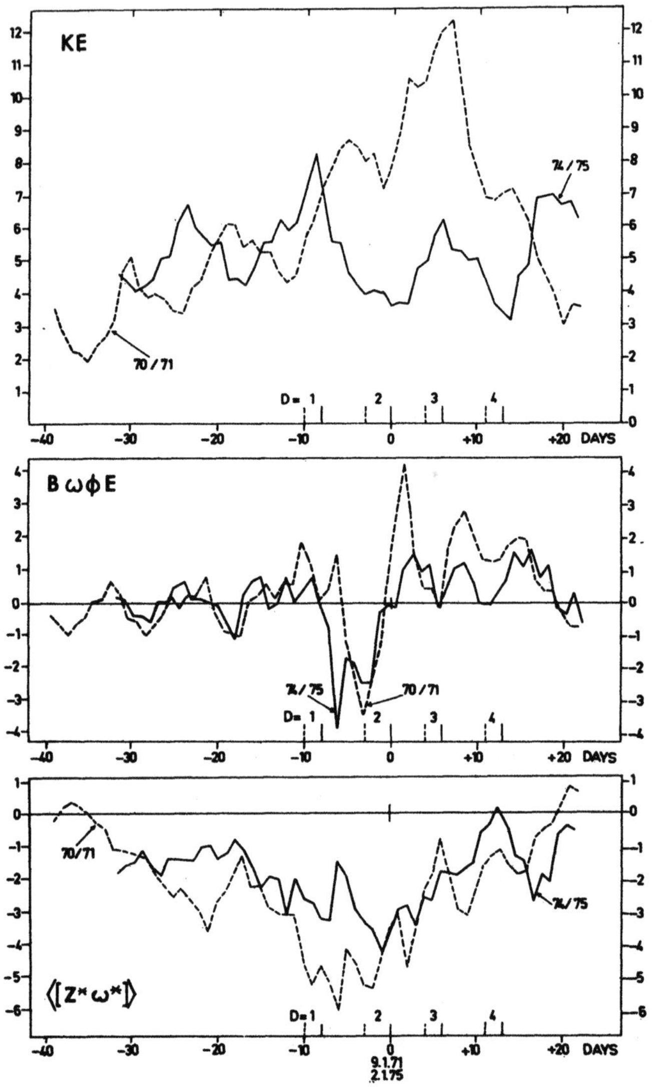

Figure 8b Daily values of KE (10^7 erg cm^{-2}) convergence of the vertical flux of geopotential (10^2 erg cm^{-2} sec) for 50-10 mb and 90°N-15°N. Daily values of the vertical flux of geopotential (10^2 erg cm^{-2} sec^{-1}) through 50 mb, 90°N - 15°N.

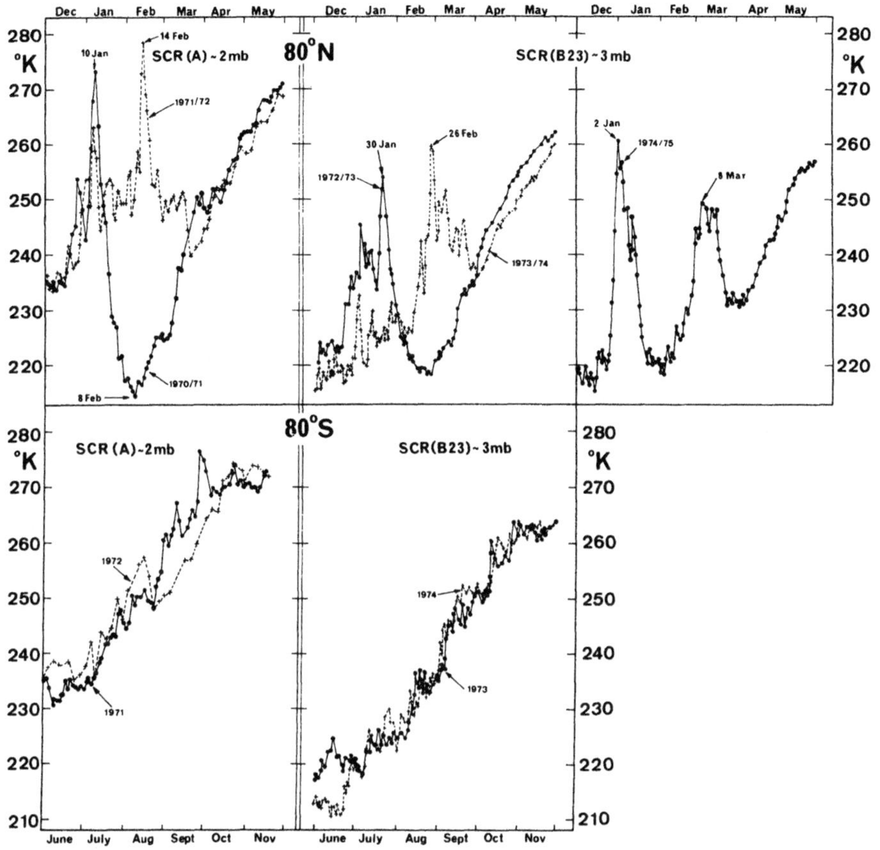

Figure 9 Daily zonal means at 80 deg. latitude of radiances of upper stratospheric channels of the SCR, flown on Nimbus 4 and 5. (The radiances are converted into equiv. black body temperatures (°K).)

ENERGETICS DURING MIDWINTER WARMINGS

Calculations of the energetics reflect the described synoptics. In Fig. 8, the geopotential flux is shown for selected days during the winter just described, i.e., a major warming. For comparison, the values for the winter 1974/75 are shown which was a winter without breakdown in the middle stratosphere (from Klinker, 1976). Most of the time, the geopotential flux and the eddy kinetic energy was much larger during the major event.

YEAR TO YEAR VARIABILITY

Over the N.H., winters with major warmings are followed by winters with minor warmings in the upper stratosphere and a relatively undisturbed, cold middle stratosphere. This variability from year to year is unexplained and of great interest, as it distinguishes the northern from the southern winter. In Fig. 9, daily zonal means of radiances of different upper stratospheric channels of the SCR are used to compare at 80° latitude 5 northern with 4 southern winters. The difference between major and minor warmings over the N.H. as well as the difference between both hemispheres is obvious. The variability during the northern winter is already very large in the middle stratosphere, where cold (undisturbed) and warm (disturbed) winter vary. The monthly mean 30-mb temperature at the North Pole (grid point value) is plotted in Fig. 10 for the last 14 years and demonstrates clearly the variability. - For discussion the monthly values of the sunspot numbers are entered as the bottom curve.

The transition from winter to spring varies according to the developments during midwinter, Often, dynamic final warmings may be observed after relatively undisturbed winter, leading to an early temperature reversal over the polar region (e.g., spring 1976). These are the so-called "early final warmings".

Over the S.H., the spring warmings (September-November) are the dominating events in the middle stratosphere, while in the upper stratosphere the antarctic starts to warm already during midwinter.

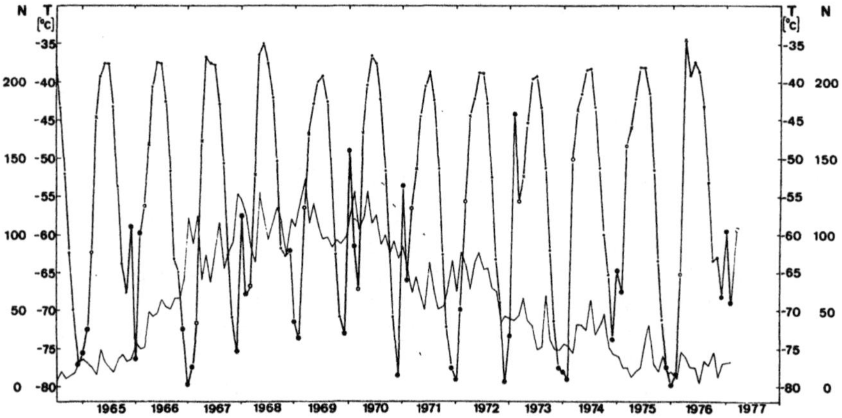

Fig. 10 ——— 30-MB MEAN MONTHLY TEMPERATURE AT THE NORTHPOLE

——— MEAN MONTHLY SUNSPOT NUMBERS

REFERENCES

J.J. Barnett et al., Nature 230, 47 (1971).

R.E. Dickinson, J. Atmos, Terr, Phys. 37, 855 (1975).

S. Fritz and S.D. Soules, J. Atmos. Sci.27, 1091 (1970).

R.S. Harwood, Quart. J.Roy. Met. Soc. 101, 75 (1975).

E. Klinker, Paper presented at COSPAR XIX (1976).

K. Labitzke, J.Atmos. Sci. 29, 756 (1972a).

K. Labitzke, J.Atmos. Sci. 29, 1395 (1972b).

K. Labitzke and H. van Loon, Meteor. Monogr. 13,No.35, 113-138 (1972).

K. Labitzke and J.J. Barnett, J.Geophys. Res. 78, 483 (1973).

C.B. Leovy, J.Atmos. Sci. 21, 238 (1964).

T. Matsuno, J.Atmos. Sci. 28, 1479 (1971)

K. Petzoldt, Paper presented at COSPAR XVI (1973).

R.S. Quiroz, Mon. Weath. Rev. 97, 491 (1969).

R.S. Quiroz, J. Appl. Met. 10, 555 (1971).

R.S. Quiroz, UAPRC-VII-3 (IAMAP/IAPSO Assemblies, Melbourne) (1974a).

R.S. Quiroz, Mon. Weath. Rev. 102, 313 (1974b).

R.S. Quiroz, J. Atmos. Sci. 32, 211 (1975).

R.S. Quiroz et al., J. Atmos. Sci. 32, 1723 (1975).

R. Scherhag, Ber. Deut. Wetterdienstes (US Zone) 38, 51 (1952).

DISCUSSION

Hargreaves: How do you define when a warming is at a maximum ? Can it be done by temperature measurements at some particular latitude and height; or is it best recognized from a global radiance map and, if so, how would the effect be quantified ?

Labitzke: The maximum is defined as maximum temperature observed during the course of the warming at any location and any height. In most cases it will be at the stratopause. If radiance maps are available, they provide the best tool for defining the peak (i.e., maximum radiance value).

Schmerling: In searching for a physical mechanism which could explain the winter warmings, are we yet at the stage where we can measure atmospheric composition ? I am particularly struck by the NO variations at high latitudes (near 90 km) seen by Atmosphere Explorer C (AE-C), presumably resulting from particle fluxes. These changes in chemistry will undoubtedly result in temperature changes.

Labitzke: The NIMBUS G satellite which is due to be launched next year (1978) should go a long way towards producing a climatology of atmospheric composition in the stratosphere and lower mesosphere at least. The SAMS instrument will measure NO, N_2O, CO, CH_4, H_2O and temperature, while the LIMS instrument will measure O_3, NO_2, HNO_3, H_2O and temperature. Both of these instruments are infrared Limb sounders. There is also an instrument (BUV-TOMS) which will measure O_3 using ultraviolet backscatter.

STRATOSPHERIC AND MESOSPHERIC TURBULENCE

S. P. Zimmerman and E. A. Murphy

Air Force Geophysics Laboratory,
Hanscom AFB, Mass., U.S.A.

INTRODUCTION. One of the more intractable atmospheric problems is the measurement of turbulence in the upper stratosphere and mesosphere. Balloons and instrumented aircraft have partially investigated the lower stratosphere to 20 km (Kellogg 1956; Gifford 1957; Zimmerman 1965; Lilly et al, 1975; Zimmerman and Loving 1975). Radio meteor (Roper and Elford 1963; Roper 1966; Zimmerman 1973, 1974) have been utilized to determine some bulk turbulent viscosities and rates of dissipation in the altitude region 81 to 97 km. Chemical releases (Blamont and de Jager 1961; Zimmerman and Champion 1963; Justus 1969; Zimmerman and Trowbridge 1973) allowed us to determine turbulent diffusion rates and turbulent dissipation by both diffusion growth rates and turbulent spectra observed in the tracer chemicals.

The altitude region from the upper stratosphere to the mesopause, however, still remains basically unknown. Attempts at understanding the fine scale structure in this region (Zimmerman et al. 1972; Cunnold 1975) were marginal at best. The first method above was a Richardson number analysis of grenade data (Smith et al. 1961-1967) which demonstrated altitude regions where turbulence, based upon a Richardson criterion equal to or less than unity, would occur. By using the turbulent energy equation and an estimated length scale, the turbulent rates of dissipation were determined within the limits imposed by this assumed length scale. The second approach (Cunnold) was to apply turbulence index of refraction theory to the Jicamarca radar returns (Woodman and Guillen 1974) which presumably show a turbulent layer in the mesosphere. The values of turbulent diffusion determined, however, were based upon an assumed "average" electron

B. Grandal and J. A. Holtet (eds.), Dynamical and Chemical Coupling, 35-47.

density distribution. Even though the values arrived at were not unreasonable, both of these analyses suffer from assumptions which were not verifiable from simultaneous measurements.

THEORY AND RESULTS

It is the purpose of this research to arrive at a more quantitative value of turbulent parameters in this altitude region (30 to 90 km) by applying a different analytic tool to the same grenade data as previously used by Zimmerman et al.

The basic logic is still the determination of the Richardson number from the temperature and wind measurements.

$$R_i = \left[\frac{g}{T} \left(\frac{\partial T}{\partial Z} + \Gamma \right/ \left(\frac{\partial V}{\partial Z} \right)^2 \right) \right] \qquad (1)$$

where g is the acceleration of gravity

T is the kinetic temperature (K)

$\frac{\partial T}{\partial Z}$ is the vertical temperature gradient,

Γ is the adiabatic lapse rate (= 10 K/km), and

$\left(\frac{\partial V}{\partial Z}\right)^2$ is the square of the vertical shear of the horizontal winds.

Given these values of R_i, there is still the problem of uniquely identifying the dominant turbulent length scale or the turbulent energy density. Fortunately, there have been some studies, Deacon (1959) and Islitzer (unpublished, C. F. Lumley and Panofsky 1967), that have shown that the ratio of the vertical turbulent wind fluctuation to the horizontal wind amplitude follows a square root relation of the Richardson number, i.e.

$$\left(\frac{w}{V}\right) \sim \sqrt{R_i} \qquad (2)$$

$$\text{Specifically} \left(\frac{w}{V}\right) = \begin{Bmatrix} -.15(R_i \ .>0) \\ +.15(R_i \ \ <0) \end{Bmatrix} \sqrt{R_i} \quad + \quad .08$$

Thus given the Richardson number and the horizontal wind we may uniquely determine the amplitude of the vertical turbulent wind, w.

Since the vertical motions will be limited by buoyancy in a stably stratified medium, the scale lengths of the turbulent motions will be functions of the atmospheric stability and the rate of turbulent dissipation. Given the above dynamical parameters, the mean wind (V), the vertical turbulent intensity

$\langle w^2 \rangle$ and atmospheric stability; it is simple to estimate the unknown turbulent parameters.

The assumptions used in the following analysis is that once we are given the turbulent intensities w^2 (eq. 2) we assume the spectrum of the turbulent motions to be inertial and, as mentioned above, the vertical length scale will be limited by this buoyancy. Thus given inertiality and one dimensionality, we have

$$v_B{}^2 = \tfrac{1}{2}\int_{k_B}^{\infty} E(k)\,dk \qquad (3)$$

and $E(k) = \alpha\varepsilon^{2/3}k^{-5/3}$

where $v_B{}^2 \equiv \langle w^2 \rangle$ is the buoyant limited turbulent intensity

set identically equal to $\langle w^2 \rangle$ as given in (2), E (k) is the turbulent energy spectrum, α is the Kolmagoroff constant; for the vertical dimension it has the value ~ 1.5, k is the wavenumber, and ε is the rate of turbulent dissipation. Thus from (3) we derive

$$v_B{}^2 \cong \varepsilon^{2/3} k_B{}^{-2/3} \qquad (4)$$

It has been discussed by Phillips (1967), in a review of the Lumley-Shur theory of turbulence in a buoyant stably stratified medium, that the turbulent spectrum will be inertial in the high wavenumber region, and that the "buoyant" subrange will strictly be a function of atmospheric stability in the form of the Brunt Väissälä frequency (N) and wavenumber (k). As a consequence of this ε, which is conserved in the inertial region, is a function of wavenumber in the buoyant region. Thus the spectral energy is given by

$$E(k) = C_1 N^2 k^{-3} + \alpha\varepsilon^{2/3}k^{-5/3} \qquad (5)$$
$$= \alpha\varepsilon_o{}^{2/3}\left[1 + \left(\frac{k_B}{k}\right)^{4/3}\right]k^{-5/3}$$

where k_B is the transition wavenumber $k_B = \left(\dfrac{C_1 N^2}{\alpha\varepsilon_o{}^{2/3}}\right)^{3/4}$

between the inertial and buoyant subranges. In this article we shall presume that the vertical turbulent spectrum is "inertial" and limited in spectral extent to the transition wavenumber (k_B), and thus the vertical turbulent kinetic energy as given by (4) is

$$\langle w^2 \rangle = 1.5\left(\frac{\alpha^{3/2}\varepsilon_o}{C^{1/2} N}\right) \qquad (6)$$

For $\alpha_v \simeq 1.5$ and $C \to 1$ to 1.5, for simplicity say 1.5, we have

$$\langle w^2 \rangle = 3.4\ \varepsilon_o/N \qquad (7)$$

or

$$\varepsilon_o \cong 1/3\ \langle w^2 \rangle\ N$$

Heisenberg (1948) had proposed an effective diffusivity (K) in explaining the establishment of an inertial subrange (C. F. Hinze 1959) as

$$K = C^1 \int_{k_B}^{\infty} \left[\frac{E(k)}{k^3}\right]^{\frac{1}{2}} dk, \text{ where } C^1 \text{ is assumed unity.} \qquad (8)$$

This reduces to $K_B = \frac{1}{2}\ \frac{\langle w^2 \rangle}{N}$ (9)

It is interesting to note here that there is a factor of ½ introduced into the amplitude of the turbulent parameters ε_o and K via the definition of the turbulent spectrum (5). If instead we had used the rationale that the one dimensional spectrum is inertial to the buoyant limited scale, then (5) would read

$E(k) = \alpha\varepsilon^{2/3}k^{-5/3}$, and 7 and 9 would follow as

$\varepsilon = \frac{1}{1.5}\ \langle w^2 \rangle\ N$ and $K = \frac{\langle w^2 \rangle}{N}$, respectively.

DATA AND RESULTS

The data used in this analysis is that reported in NASA publications (Smith et al. 1961-1967) that derives the winds and temperatures from rocket grenade experiments for the years 1961-1967. The data, whose resolution varies from ∿ 3 to 6 km, is then spline fitted and interpolated at 1 km intervals, where the interpolation routine guarantees that there exists no more than one inflection point between data points. These data are then differentiated and Richardson numbers determined to a one km interval.

For this calculation we will examine the annual and the 6 months summer/winter averages of temperature, winds and Richardson number (eq. 1). Also the turbulent parameters will include (1) the occurrence rate of turbulence, based upon the occurrence of R_i less than or equal to ¼; (2) the rate of dissipation of turbulent kinetic energy (ε); (3) vertical turbulent intensity $\langle w^2 \rangle$; and (4) vertical turbulent diffusivity (K) within the limits set forth in the theory. To repeat, these limits are the minimal values of dissipation rate and diffusivity for the given intensity, since they are based upon a length scale which marks the termination of the inertial subrange and does not consider the contribution of the "buoyant" range to these parameters.

Given the interpolated experimental data, the Richardson number for each experiment is calculated. Then the results of wind, temperature, R_i, occurrence rates of $R_i \leq .25$, $\langle w^2 \rangle$ determined from Eq. 2, turbulent diffusivity and rate of turbulent dissipation are averaged over the number of experiments. In the averaging process the zeros of the occurrence rate, that we use when $R_i > .25$, are added into the total of these experiments. The data, as displayed, shows the standard deviation of the mean. For the cases where there is a single occurrence of turbulence (i.e., $R_i \leq .25$ for one case out of the total number of experiments) the standard deviation is meaningless and is not shown in the illustrations. Once we have determined the annual average of the data base we then examine the summer and winter averages to ascertain the seasonal differences.

RESULTS

The following figures (1-4) show the parameters arrived at as a result of the analyses that are divided into the four latitude points (70°N Pt. Barrow, 60°N Churchill, 38°N Wallops Island and 6°S Natal, Brazil). In each grouping we display the occurrence rate of turbulence defined by the fraction of occurrence of $R_i \leq .25$, the turbulent diffusivity (K), the turbulent intensity $\langle w^2 \rangle$, and the rate of turbulent dissipation (ε) for the average of the data base and for the winter and summer averages.

These analyses show many expected features, but also some facets that were not expected. First let us examine the occurrence rates of $R_i \leq .25$. Generally the results are almost exactly what we would expect; a low percentage occurrence of turbulence in the upper stratosphere and a gradual increase in an approximately exponential manner from the lower mesosphere upward to 90 km. This behavior is indicative of an ensemble of freely propagating internal gravity waves in an exponentially decreasing atmosphere. Thus we would expect to see breakdown more often at the higher altitudes where the waves would intensify and break. This reasoning quite obviously does not apply to the tropical data (Natal 6°S lat.) where we observe no Richardson numbers less than $\leq .25$ until $\sim$ 75 km. Then there is an extremely rapid increase with altitude to the upper limit of $\sim$ 90 km. This behavior is so different than that of the other three latitudes that we are led to the conclusion that the underlying mechanism for equatorial turbulence is not the internal gravity wave system, evident at the mid and northern latitudes, but the tidal oscillations discussed by Lindzen (1968). The altitude of onset of turbulence occurring at $\sim$ 75 km and the very rapid increase of occurrence rate with altitude resembles, albeit roughly, Lindzen's theoretical distribution of Richardson number based upon diurnal tidal theory. The other aspect not expected is mid and northern latitude winter

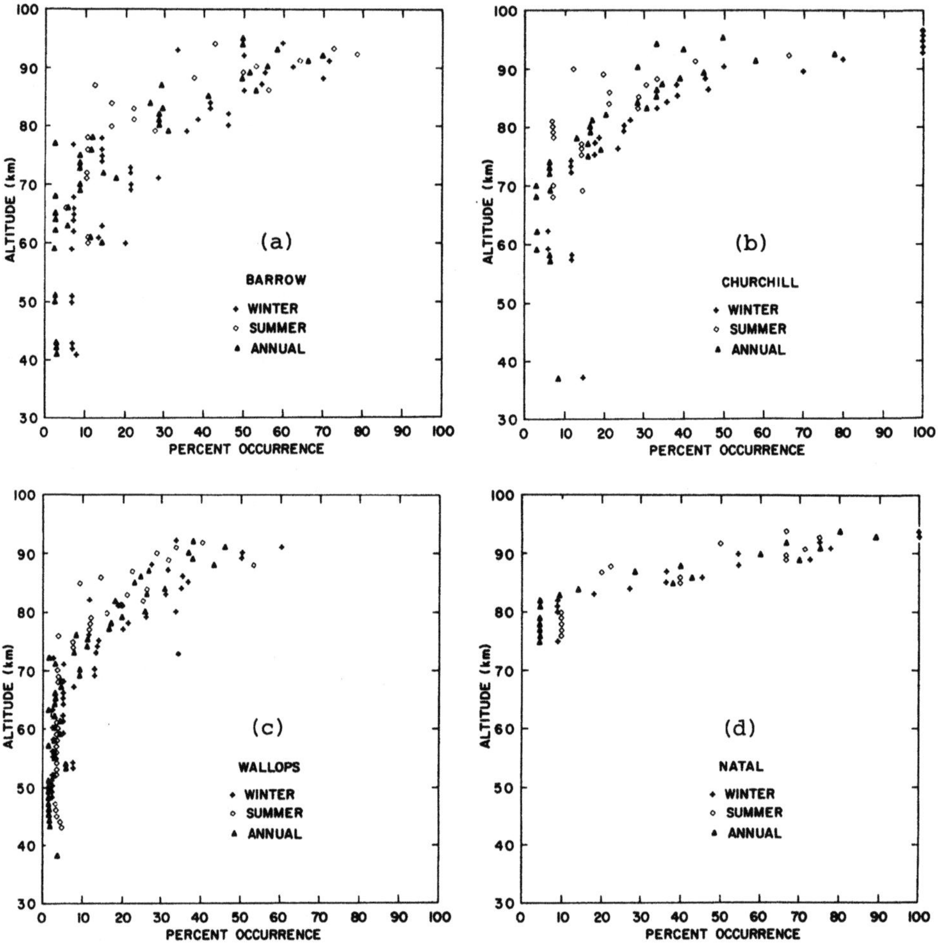

Figure 1

a) The turbulence occurrence rate for $R_i \leq .25$ at Point Barrow, Alaska (70°N Lat.) for the years 1965-1967.

b) The same as a) except at Churchill, Can. (60°N Lat.) and for the years 1962-1967.

c) The same as a) except at Wallops Island, Va. (38°N Lat.) and for the years 1961-1967.

d) The same as a) except at Natal, Brazil (5°S Lat.) and for the years 1966-1967.

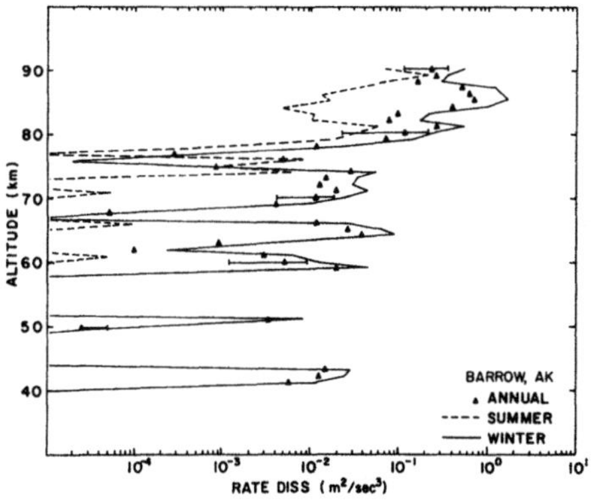

a

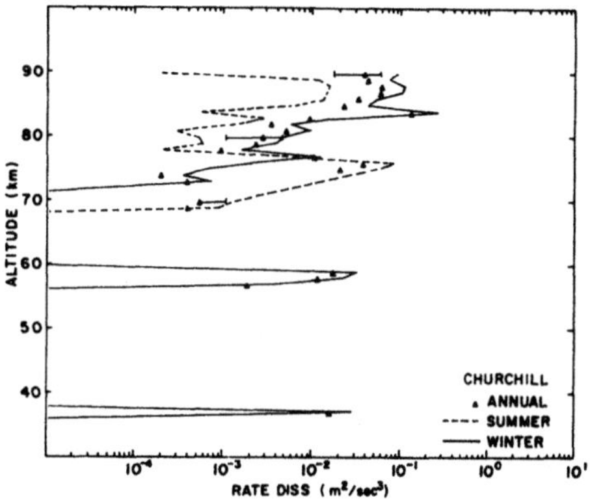

b

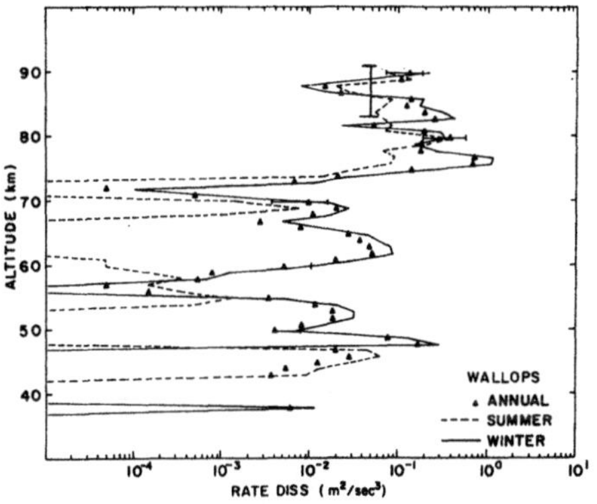

c

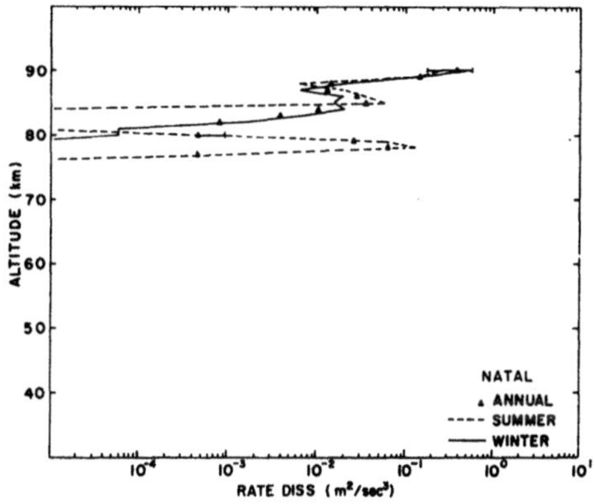

d

Figure 2

The minimum vertical turbulent diffusivity (K). The sections a,b,c and d are the same as in figure 1.

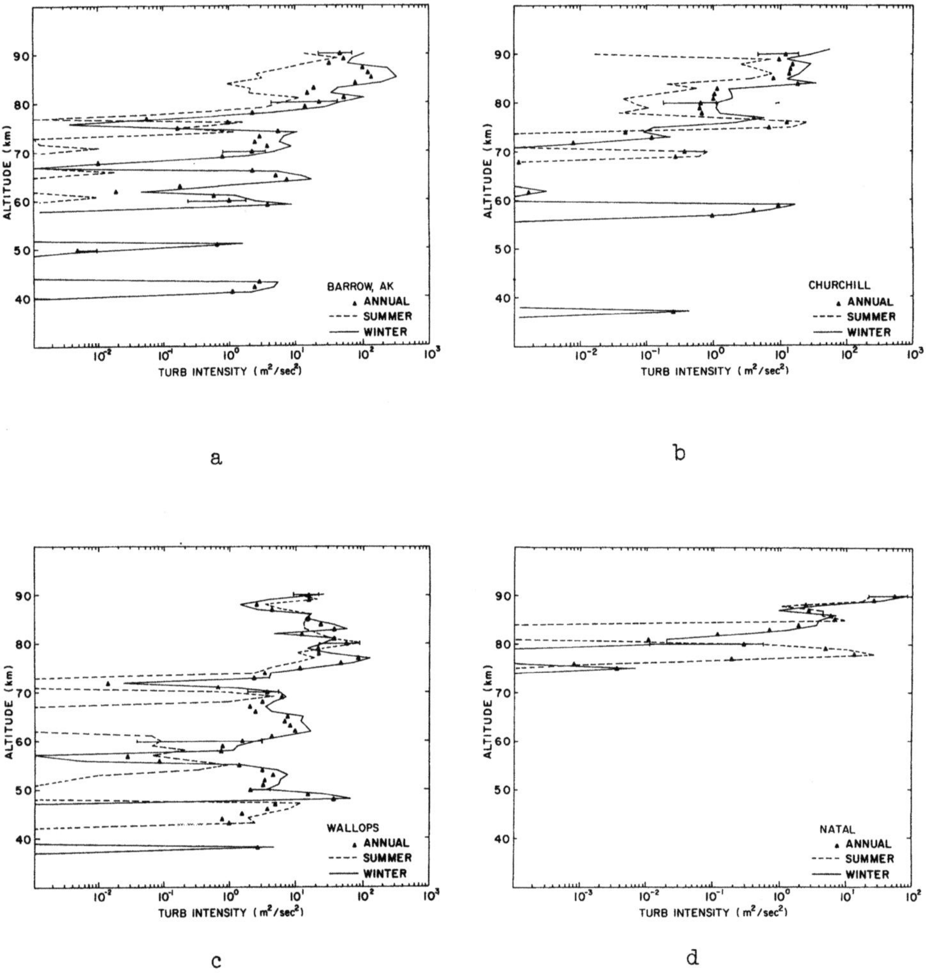

Figure 3

The turbulent intensity $\langle w^2 \rangle$. The sections a,b,c, and d are the same as in figure 1.

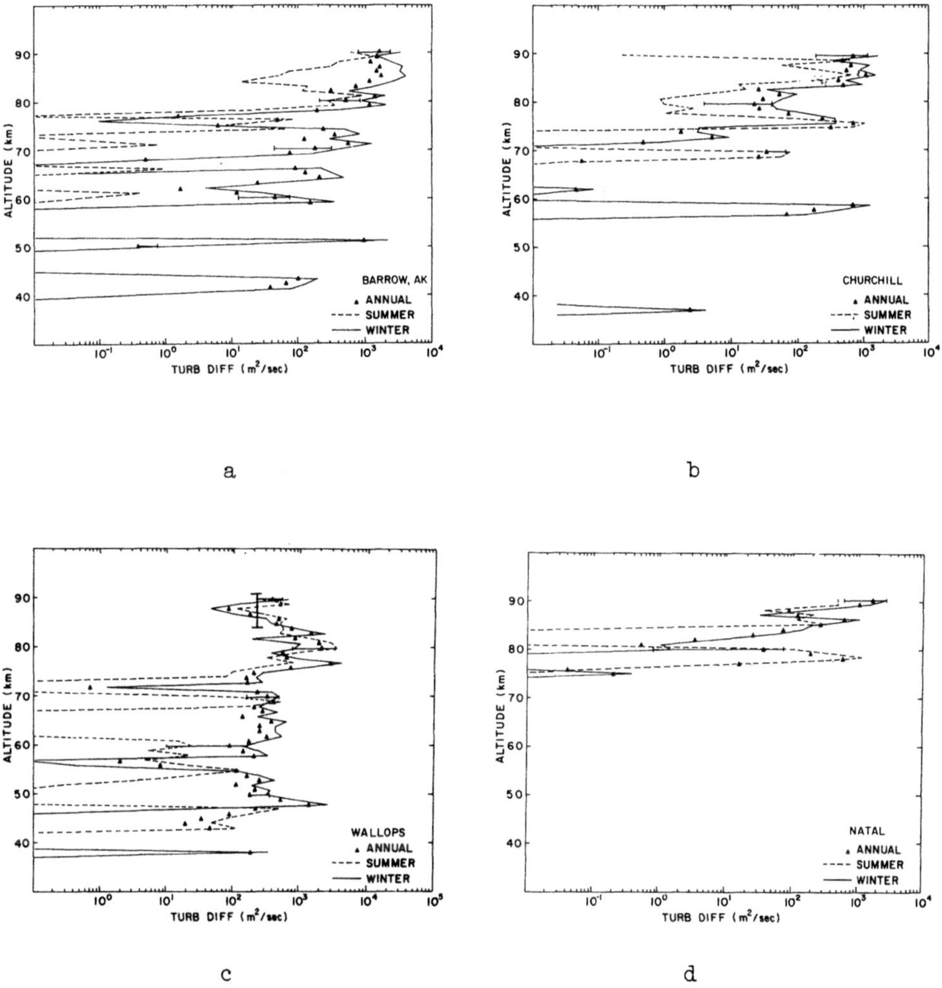

Figure 4

The rate of turbulent dissipation (ε). The sections a,b,c and d are the same as in figure 1.

phenomenon. In the figures of the winter occurrence rate we observe a ledge or peak, denoting an increased occurrence of turbulence at the stratopause region. This increase, while not huge is still significant, with an occurrence of $\sim$ 10% - 20%, while the summer data shows no such increase. As mentioned above the altitudes where these layers occur are around the stratopause being respectively 53 km at 38°N, 57 km at 60°N and 60 km at 70°N latitude. Comparison with the 1966 U. S. Standard Atmosphere Supplements does show some area of agreement. The stratopause being at 53 km for the 30°N winter model and 55 km for the 60°N winter model. However, the identification of turbulence via the Richardson number criterion is, in a stably stratified atmosphere, more dependent upon wind shear as the destabilizing mechanism than upon temperature gradient. Thus we have examined synoptic data (Groves 1971) for correlation and/or confirmation with our observations. Here the results are marginal, at best. Excellent correspondence with our findings is observed, however, in some data analysis performed by Cole (private communication 1977, Figure 1) our Figure 5. This shows cross-sections of scalar wind speed (m sec^{-1}) equal or exceeded 10 percent of the time between 30 and 65 km. Indicated also on the figure are the winter time turbulence occurrence of the layers at 38°, 60° and 70°N latitudes. The comparison is quite good in that it demonstrates that regions of high wind shear occur on the average 10% of the time at the latitudes and altitudes where instability measurements show similar occurrence rates.

The end results of the calculations are the determination of the following turbulent parameters (1) the rate of turbulent dissipation, (2) the minimum vertical turbulent diffusivity and (3) the turbulent intensity. These demonstrate marked variability as a function of latitude, season and altitude. The results are self-explanatory, but some discussion and comparison with other analysis would be appropriate. The altitude and latitude region where there is greatest comparison would be the mid-latitude 30° to 40° latitude and around the 80 to 90 km region. Thus we look at the Wallops Island data from 80 to 90 km and compare with the meteor wind analysis (Zimmerman 1973-1974) for the annual average turbulent rate of dissipation and diffusivity at 30°S latitude. This meteor data comprises the bulk diffusivity over the altitude region 83 to 91 km and is shown as the large bar in figures 2c and 4c. It is quite apparent that there appears to be good quantitative agreement between the meteor measurements and the average over the 80 to 90 km region in these data, even though the meteor data is southern hemisphere data.

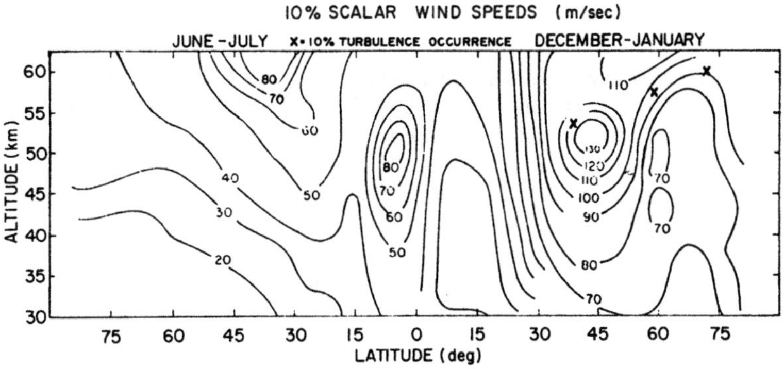

Figure 5

The scalar windspeeds that are equal or exceeded 10% of the time. The x's mark the altitude region where the dominantly winter occurrence rate appears as a ledge.

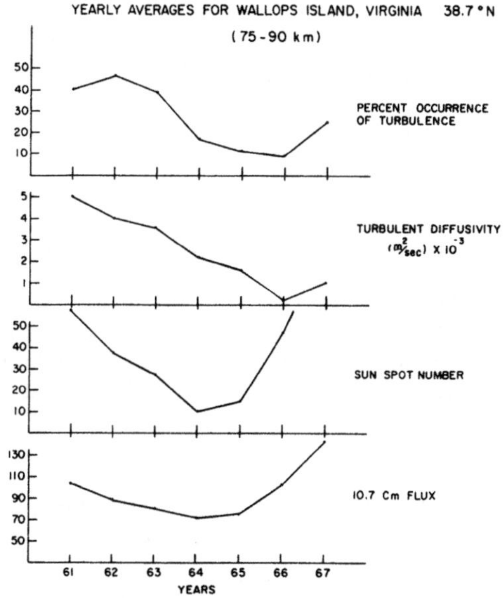

Figure 6

The Wallops Island yearly averaged occurrence rate and turbulent diffusivity summed over the altitude region 70 to 90 km and compared to the 10.7 cm flux and sunspot number.

SUMMARY AND CONCLUSIONS

These results show the determination of turbulent occurrence rates, vertical dissipation, momentum transfer coefficients (or diffusivity) and intensities from reported rocket grenade data at 4 latitudes (70°N, 60°N, 38°N and 6°S) and for the altitude region ∿ 30 to 90 km over the American continent. There is surprising agreement with estimates used in model aeronomic calculations (for example, Keneshea and Zimmerman 1970 and many others) and those deduced from mesospheric heat transfer requirements (Johnson and Wilkins 1965). Also there are regions of not so good agreement, particularly the large amplitude diffusivity in the middle mesosphere, mid-latitude results. A surprising feature is the parallel behavior of the 70 to 90 km yearly averaged occurrence rates and turbulent diffusivity with (Figure 6) the 10.7 cm flux. This suggests a strong solar control, most probably in the lower stratosphere or troposphere, but nonetheless manifested in the turbulent mesospheric motions. Indeed these turbulent motions may be a segment of the linkage observed between the lower and upper atmosphere in correlation studies.

The equatorial turbulence is considered to be a manifestation of only tidal wind breakdown (Lindzen 1968) as contrasted to the higher latitude results, in that the occurrence of $R_i \leq .25$ is absent below ∿ 65 km and appears to begin about 75 km. However, this result could also be due to the large vertical spacing of the data points. A more finely detailed result could conceivably show the existence of turbulence over scales small compared to the spacing between data points in this analyses. This same reasoning could hold in the mid-latitude data in the lower mesosphere-upper stratosphere, where there appears a null region of turbulence. More experiments are truly needed to provide more definitive results than presented here, if the fine scale structure transfer coefficients of the atmosphere are to be determined.

REFERENCES

Blamont, J. E. and deJager, C. 1961, Am. Geophys. 17, 134.
Cole, A. E. 1977, Private communication.
Cunnold, D. 1975, J. Atmos. Sci. 32, 2191.
Deacon, E. L. 1959, Intern. J. Air Pollution 2, 92.
Gifford, F. Jr. 1957, J. Meteor 14, 410.
Groves, G. V. 1971, AFCRL Envir. Res. Papers 368 #71-041.
Heisenberg, W. 1948, Z. Physik 124, 628.
Hinze, J. O. 1959, Turbulence, McGraw-Hill, New York.
Johnson, F. S. and Wilkins, E. M. 1965, J. Geophys. Res. 70, 4063.
Justus, C. G. 1969, J. Atmos. Sci. 26, 1137.
Kellogg, W. W. 1956, J. Meteor 13, 241.

Keneshea, T. J. and Zimmerman S. P. 1970, J. Atmos. Sci. 27, 831.
Lilly, D. K., Waco D. E. and Adelfang, D. I. 1975, The Natural Stratosphere of 1974, Dept. of Transportation CIAP Monograph 1.
Lindzen, R. S. 1968, Proc. Roy. Soc. A303, 299.
Lumley, J. L. and Panosky, H. A. 1964, The Structure of Atmospheric Turbulence, Interscience Publisher, New York.
Phillips, O. M. 1967, Atmospheric Turbulence and Radio Wave Propagation, Proc. Int. Coll., Moscow, 1965, Nauka, Moscow.
Roper, R. G. 1966, J. Geophys. Res. 71, 5785.
Roper, R. G. and Elford, W. G. 1963, Nature 197, 963.
Smith, W. S., Theon J. S., Swartz, D. C., Katchen, L. B. and Horvath, T. T. 1961-1967, NASA Tech. Rep, TR R-211,245, 263 and 288.
Woodman, R. and Guillen, A. 1974, J. Atmos. Sci. 31, 493.
Zimmerman, S. P. 1965, J. App. Meteor. 4, 279.
Zimmerman, S. P. 1973, J. Geophys. Res. 78, 3927.
Zimmerman, S. P. 1974, J. Geophys. Res. 79, 1095.
Zimmerman, S. P. and Champion, K. S. W. 1963, J. Geophys. Res. 12, 623.
Zimmerman, S. P. and Trowbridge, C. A. 1973, Space Res. XIII, 203.
Zimmerman, S. P., Faire, A. C. and Murphy, E. A. 1972, Space Res. XII, 615.

ACKNOWLEDGEMENT

The authors wish to take this opportunity to acknowledge the efforts and expertise of Ms. Muriel G. Hervey in the programming and analysis of the data.

DISCUSSION

Thrane: Trying to relate the occurrence of turbulence to the occurrence of partial reflections, I note that you see turbulence only part of the time, typically 10-20% near 60 km, whereas one has partial reflections at all times. Do your results indicate that there is no turbulence at all during certain periods ?

Zimmerman: My small percentage occurrence may be due to the gross dimensions between data points. However, as shown by Cole, shears that may cause destabilization over the N. American sites are present ~ 10-20% of the time. So, this is a fair confirmation of our analysis. If this layer of reflection is present all the time in your data, it may suggest a different mechanism to create the reflection. Alternatively it may suggest that turbulence is present all the time at your measuring site.

INERT GAS ABUNDANCES AS INDICATORS FOR THE STRENGTH OF EDDY DIFFUSION AT TURBOPAUSE ALTITUDES

U. von Zahn and T. Herwig

Physikalisches Institut der Universität Bonn, FRG

At the present time there is no possibility for calculating the strength of turbulence in the 100 km altitude region from first principles. We can, however, describe the naturally observed turbulence using the concept of eddy diffusion, in which the strength of turbulence is parameterized with a fudge factor called eddy diffusion coefficient K_e. K_e is related to the net particle flux Φ of the tracer gas in the (vertical) direction z through equation (1) (Hunten 1975):

$$\Phi = -K_e\, n_a \frac{dF}{dz} - D\, n_q \frac{dr}{dz} \qquad (1)$$

where n_a ambient density
n_q diffusive equilibrium solution
n tracer density
$F = n/n_a$
$r = n/n_q$
D molecular diffusion coefficient.

Even though the concept of eddy diffusion is a very simple one, there remains considerable uncertainty about the altitude dependence and the temporal as well as spatial variations in the eddy diffusion coefficient. Figure 1 shall exemplify this fact. We realize, that most authors put the eddy diffusion coefficient at turbopause heights in the range of 10^2 to 10^3 m^2/s. We also note, that there is hardly any agreement as to the altitude dependence of K_e about 100 km.

B. Grandal and J. A. Holtet (eds.), Dynamical and Chemical Coupling, 49-52.

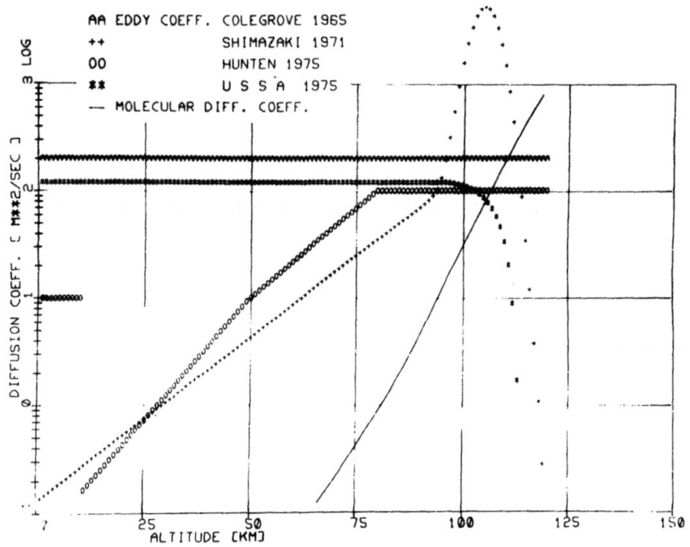

Fig. 1: Eddy diffusion coefficients (in m^2/s) vs. altitude from Colegrove et al. (1965), Shimazaki (1971), Hunten (1975), the U.S. Standard Atmosphere 1975, and the molecular diffusion coefficient Ar in N_2 from Banks and Kockarts (1973).

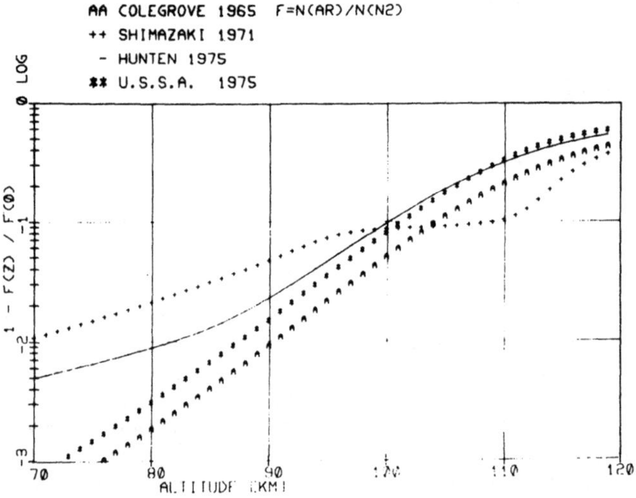

Fig. 2: Deviation of the $n(Ar)/n(N_2)$ ratio from its ground value calculated for different K_e-profiles. A value of 10^{-1} means a 10% deviation, 10^{-2} means a 1% deviation from the ratio at the ground level.

The main purpose of this paper is to emphasize that the various assumed K_e-profiles shown in figure 1 will lead to significantly different altitude profiles of the noble gas abundances at and even below the 100 km region.

Let us first consider the ratio of argon/nitrogen in the altitude region 70 to 120 km. We have calculated the altitude profiles of the number densities of nitrogen and the noble gases by a numerical integration using different K_e models and assuming $\Phi = 0$, as follows:

$$\frac{dn}{dz} = -n\left[\frac{1}{1+\Lambda}\left(\frac{1}{H}+\frac{\Lambda}{H_a}\right)+\left(1+\frac{\alpha_T}{1+\Lambda}\right)\frac{1}{T}\frac{dT}{dz}\right] \qquad (2)$$

where Λ = K_e/D
H tracer gas scale height
H_a ambient gas scale height
α_T thermal diffusion factor
T temperature

For that purpose we took the profiles of temperature T(z), gravitational acceleration g(z), and mean molecular mass m(z) from CIRA 1972. Values for D(z) and α_T were taken from Banks and Kockarts (1973). For Ar we have used either $\alpha_T = 0$ or $\alpha_T = 0.167$ which, however, did not produce any noticeable differences in the altitude region under study. Here we present results assuming $\alpha_T = 0$.

In figure 2 the ratio $n(Ar)/n(N_2)$ is shown using four K_e profiles referenced in figure 1. We note that at 100 km altitude all four models yield approximately the same Ar/N_2 ratio at 100 km altitude which, however, is depressed by 8% with respect to the ground value ratio.

We also see, that above and below 100 km the various assumptions for the eddy diffusion coefficient yield notably different ratios of Ar/N_2. The model of Shimazaki exhibits the lowest K_e values in the mesosphere yielding the largest diffusive separation effects below 100 km. Figure 3 shows you the calculated magnitude of diffusive separation of He, Ar, Kr for the K_e profile of Shimazaki. As the atomic weight of He and Kr is more different from the mean molecular weight of air than of Ar, their diffusive separation is stronger than in the argon case. At 100 km the deviation from the ground value is about 50% for He and 20% for Kr.

Although we do not want to go into experimental details here, we wish to mention, that our recently developed neutral mass spectrometers should be able to measure the Ar/N_2 ratio to a precision of about 1% down to around 85 km. Thus we conclude, that means have become available to distinguish between the various proposed K_e profile shapes by in situ composition measurements.

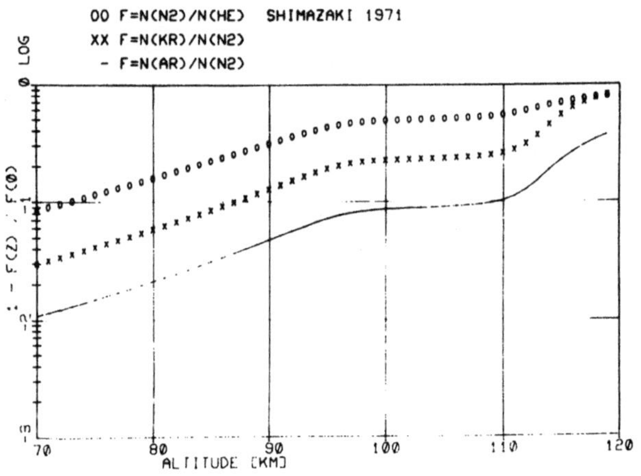

Fig. 3: Deviation from the ground value of the density ratios of different tracer gases to N_2, using the K_e-profile of Shimazaki.

Beyond the goal of establishing more firmly the gross shape of the mean K_e-profile we can see a more sophistidated application for high precision composition measurements at turbopause altitudes. Layers of strong turbulence have been frequently observed in this altitude region during rocket grenade experiments and with partial reflection sounders (see for example the paper by Zimmerman and Murphy).

These layers should be characterized also by a gradient of the ratio Ar/N_2 which is more shallow than directly above and below the turbulent layer. These compositional variations will depend on both the strength of turbulence in the layer and its life time. Thus precision measurements of local compositional variations may contribute to the investigations of small scale dynamic processes at turbopause altitudes.

REFERENCES

Banks, P.M., and Kockarts, G., 1973, Aeronomy, Part B. Academic press New York.

CIRA 1972, COSPAR International Reference Atmosphere 1972.

Colegrove, D.F., Hanson, W.B., and Johnson, F.S., 1965, J. Geophys. Res., 70, 4431-4441.

Hunten, D.M., 1975, "Vertical Transport in Atmosphere" in Atmospheres of the Earth and the Planets (ed.McCormac), Reidel Dordrecht.

Shimazaki, T., 1971, J. Atm. Terr. Phys., 33, 1383.

U.S.S.A. 1975, United States Standard Atmosphere 1975.

ATMOSPHERIC WAVES

M.W. Moncrieff

Atmospheric Physics Group, Department of Physics,
Imperial College, London.

1. INTRODUCTION

The basic aim of this paper will be to outline physical and dynamical principles with a minimum of mathematical complexity, to communicate theoretical aspects of the subject to a range of physical disciplines. The intention is not, of course, to deprecate the elegance of mathematical research on wave dynamics, which is necessary to provide a rigorous theoretical basis. An extensive bibliography exists on the mathematical aspects and this will be referred to in context.

It is useful to put the subject of this paper, which is mainly on gravity waves, into the context of work described in more detail by other speakers, namely tidal and acoustic waves. The atmospheric wave spectrum can be concisely summarised by reference to a dispersion diagram, in the form of a graph of horizontal phase speed against wavelength in isopleths of vertical wavelength, shown in Fig. 1. The mathematical derivations may be found in an unpublished set of notes by Green and Moncrieff. It is clear from this diagram that three distinct classes of atmospheric waves exist:

(a) Acoustic waves with high propagation speed and frequencies greater than $\sqrt{gB}$, the Brunt-Vaisala frequency.
(b) Gravity-inertial waves with frequency between $\sqrt{gB}$ and f, the Coriolos parameter. The short period waves oscillate with the Brunt-Vaisala frequency and there is a class of gravity waves of long vertical wavelength which are not greatly influenced by inertial effects.

B. Grandal and J. A. Holtet (eds.), Dynamical and Chemical Coupling, 53-62.

(c) Planetary scale waves of frequency less than f. These include the geostrophic Rossby waves, where the β-effect is dominant, and baroclinic waves.

Also included in this diagram are the so-called surface waves, which are distinguished by the fact that they do not transfer energy in the vertical direction. Their mechanics depend on the top and bottom boundary conditions and there are two distinct types: a Lamb wave of acoustic origin which can exist with lid conditions and free surface gravity waves. The three main wave types have completely distinctive dynamics and structure. Each may be independently eliminated from the equations of motion. Acoustic waves are eliminated by neglecting the local rate of change of density in the continuity equation (elastic compressibility). Gravity waves are eliminated by neglecting the rate of change of horizontal divergence in the balance equation, and Rossby waves by setting $\beta=0$. In terms of energy transport, acoustic waves are relatively unimportant in atmospheric dynamics. Geostrophic waves are important to the dynamics of the troposphere and stratosphere. Since both these regions are baroclinically unstable, transient waves are generated in situ. Long waves may propagate energy upwards. The hypothesis that waves generated in the lower atmosphere influence the upper atmosphere was made by Hines (1960) and in more detail by Eliassen and Palm (1960). No specific reference will be made to the upward propagation of geostrophic wave energy, although some of the physical principles of reflection, transmission and absorption also apply in a modified form to planetary waves. The gravity-inertial waves will now be considered in more detail. Low frequency gravity-inertial waves are important in the atmosphere since they represent the atmospheric tides.

2. THE ATMOSPHERIC TIDES

Tides will be treated in more detail by another speaker, but in the context of physical principles, some discussion of salient features is useful. Much of the basic mathematics of tidal motion was due to Wilkes (1949), and a paper by Green (1965) extended some of these principles, as well as giving useful dynamical interpretations. Since the tides are of planetary scale, it is of some advantage to allow for spherical geometry and the relevant mathematics defines a series of orthogonal (Hough) functions which are the natural eigenfunctions for the problem. The related eigenvalues define a characteristic depth - either a vertical wavenumber if the motion is oscillatory, or a scale height if it is exponential. For constant values of the Coriolis parameter, this characteristic depth can be represented in closed form by

$$h_{sn} = \frac{w^2a^2}{g} \frac{(s^2-s_o{}^2)}{n(n+1)} \tag{1}$$

where n and s are integers, $\omega = 2\pi/(1 \text{ day})$, $s_o = f/w$, where f is the Coriolis parameter and a = Earth radius. The diurnal tide is given by $s=1$ and the semi-diurnal tide by $s=2$. It can be shown that $s_o{}^2 = 0.95$ for the diurnal and $s_o{}^2 \simeq 2$ for the semi-diurnal tide. If H_o is the density scale height ($\simeq 7$ km), the phase speed (c) of long gravity waves is given by:

$$\frac{gB}{c^2} = m^2 + \frac{1}{4H_o{}^2} \tag{2}$$

where m is the vertical wavenumber. It should be stressed that for waves in general (acoustic, gravity and Rossby), the gross variation of density with height appears in conjunction with the vertical wavenumber as $m^2 + 1/4\, H_o{}^2$, and it is implied that the flow perturbations (but not the phase speed) increase with height proportional to $\rho^{-\frac{1}{2}}$. This maintains the specific kinetic energy approximately constant. This height amplification factor may be large for gravity waves, since the density changes by some 14 density scale factors between the surface and 100 km. A perturbation originating in the troposphere and propagating to 100 km, would be amplified by a factor of $\sim 10^3$. Apart from this consideration, the atmosphere can be treated as virtually incompressible for wave motion provided Eq.(2) is used. By equating the phase speed c in Eq.(2) to the speed of the forcing function $(g\, h_{sn})^{\frac{1}{2}}$ from Eq.(1), Green deduced the results shown in Table 1. The important conclusion is that the principal diurnal wave (1,1) has a small vertical wavelength (4 km) whereas the principal semi-diurnal wave (2,2) has a large vertical wavelength (70 km). Thus the diurnal wave probably responds to local forcing whereas the semi-diurnal wave responds to an integrated effect over a deep layer.

TABLE 1

s	n	h_{sn}(km)	$L_z=2\pi/m_{sn}$
1	1	0.63	4.0
1	2	0.24	2.5
2	2	7.85	70.0
2	3	3.66	13.0
2	4	2.12	8.5

There are two possible forcing functions for tidal waves, gravitational forces and the diurnal variation of heating. If Ω is the gravitational potential then under equilibrium conditions a pressure deviation of

$$\delta p = -\rho\Omega \tag{3}$$

is generated. A temperature deviation ΔT over Δz, will, by hydrostatic considerations, generate

$$\delta p = -\rho g \Delta z \, \Delta T/T \tag{4}$$

On comparing Eqs.(3),(4), since $\Omega \simeq 2\ m^2s^{-2}$, the effect of solar gravitational potential is equivalent to a temperature change of $10^{-2}\ ^\circ K$ over 4 km. Therefore, since the net diurnal temperature changes by water-vapour absorption in the troposphere, ozone heating between 40–55 km and molecular oxygen above 100 km are respectively, $0.5^\circ K$, $2^\circ K$ and $5^\circ K$, evidently the gravitational forcing is negligible by comparison to heating.

(a) The semi-diurnal tide

Since the height scale of this heating is of the order of 10 km and the vertical scale of the principal component of the semi-diurnal tide is 70 km, the distribution of heating with height is likely to be unimportant. By making the approximation of an infinite equivalent depth, total heating energy Q concentrated within a small distance ε of the level $z=z_*$, assuming zero vertical velocity at the earth's surface $z=0$, and an exponential boundary condition as $z\to\infty$, Green obtained a solution for the pressure deviation:

$$\frac{\delta p}{\rho} = \begin{cases} \frac{1}{2}(z-4H_0)e^{\frac{1}{2}(z-z_*)/H_0} \displaystyle\int_0^\infty Q\,dz & z < z_* \\ \frac{1}{2}(z-2H_0)e^{\frac{1}{2}(z-z_*)/H_0} \displaystyle\int_0^\infty Q\,dz & z > z_* \end{cases} \tag{5}$$

The main features of this solution are that the velocities (proportional to $\delta p/\rho$) increase exponentially upwards, are proportional to the total heating and have a change of phase at $z = 4H_0$ ($\simeq 27$ km) which is independent of the heating. This solution compares well with the more complete solution of Butler and Small (1963). By superposition of the above heating regimes it follows that ozone heating is the dominant contributor to the diurnal change in surface pressure with a $\simeq 25\%$ contribution from tropospheric water-vapour heating and a negligible contribution from atomic oxygen. Ozone heating also appears to dominate at levels up to 120 km.

(b) The diurnal tide

Tidal theory for small values of equivalent depth (h) was treated by Wilkes (1949) and in a more general form by Green (1965). It was shown that for small values of equivalent depth,

$$\frac{\delta p}{\rho} = gh^2\left(\frac{\partial Q}{\partial z} - \frac{Q}{H_o}\right) \qquad (6)$$

The motion generated in each heating zone is of interest. Green showed that heating rate in the lower troposphere that gives a 0.5°K temperature wave produces a pressure deviation of about 0.3 mb at the surface and an airspeed amplitude of $\simeq 30$ cms^{-1}. Diurnal heating in the ozone layer is equivalent to a temperature amplitude of 2°K. Since the induced wind speeds are proportional to $\delta p/\rho$ and by Eq.(6) proportional to the heating rate, it follows that these will only be a factor of four greater than the velocities effected by tropospheric heating i.e. around 1 ms^{-1} in the ozone layer; at least an order of magnitude smaller than those observed. Likewise, the local heating around 90 km due to atomic oxygen will generate around 3 ms^{-1} rather than the observed 30 ms^{-1}. The important conclusion is that direct local heating does not generate the observed diurnal wind oscillations. It was suggested by Green (1965) that the observed structure may be explained in terms of an upward propagation of a free wave generated by tropospheric heating. The large amplitude at higher levels is likely to be a feature of the $\rho^{-\frac{1}{2}}$ amplification effect.

3. GRAVITY WAVES

For gravity waves of periods up to (say) 200 min, the earth's rotation is relatively unimportant. Gravity waves frequently occur in the atmosphere, but their associated heat and momentum transfer is complicated. This is the subject of much detailed theoretical research e.g. Scorer (1949), Eliassen and Palm (1960), Bretherton (1966), Booker and Bretherton (1967).
Stationary gravity waves generated by mountains extend into the upper atmosphere, where they are occasionally observed as mother-of-pearl clouds around 25 km. The distortion of meteor trails (Greenhow and Neufeld (1960)) and rocket soundings indicate that between $80-100$ km oscillation of the order of 10 km vertical wavelength and 100 km horizontal wavelength and amplitudes of some 30 ms^{-1} exist. Hines (1960) interpreted these oscillations as gravity waves of periods $10-200$ minutes. The dispersion relation for neutral gravity waves in the presence of a zonal current U is

$$(U-\sigma/K)^2 = \frac{gB}{K^2 + m^2 + \frac{1}{4H_o^2}} \qquad (7)$$

where $K = 2\pi/L_x$ and $m = 2\pi/L_z$ are the horizontal and vertical wavenumbers respectively.

An important feature distinguishing internal gravity waves from say surface waves is their ability to transport energy and momentum in the vertical. Also the vertical component of the group velocity $\partial\sigma/\partial m$ is in the opposite sense from the vertical component of phase velocity σ/m, a condition readily obtained by differentiating Eq.(7). Thus in order to transport energy upwards, the vertical component of the phase speed must be downwards. A related effect is the ability of waves to transfer momentum in the vertical, a process determined by the phase variation with height. This is an example of how momentum can be transported without mixing of fluid particles being involved.

Taking stationary waves as an example ($\sigma/K = 0$) the characteristic depth of gravity waves is given by the formula

$$m^2 = \frac{gB}{U^2} - \frac{1}{4H_o^2} - K^2 \qquad (8)$$

since $gB/U^2 \sim 10^{-6}\ \mathrm{m}^{-2}$ and $1/4H_o^2 \sim 10^{-8}\ \mathrm{m}^{-2}$, $m^2 \simeq gB/U^2 - K^2$. The vertical variation which satisfies the equations of motion is of the form $e^{\pm imz}$, and clearly these can only be wavelike if m is real, that is $K^2 < gB/U^2$, thus it follows that waves of horizontal wavelength greater than ~ 10 km have oscillatory variation in the vertical, whereas the shorter waves are of exponential form. Thus, in constant wind and constant static stability, there is a broad spectrum of stationary waves which are capable of vertical propagation. An extra physical consideration is however necessary. The general solution to the simplest stationary gravity wave problem are those of the form $\mathrm{Re}\{A \exp i(Kx+mz) + B \exp i(Kx-mz)\}$ where A and B are constants. The distinction between these two components is in the vertical phase. Since the vertical component of group velocity of gravity waves is in the opposite sense to the vertical component of phase speed, if only upward energy propagation is physically acceptable, it follows that the solutions must be of the form $\exp i(Kx+mz)$. That is, the phase of the waves slopes into the flow with height, consistent with a downward transport of horizontal momentum.

If either the stratification or the wind speed varies with height then additional effects exist. These are associated with vertical variation of the reflection, transmission and absorption of wave energy. In general, this is a complicated theoretical problem and only the simplest examples will be considered here. Further information may be obtained by referring to Booker and Bretherton (1967), Bretherton (1966), Scorer (1949), Hines (1960).

(a) The variation of static stability

The basic effect of the variation of static stability with height may be obtained by considering a two layer model. First, suppose that a layer of large static stability underlies a layer of smaller static stability with U constant throughout. If $K^2 < gB_1/U^2$ then vertically propagating waves can exist in the lower regions, but if $K^2 > gB_2/U^2$ then the solutions vary exponentially in the upper layer. If the upper layer is sufficiently deep then there will be total reflection. If the lower layer is of depth H then from Scorer (1949) it is possible to have resonance if $gB_2/U^2 < K^2 < gB_1/U^2 - \pi^2/4H^2$. Shorter waves are refracted towards the horizontal.

Second, suppose that the layer of larger static stability is uppermost. Then $K^2 + m_2{}^2 = gB_2/U^2 > gB_1/U^2 = K^2 + m_1{}^2$, so both $K^2 + m_2{}^2$ and $m_2{}^2$ are larger in the upper layer. The wave fronts in the upper layer are therefore more nearly horizontal. The waves are refracted upwards and some energy is reflected back down.

Continuity of displacement and pressure must be satisfied at the interface. If $\psi_2 = \mathrm{Re}\, A\, e^{\pm i(Kx+m_2z)}$ giving upward energy propagation in the upper layer, then it can be shown that in the lower layer the streamfunction is

$$\psi_1 = \mathrm{Re}\ \{A\left[\frac{m_1+m_2}{2m_1}\exp i(Kx+m_1z) - \frac{m_2-m_1}{2m_1}\exp i(Kx-m_1z)\right]\}$$

which is the sum of an incident and a reflected component. Thus for any incident wave, there is a nonzero transmitted wave. Some energy escapes and resonance is not possible.

When the wind is sheared in the vertical but $U \neq 0$ anywhere then the analysis is analogous except that gB/U^2 is replaced by $gB/U^2 - 1/U\ d^2U/dz^2$. The second term is normally substantially smaller than the first. For waves travelling at speed c the above analysis follows through except that U is replaced by $(U-c)$.

(b) Critical levels

If there exists a critical level where the horizontal phase speed of the waves coincides with the flow speed at that level ($U=c$ or in the case of stationary waves $U=0$) then the dynamics is drastically modified. Such a level is referred to as a critical level. A considerable amount of research, mostly of a mathematical nature, exists on this aspect i.e. Bretherton (1966), Booker and Bretherton (1967). From Eq.(7) we see that as the fluid speed U approaches the horizontal phase speed c, the vertical wavenumber (m) becomes large. The wavefronts are therefore nearly horizontal and closer together. Moreover, the group velocity relative to the flow at that level approaches zero and wave energy is absorbed by the flow. Booker and Bretherton show that when the gradient Richardson Number Ri = $(g/\theta)(\partial\theta/\partial z)/(\partial U/\partial z)^2$ exceeds units in a shear layer then most of the energy is absorbed. Wave momentum is given up to the mean flow and this may cause instability and local turbulence.

Concentration of energy on a diminishing scale contributes to non-linearity, analogous to a surface gravity wave approaching a shelving beach. Critical level absorption is physically important since it acts to absorb energy and consequently removes parts of the wave spectrum, which would otherwise propagate upwards to high levels. It is likely that critical level absorption is important in explaining the existence of gaps in the gravity wave spectrum in the high atmosphere.

(c) Shearing instability

The energetics of neutral gravity waves is fundamentally different from that of unstable waves. Neutral gravity waves are generated by external forcing and there is an oscillatory exchange between potential and kinetic energy produced by the pressure field. The interchange is affected only a little by vertical shear. Dynamically unstable waves rely on the vertical shear of the mean flow, from which energy is extracted to supply the energy of the amplifying perturbations.

In the previous discussion, the frequencies of the oscillations were real. However, when the gradient Richardson number Ri is less than $1/4$, it is likely for the frequency to have an imaginary part. In physical terms, this means that the waves amplify exponentially, instead of being neutral as they are when Ri $> 1/4$. There is extensive literature on this subject. After preliminary work by Rayleigh (1879), Goldstein (1931), Drazin (1958), Miles (1963), and others showed that when the flow is unstratified above and below a shear layer, waves of horizontal wavelength long by comparison with the depth of the shear layer are unstable; the short waves behave very much like neutral gravity waves. An important aspect is that there exists a wavelength of maximum growth rate, that is a selection principle exists. It was also shown that an additional condition for instability exists, namely there must exist a point of inflection in the velocity profile ($\partial^2 U/\partial z^2 = 0$ somewhere) as well as that Ri $< 1/4$. Stratification above or below the shear layer is a stabilising influence since wave energy can then be radiated away from the shear layer by neutral gravity waves.

There probably exists a close relationship between critical level absorption and shearing instability in stratified flow. A satisfactory theory explaining this aspect does not, however, yet exist. Shearing instability is an important process for producing areas of local turbulence and transfer, in which the kinetic energy of the mean flow is dissipated as wave energy, a process distinct in effect from critical level absorption.

4. CONCLUSIONS

Some of the more important properties of atmospheric wave transfer have been outlined in a simplified manner, with reference to more complex analyses. Both large and small scale gravity waves are probably important as transfer processes in the upper atmosphere in contrast to the troposphere, their effects are not dominated by the presence of large-scale baroclinic instability and small-scale convective processes. It is by no means clear at present how wave theories fit into the feedback system in order to explain the observed structure of the upper atmosphere. It may indeed be that the situation is even more complicated than the dynamics of the troposphere, where baroclinic instability of the zonal flow and its secondary effects dominate the global scale circulation, thus acting as a controlling influence on the smaller scale.

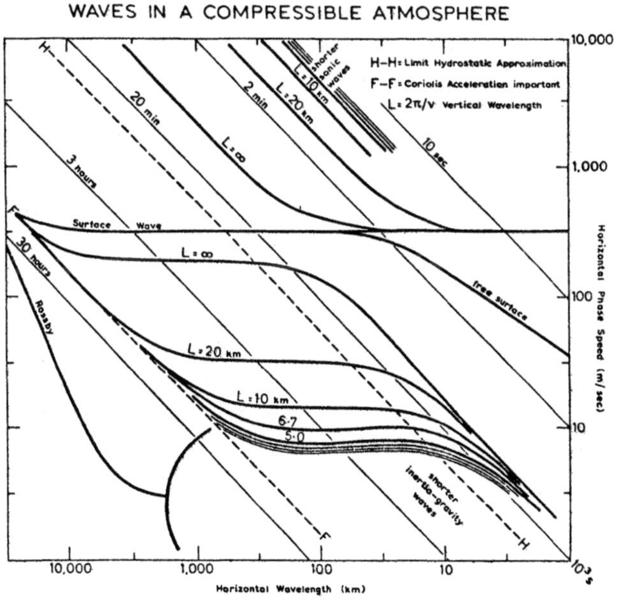

Figure 1

REFERENCES

Booker, J.R. and Bretherton, F.P., 1967, J.F.M., 27, 513.
Bretherton, F.P., 1966, Quart.J.R. Met.Soc., 92, 466.
Butler, S.T. and Small, K.A., 1963, Proc.Roy.Soc., A274, 91.
Chapman, S. and Lindzen, R.S., D. Reidel Publishing Company.
Drazin, P.G., 1958, J.F.M., 4, 214.
Eliassen, A. and Palm, E., 1960, Geof. Publ., 22, 3.
Goldstein, S., 1931, Proc. Roy. Soc., A132, 524.
Green, J.S.A., 1970, Phys.Education, 5, 37.
Green, J.S.A., 1965, Proc. Roy. Soc., A288, 564.
Green, J.S.A. and Moncrieff, M.W., 1970, unpublished notes.
Greenhow, J.S. and Neufeld, E.L., 1960, Proc.Phys.Soc., 75, 228.
Hines, C.O., 1960, Canad.J.Phys., 38, 144.
Miles, J.W., 1963, J.F.M., 16, 209.
Rayleigh, Lord, 1879, Proc.Lond.Math.Soc.
Scorer,R.S., 1949, Quart.J.R.Met.Soc., 87, 24.
Siebart, M., 1961, Adv. Geophys., 7, 105.
Wilkes, M.V., 1949, Cambridge University Press.

DISCUSSION

Rastogi: What is the order of vertical scale associated with gravity waves which do not propagate vertically (exponentially damped) ?

Moncrieff: For a given Brunt-Vaisala frequency, the vertical scale height of exponentially decaying gravity waves is a strongly varying function of frequency and horizontal wavelength. If $L_x \sim 100$ km, and period ~ 30 min, then the vertical height scale is some 50 km. However, no wave energy or momentum can be transported vertically.

Rastogi: I would like to point out that of the many different gravity waves with associated wave number K and frequency σ, wind shear would cause only one to become critical at any given level.

Moncrieff: Since the horizontal phase speed of gravity waves depends on m and K, the critical level is not unique. However, gravity waves with $L_x \sim 100$ km are virtually non-dispersive in the horizontal. Note that waves of large vertical wavelength do not have a critical level since the horizontal phase speed is too large. For critical level analysis of wave-packets refer to Bretherton (1966).

INFRASONIC WAVES - INTERACTION WITH THE ATMOSPHERE

L. Liszka

National Board of Occupational Safety and Health, Umeå
and Kiruna Geophysical Institute, Kiruna, Sweden

The purpose of the present paper is to review known facts about common types of high intensity infrasonic sources and discuss their possible effects on the atmosphere. The discussion will be limited to frequencies above 0,01 Hz. Infrasonic waves with frequencies below 4 Hz, if launched at sufficiently large angle from the vertical direction will be reflected and will propagate to large distances in the waveguide created by the reflecting layer and the earth surface, (see Fig. 1). Waves propagated in directions close to the vertical direction will be dissipated in the atmosphere at heights where the wave frequency is close to the collision frequency. The wave energy will there be converted into heat and will thus rise the local temperature. At distances large in comparison with the height of the waveguide, the sound intensity on the ground will be proportional to 1/R (R-distance to the source).

B. Grandal and J. A. Holtet (eds.), Dynamical and Chemical Coupling, 63-71.

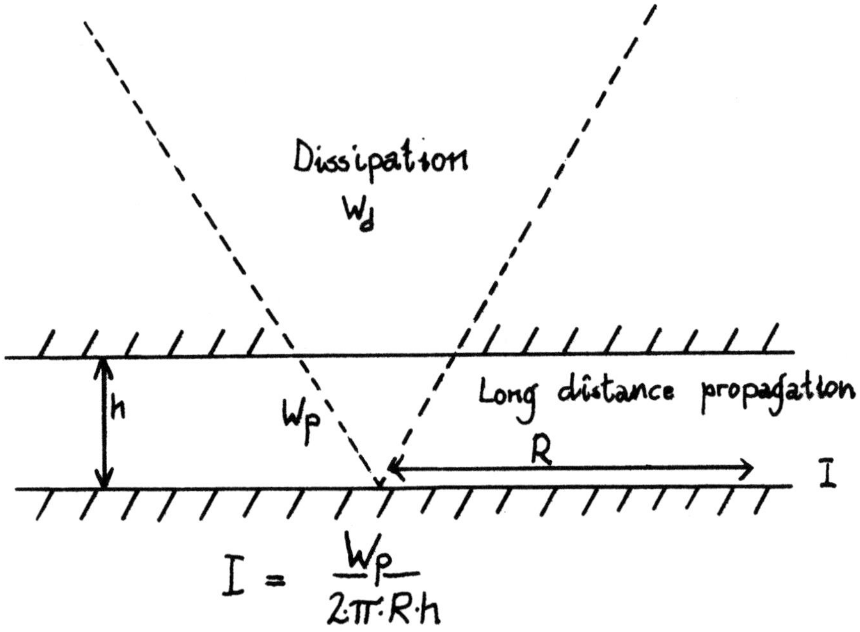

Fig. 1

When the amplitude of infrasonic waves is large (that is when particle velocity in the wave is large in comparison with the sound velocity) the atmosphere will be influenced by the waves along the large portions of the propagation path.
In particular, when infrasound propagates upwards, its amplitude grows and the wave will convert into an N-shaped wave after $0{,}2\ (\frac{c}{u})\ \lambda$ (cf eg Morse and Ingard, 1968) where c is sound velocity, u is particle velocity in the wave with the wavelength λ.
In each compression of the gas due to the wave there will be a temperature increase θ:

$$\theta = \frac{u}{c}\,(\gamma - 1)\ T_o \tag{1}$$

where γ is the specific heat ratio and To is the ambient temperature. Each particle in the atmosphere will there experience a maximum displacement ξ given by:

$$\xi = \frac{u}{2\pi f} \tag{2}$$

where f is the frequency of the wave.
For a plane infrasonic wave the sound intensity I is constant. As

$$I = \rho c \, |u|^2 \tag{3}$$

the particle velocity will increase with height as $\rho^{-1/2}$.
In an N-shaped wave travelling upwards the energy is steadily shifted to high frequency components of the wave spectrum. These high frequency components can propagate only up to heights where the collision frequency in the atmosphere is equal to frequency of the particular wave component. The high intensity infrasonic wave will thus heat the atmosphere at much lower heights than a weak wave.

Atmospheric processes may be influenced by propagation of infrasonic waves due to density fluctuations, particle velocity and displacement field imposed on the atmosphere by the waves.
In the following sections above wave parameters for infrasound produced by more important types of sources are discussed.

1. AURORAL INFRASONIC WAVES

Auroral infrasonic waves (AIW) were recorded in northern Scandinavia both at 2 Hz and at lower frequencies (0,1 - 0,01 Hz). A review of results may be found in a recent publication (Westin, 1976). Infrasonic signals are observed when auroral arcs are moving southward with supersonic velocities. The hitherto most complete theory for the generation of AIW by supersonic motion of an auroral electrojet has been developed by Swift (1973).

Here, possible effects of AIW on the lower atmosphere will be discussed. Typical AIW have at the ground level amplitudes of about 0,6 N/m^2 (peak-to-peak value) which corresponds to a sound pressure level (rm s value) of 81 dB re 20 $\mu N/m^2$ and to a sound intensity of 1,1 x 10^{-4} W/m^2. The frequency is assumed to be of the order of 0,05 Hz. As the source may be, at large distance from it, considered as having cylindrical symmetry, it will be assumed that the intensity is proportional to $\frac{1}{R}$. The wave intensity will thus be larger closer to the auroral arc. Calculations of different quantities in the atmosphere have been made for above typical AIW parameters. The particle velocity has been calculated from expression (3) using the characteristic acoustic impedance of the atmosphere $Z = \rho c$ obtained from a standard atmosphere.
Table I shows results of these calculations: particle velocity u, particle displacement ξ, pressure deviation in the wave p, relative pressure deviation and the temperature increase θ due to

each pressure maximum of the wave are shown as a function of height.

Table I

H (km)	u_{rms} (m/s)	rms (m)	P (N/m^2)	P/Po	θ (^oK)
10	$1,04x10^{-3}$	$3,31x10^{-3}$	$1,29x10^{-1}$	$4,88x10^{-6}$	$3,10x10^{-4}$
20	$2,55x10^{-3}$	$8,12x10^{-3}$	$6,69x10^{-2}$	$1,21x10^{-5}$	$7,51x10^{-4}$
30	$6,34x10^{-3}$	$2,02x10^{-2}$	$3,51x10^{-2}$	$2,94x10^{-5}$	$1,91x10^{-3}$
40	$1,55x10^{-2}$	$4,92x10^{-2}$	$1,96x10^{-2}$	$6,83x10^{-5}$	$4,89x10^{-3}$
50	$3,59x10^{-2}$	$1,14x10^{-1}$	$1,22x10^{-2}$	$1,52x10^{-4}$	$1,18x10^{-2}$
60	$8,34x10^{-2}$	$2,66x10^{-1}$	$8,18x10^{-3}$	$3,72x10^{-4}$	$2,66x10^{-2}$
70	$2,16x10^{-1}$	$6,88x10^{-1}$	$5,61x10^{-3}$	$1,02x10^{-3}$	$6,40x10^{-2}$
80	$7,12x10^{-1}$	2,27	$3,83x10^{-3}$	$3,68x10^{-3}$	$1,92x10^{-1}$
90	1,78	5,66	$1,52x10^{-3}$	$9,49x10^{-3}$	$4,79x10^{-1}$

It may be seen that the particle velocity and the displacement reach considerable values first at heights above 80 km. At lower stratosphere the disturbance due to AIW will be very small. It is of considerable interest to know whether the magnitude of the disturbance is large enough in order to influence the water vapour condensation in the atmosphere and thus the cloud formation. It is known that it is the case for internal gravity waves and mountain waves.

2. INFRASOUND PRODUCED BY SEVERE WEATHER

During the last 15 years a number of observations of low frequency infrasound has been reported (cf eg Bowman and Bedard, 1971). Periods of these infrasonic waves are distributed between 10 and 80 seconds. Reported amplitudes indicate that the total acoustic power radiated by a severe weather is of the order of 1-10 MW. Although there is no information about the directivity of such a source it is plausible to assume that the amount of energy propagated upwards and dissipated in the upper atmosphere is of the same order of magnitude as that propagated in the atmospheric waveguide.

At higher frequencies, for example at 2 Hz, where observations were carried on in Northern Sweden since 1971, it has been found that the radiated acoustic energy is considerably smaller. Since 1971 about 20 events were observed, where the source could be traced for at least 6 hours and which was identified as a low pressure area with adjacent front system. An example of such low pressure area is shown in Fig. 2.

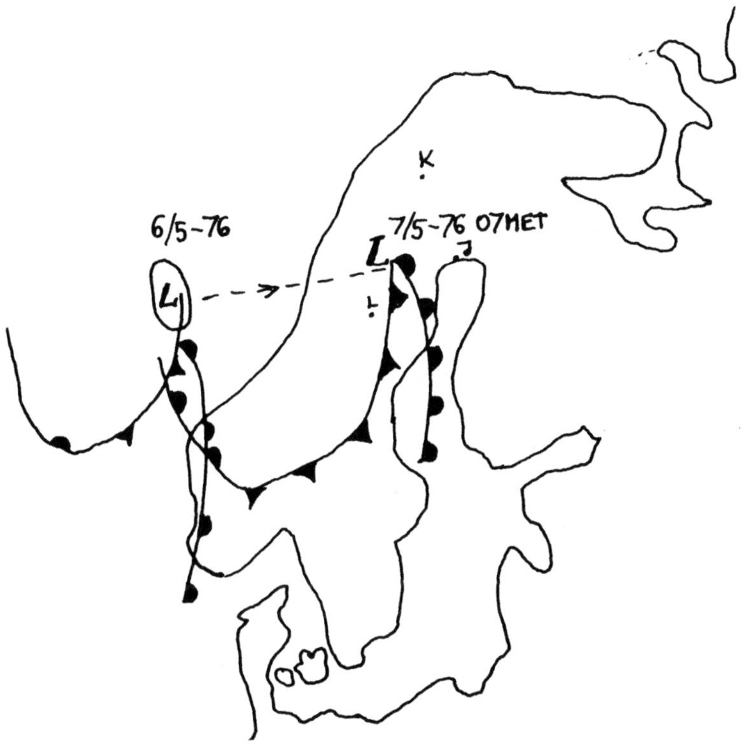

Fig. 2

The corresponding recordings of azimuth and pressure spectrum level at 2 Hz as a function of time are shown in Fig. 3. It may be calculated for the above example that the total acoustic power at 2 Hz propagated in the atmosphere was of the order of 10 kW.

It may be concluded that meteorological phenomena, as low pressure areas connected with intense air circulation systems produce large amount of acoustical energy. The major part of the energy is located in the lowest part of the infrasound spectrum. These low frequency infrasonic waves propagate up to the ionosphere and may be detected using radio methods, for example HF Doppler ionospheric soundings (Georges, 1968).

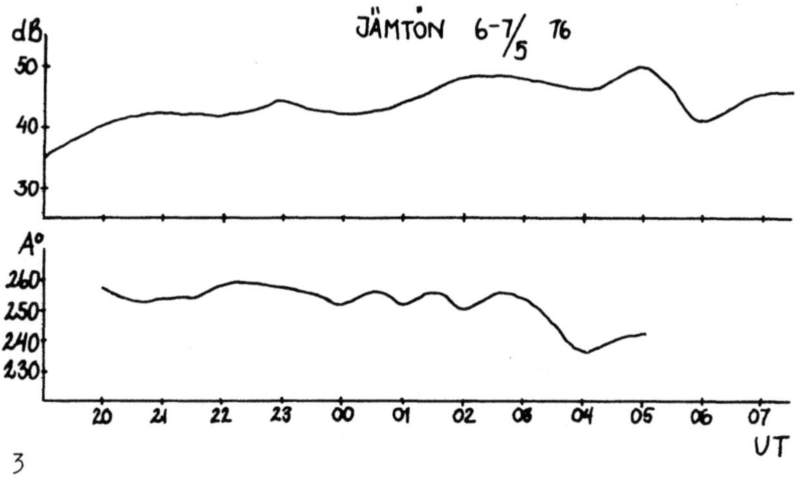

Fig. 3

3. MAN-MADE INFRASONIC SOURCES

The most intense man-made infrasound is produced by nuclear explosions. There are however numerous sources of infrasound, which although much less powerful, may still influence the atmosphere. Man-made infrasonic sources are dominated by frequencies above 1 Hz which influence mostly the stratosphere and the lowest termosphere. Recording of infrasound at 2 Hz shows that strong emissions from man-made sources are very frequent (Liszka, 1974). Besides such spectacular sources as large rockets and large supersonic aircraft there are more common industrial sources of infrasound. A source which deserves most attention is an oscillating hydroelectric power plant (see Fig. 4).

Fig. 4

At certain flow condition oscillations in the water flow are produced. These will result in a surface wave in the water magazine. The surface of the magazine will radiate acoustic energy into the atmosphere. A large power plant, producing more than 100 MW of electric power, may during periods of power oscillation radiate 1-10 MW of acoustic power. During an emergency stop a large wave in the water magazine is produced and the amount of acoustic power which is radiated during a short time into the atmosphere may be an order of magnitude larger than in case of persistent power oscillations. Due to the fact that the acoustic energy is radiated by a large water surface, the radiation pattern is highly directive, with its maximum at the direction normal to the oscillating surface. It is why only a small fraction of the energy may be detected in the plane of the water surface. The typical oscillation frequency is between 0,1 and 2 Hz. At few occasions emergency stops have been arranged at the power plant Tuggen, about 20 km from the ionosounder station in Lycksele, Sweden. Travelling disturbances were observed on F-layer traces 15-20 minutes after the stop. Assuming that the acoustic power is radiated in a cone 30° around the normal to the surface, the acoustic intensity, particle velocity and the pressure deviation in the wave have been calculated for different heights in the atmosphere (see Table II). Calculations have been made for total radiated power of 3 MW. It may be seen that considerable pressure variations will take place in the troposphere and that particle velocity and thus displacement will be important first at E-layer heights. In the lower part of Table II calculations are made for total radiated power of 30 MW, which should be representative for an emergency stop. Atmospheric effects of these waves could be studied besides of their ionospheric effects, for example by observations of high attitude clouds.

The same is aplicable to large supersonic aircraft. Only chemical effects of fuel combustion products on the stratospheric composition have been discussed. It seems however that also effects of acoustic energy radiated by the aircraft should be taken into account.

Above remarks are concluded by stating that acoustic energy of infrasonic waves of both natural and man-made origin must be taken into account in the atmospheric energy balance. A large part of energy waste produced by the modern industrial civilisation is radiated into the atmosphere in the form of acoustic energy. The question is to which extent this form may influence the atmospheric energy balance.

Table II

For $W_d = 3x10^6$ Watt

h (km)	I (W/m^2)	u_{rms} (m/s)	P (N/m^2)
10	$2,73x10^{-2}$	$1,48x10^{-2}$	1,83
20	$6,84x10^{-3}$	1,62	$4,23x10^{-1}$
30	$3,04x10^{-3}$	2,34	1,30
40	$1,71x10^{-3}$	3,67	$4,65x10^{-2}$
50	$1,09x10^{-3}$	5,67	1,92
60	$7,59x10^{-4}$	8,80	$8,63x10^{-3}$
70	$5,59x10^{-4}$	$1,46x10^{-1}$	$3,81x10^{-3}$
80	$4,27x10^{-4}$	2,82	1,52
90	$3,38x10^{-4}$	$6,29x10^{-1}$	$5,37x10^{-4}$
100	$2,73x10^{-4}$	1,38	1,98

For $W_d = 3x10^7$ Watt

h (km)	I (W/m^2)	u_{rms} (m/s)	P (N/m^2)
10	$2,73x10^{-1}$	$4,69x10^{-2}$	5,82
100	$2,73x10^{-3}$	4,35	$6,27x10^{-4}$

REFERENCES

Bowman, H.S. and Bedard, A.J.: Geophys. J. R. Astr. Soc. 26, 215, 1971.

Georges, T.M.: Proc. ESSA/ARPA Symposium on Acoustic-Gravity Waves in the Atmosphere, Boulder, Colorado, July 1968.

Liszka, L.: J. Acoust. Soc. Am. 56, 1383, 1974.

Morse, P.M. and Ingard, K.U.: Theoretical Acoustics. McGraw Hill Book Comp., 1968.

Swift, D.W.: J. Geophys. Res. 78, 8205, 1973.

Westin, H.: Studies of infrasonic waves in the auroral zone. Kiruna Geophysical Institute Report No. KGI 76:5, November 1976.

DISCUSSION

Inan: Is there a special propagation characteristic associated with infra-sonic frequencies as opposed to sonic and ultra-sonic frequencies ?

Liszka: The main difference in comparison with sonic waves is the extremely low attenuation (the attenuation proportional to f^2). The infrasonic waves may thus reach considerable heights in the atmosphere, where temperature gradients permanently exist which can reflect these waves downwards and produce long distance propagation.

Schlegel: Can you say something about the frequency spectrum of the auroral infrasound ?

Liszka: Frequency spectrum of the auroral infrasound is dominated by frequencies between 0.01 and 0.05 Hz. Sound pressure levels at 2 Hz associated with supersonic auroral events are at least 20 dB below those at 0.05 Hz.

Thrane: One should bear in mind that the acoustic frequency spectrum produced by an aurora will depend upon its heights simply because you cannot generate a soundwave with wavelengths shorter than the mean free path of the atmospheric molecules.

Jones: Have you considered monitoring a rocket launched from a nearby range (ESRANGE or Andøya) to provide an acoustic source of variable height? The acoustic spectrum change should yield information on the cut-off angles mentioned in the paper.

Liszka: We have recorded most of rocket launch from ESRANGE and Andøya. However, we have not used these data to study the acoustic spectrum variation with height, which is of course possible.

Jones: What fraction is the acoustic energy observed from "Concorde" of the total energy recorded from all supersonic aircraft ?

Liszka: Due to a long duration of the supersonic flight of Concorde it is responsible sometimes for more than 10% of all observed infrasonic emissions during the day.

REMOTE SENSING OF THE NEUTRAL CONSTITUENTS OF THE LOWER THERMOSPHERE*

D. R. Hickman, F. A. Morse and J. B. Pranke

The Ivan A. Getting Laboratories
The Aerospace Corporation
El Segundo, California, USA

1. INTRODUCTION

Measurement of densities of the neutral constituents in the thermosphere is of continuing interest to aeronomers. As understanding is improved of the gross response of the thermosphere to such things as changes in solar ultraviolet flux and geomagnetic activity, more and more specific questions are being asked about the detailed thermospheric behavior. The search for greater understanding requires ever more detailed measurements. In the past, measurement of the thermospheric density has been done primarily through drag studies or rocket and satellite borne instruments which directly sample the atmosphere. Only methods employing satellite borne sensors have been able to provide global scale measurements. A common limitation of almost all of the satellite borne sensor techniques used in the past is that neutral densities and compositions are measured only at the satellite location. Thus, while good latitude and longitude coverage is possible, altitude profiles above a fixed surface point have not been possible. In the lower thermosphere (below 150 km) the high drag experienced by satellites with such low perigees makes in situ measurements unattractive. One technique, the measurement of the attenuation of solar ultraviolet flux at satellite sunrise and sunset, is capable of giving vertical profiles of atmospheric density, but since only

*This work has been supported by the U.S. Air Force Space and Missile Systems Organization through Contract Number F04701-76-C-0077.

B. Grandal and J. A. Holtet (eds.), Dynamical and Chemical Coupling, 73-78.

two profiles are determined per satellite revolution, and these are either dawn or dusk times, this technique is not suitable for global thermospheric density monitoring. There is thus a clear need for an instrument which can monitor not merely the characteristics of the atmosphere through which it flies, which has been done for 15 years, but rather which uses a remote sensing technique. Such a remote sensing method would allow vertical profiles of density in the lower thermosphere to be determined from measurements made by an instrument carried on a satellite with perigee well above the high drag regime.

Based on initial work by Dr. A. B. Prag, we have recently developed a method which we feel will, for the first time, allow determination of vertical profiles of the neutral constituents of the lower thermosphere by a satellite borne instrument.

2. THEORY

Optical emissions of the thermosphere are the most obvious property suitable for remote sensing. The principal constituents of the lower thermosphere are N_2 and O, with O_2 also important at altitudes below 150 km. Simplified energy level diagrams for O and N_2 are shown in Fig. 1. In choosing lines for observation we have sought those which are strong and for which the production mechanisms are relatively simple and well understood. In addition, the necessity for complex radiative transfer calculations is to be avoided if at all possible.

The well known forbidden lines for O at 557.7 and 630.0 nm were felt to have production mechanisms more complex than desired. The 130.4 nm triplet, in addition to having complex production mechanisms, would require radiative transfer calculations more sophisticated than only consideration of absorption and single scattering. The most suitable emission is the 135.6 nm line. This line is both strong and, because it is an optically forbidden transition, multiple scattering may be ignored. In the lower thermosphere, absorption of the 135.6 nm radiation in the Schumann-Runge continuum of O_2 is important and may be viewed as either a complication or as a means of obtaining some information on O_2 densities.

Selection of the N_2 line is made easier by the lack of many emissions from which to choose. The clear choice was the second positive system and we have selected the strong (0,0) band starting at 337.1 nm. As the lower state of the transition is not the ground state, multiple scattering and self absorption may be neglected.

The excitation transition from the ground state to the upper level of the selected emission line is forbidden for both 135.6 nm and 337.1 nm and therefore fluorescence and resonant scattering contributions are absent. The only excitation mechanism is by fast photoelectron impact, with an excitation cross section peak near 15 eV. Theoretical calculation of emission rates is relatively uncomplicated as only a single mechanism needs to be considered. In computing photoelectron fluxes local steady-state calculations are justified. The lower energy electrons which might require

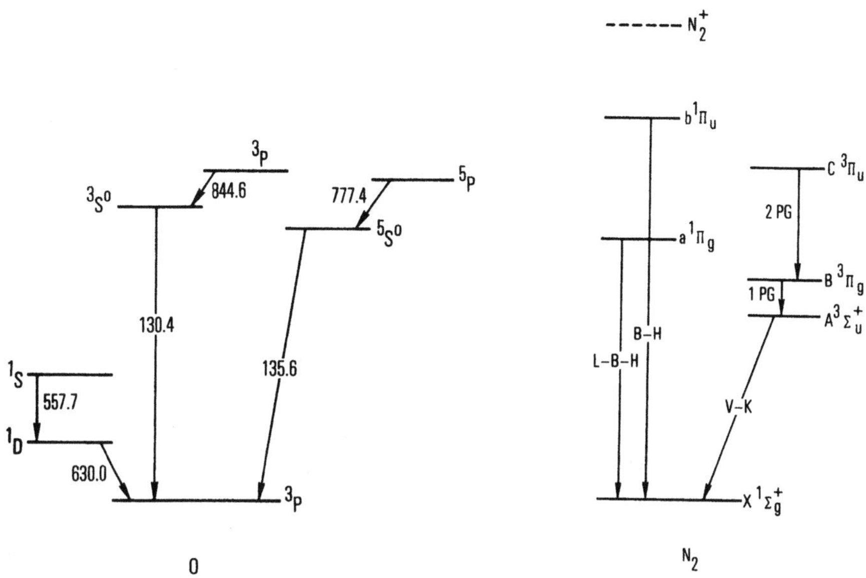

Fig. 1 Simplified energy level diagrams for O and N_2.

non steady-state calculations do not have sufficient energy to stimulate the exciting transitions and non-local calculations are not necessary because of the small number of high energy photelectrons and the small excitation cross section at high energies. The fast photoelectron excitation mechanism limits use of this technique to the dayside hemisphere.

The emissions are observed using an earth limb scanning ultraviolet spectrometer which will be described in a later section of this paper. The viewing geometry is seen in Fig. 2. The spectrometer has a vertical field of view of 0.1 degrees corresponding to a tangent height spread of about 5 km. The tangent height of the line of sight is stepped from 120 to 250 km allowing determination of the apparent brightness at 337.1 and 135.6 nm vs tangent height altitude.

The altitude profiles of the thermospheric gases are described in a parameterized form involving the densities of N_2, O, and O_2 at 120 km, a Bates-Walker temperature profile, and the assumption of diffusive equilibrium above 120 km. It might be noted that it is not actually necessary that the gases be in diffusive equilibrium, but only that the actual profiles can be well described by the chosen parametric form. In addition to the atmospheric parameters there is a parameter which describes the strength of the exoatomospheric solar EUV flux in terms of an overall multiplier to be applied to a Hinteregger standard model.

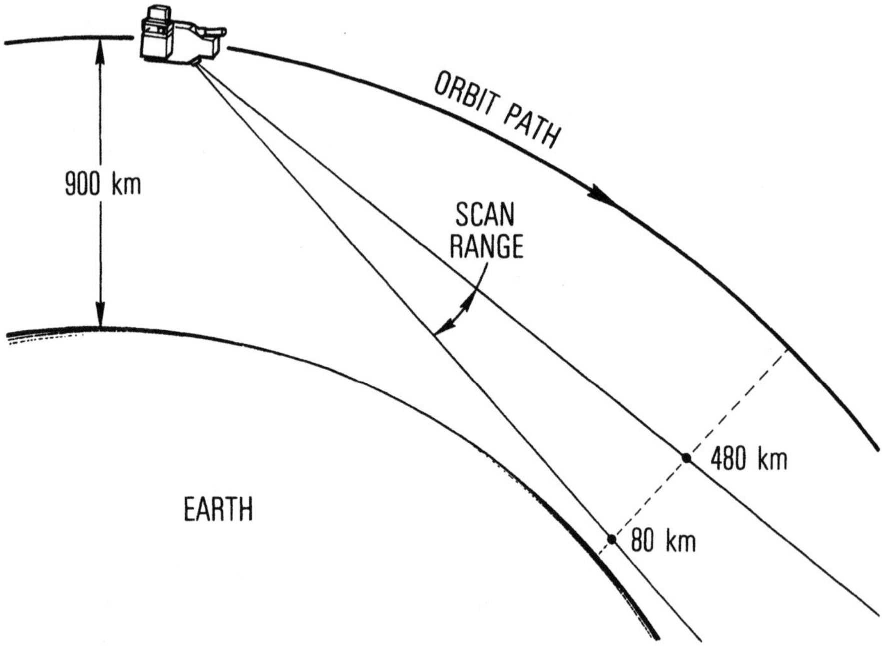

Fig. 2 Viewing geometry for remote sensing instrument.

For analysis of the data obtained from one vertical scan of the earth's limb, an initial estimate of the parameters is obtained from an atmospheric model. A theoretical calculation is then made of the intensity profiles which would be seen with the assumed densities. If the computed and observed intensity profiles do not agree, a least squares procedure is used to adjust the parameters in order to obtain agreement. The density profiles are then derived from the parameters thus determined.

In order to test the numerical accuracy and stability of the method, synthetic data have been generated and analyzed. Ideal data as well as data with random noise superimposed have been analyzed beginning with initial parameter estimates which were either reasonable or significantly perturbed. These tests indicate satisfactory stability and accuracy.

3. INSTRUMENT

The instrument to be used is an earth limb scanning ultraviolet spectrometer schematically illustrated in Fig. 3. It will be flown on a three axis stabilized spacecraft in a noon-midnight sun-synchronous orbit. Spacecraft attitude knowledge and control significantly better than 0.1 degree assure the pointing accuracy necessary for good tangent height determination.

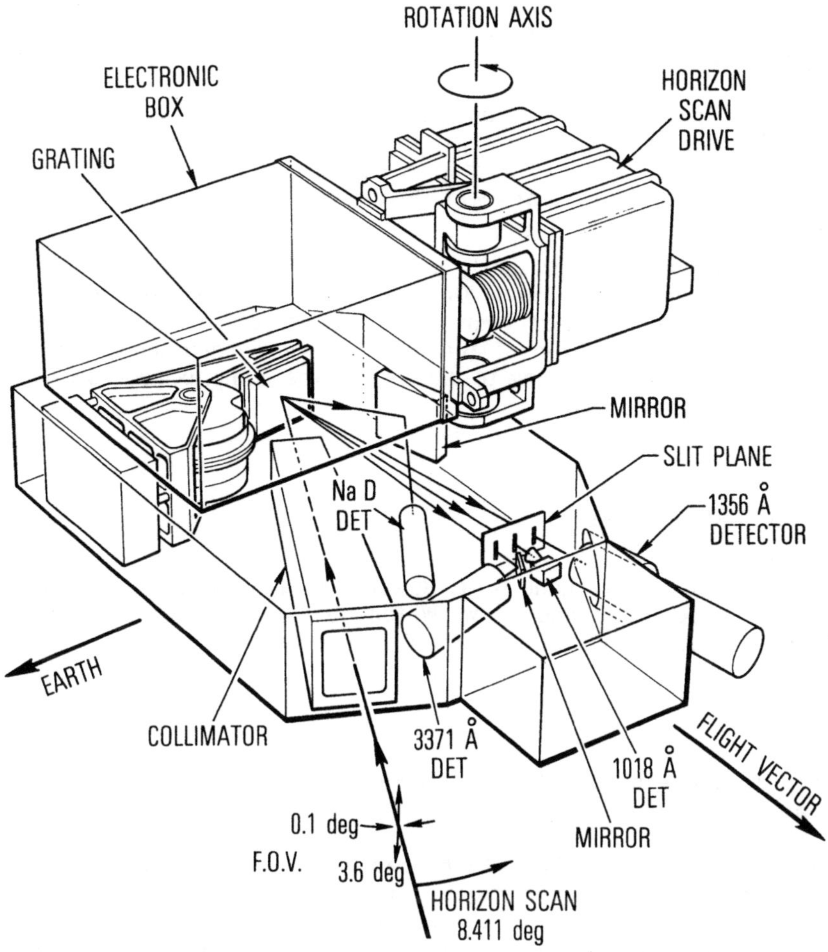

Fig. 3 Limb scanning ultraviolet spectrometer to be used for remote sensing.

The instrument consists of three major assemblies. The electronics box and horizon scan drive box are fastened to the spacecraft. The optics box rotates about an axle and this motion controls the tangent height of the line of sight. Light enters the instrument through a slat collimator which defines the field of view. After passing through the collimator, light strikes a concave diffraction grating. The dispersed light passes through a slit assembly and is detected by photomultipliers operated in a counting mode. The wavelengths observed are controlled by rotating the grating. The spectral ranges covered by the slits are 290-395 nm, 110-163 nm, and 85-120 nm with resolution of 1.2, 0.8, and 0.4 nm, respectively. A sodium D detector uses the non-dispersed zeroth order light to obtain an altitude verification from the Na D layer near 92 km. A large number of operating modes allowing both the wavelength drive and the altitude scan drive to be either continuously scanned or stepped to a small number of commandable positions give the instrument great flexibility in carrying out its investigations.

The instrument is presently undergoing final testing and calibration and is scheduled to fly on a Defense Meterological Satellite Program spacecraft within the next year. The success of this new approach to measuring the composition of the lower thermosphere should furnish an imporant new tool for trying to understand the complex processes of this region of the atmosphere.

ACKNOWLEDGEMENT

We would like to thank B. R. Baldree, W. T. Chater, C. K. Howey, E. M. Irwin, D. A. Jones and D. Y. Watanabe for their expert work in the design, fabrication and testing of the instrument.

DISCUSSION

Witt: Is there no use for the technique during the night and in the auroral zone ?

Hickman: We will not be able to derive density profiles on the night side using this technique. In the auroral zones there may be difficulties because of auroral emission at 135.6 and 337.1 nm in addition to the photoelectron produced emission. The day-side auroral zones are also regions of large solar zenith angle where the analysis may be difficult even in the absence of auroral emissions. I think we will have to wait until we have tried to analyze real data to know how well we can do at high latitudes. At night there are emissions such as 391.4 nm from ionized molecular nitrogen and Lyman alpha which we may be able to study although this is not the principal object. There has also been a suggestion to use the 135.6 nm emission due to radiative recombination of atomic oxygen ions near the F_2 peak to derive electron density profiles, but in the daytime there is competition from photoelectron excitation of neutral atomic oxygen and at night our instrument will probably not have enough sensitivity.

THE EFFECTS OF CHANGING FLUXES OF SOLAR UV RADIATION AND SOLAR COSMIC RAYS ON THE MIDDLE ATMOSPHERE TEMPERATURE AND OZONE CONCENTRATION

A.G. Theobald, R.G. Williams and M.J. Rycroft

Department of Physics, University of Southampton, U.K.

1. INTRODUCTION

The question "How constant is the solar constant ?" has been the subject of much speculation for many years. Heath (1973) made significant progress in placing this curiosity on firmer ground; using Nimbus 3 and 4 data he reported on variations of the solar constant between 120 nm and 340 nm suggesting that the variations were mainly associated with two active regions on the sun. Ruderman and Chamberlain (1975) demonstrated that cosmic rays produce nitric oxide in the atmosphere which in turn catalytically destroys odd oxygen. Since solar cosmic rays are modulated by the interplanetary magnetic field sector structure, which appears to be related to active regions (Heath et al. 1975) it is appropriate to estimate the combined effects of these variations of the solar output on the stratosphere and mesosphere.

In this calculation we consider wavelengths in the region 190 nm $\leqslant \lambda \leqslant$ 310 nm with the amplitude of the variation given by

$$\frac{\Delta I(\lambda)}{I(\lambda)} = 0.00107 \exp \frac{6656}{\lambda} \qquad (1)$$

This represents a 0.04% variation of the solar constant. The amplitude of the cosmic ray flux variation considered is based on the minor solar flare of September 1966. In order to estimate the effects, on odd oxygen and temperature profiles, of these variations. we have constructed a model with simple oxygen chemistry modulated by the production of NO by solar cosmic rays. We ignore the effects of dynamics and also assume photochemical equilibrium. The calculations have been performed for a sinusoidal variation of the solar

B. Grandal and J. A. Holtet (eds.), Dynamical and Chemical Coupling, 79-84.

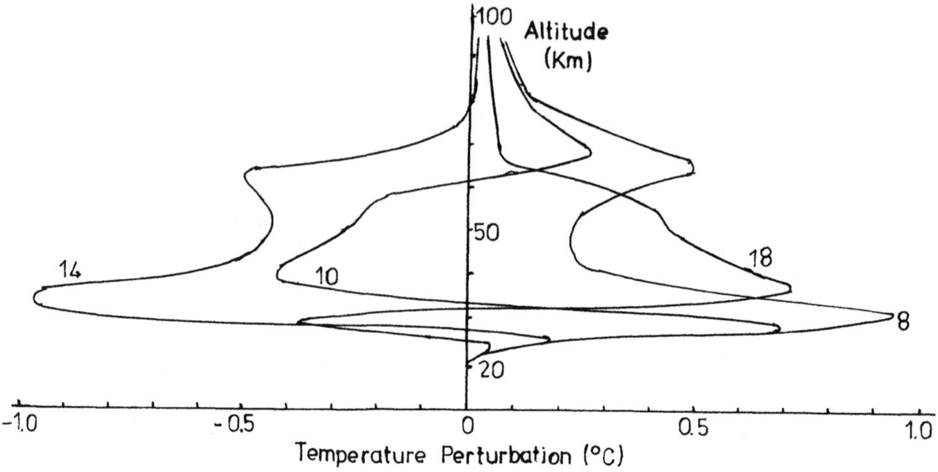

Figure 1. Temperature profiles for a sinusoidal variation (with a 14 day period) of UV radiation only. Each profile is marked with the number of days since the variation started.

output of 14 or 28 day period, and also for pulsed variation which repeat with a period of 14 days.

2. RESULTS

In this paper we present the results of using this computer program for two different forcing functions of solar variability. Firstly, we compare the response of the atmosphere to a sinusoidal variation, of period 14 days, of just the solar UV radiation with the corresponding response when we also include a sinusoidal variation of the solar cosmic ray flux. Secondly, we illustrate the longitudinal variations which occur by considering a more realistic simulation of the active region on the sun and the variation of solar cosmic rays due to the interplanetary magnetic field sector structure.

Figure 1 shows the profile from 10 km to 95 km of the perturbed temperature, for a variation of solar UV flux only, at the equator for April. We see that the main perturbation occurs in the mid-stratosphere and has an amplitude $\sim \pm 1^{\circ}$C. Above 40 km, the amplitude decreases (to $\sim \pm 0.6^{\circ}$C at 50 km), but at 70 km a shoulder or secondary maximum occurs with an amplitude of 0.5°C. The amplitude above 85 km is relatively independent of altitude, and is much smaller ($\lesssim \pm 0.1^{\circ}$C) than the stratospheric variation.

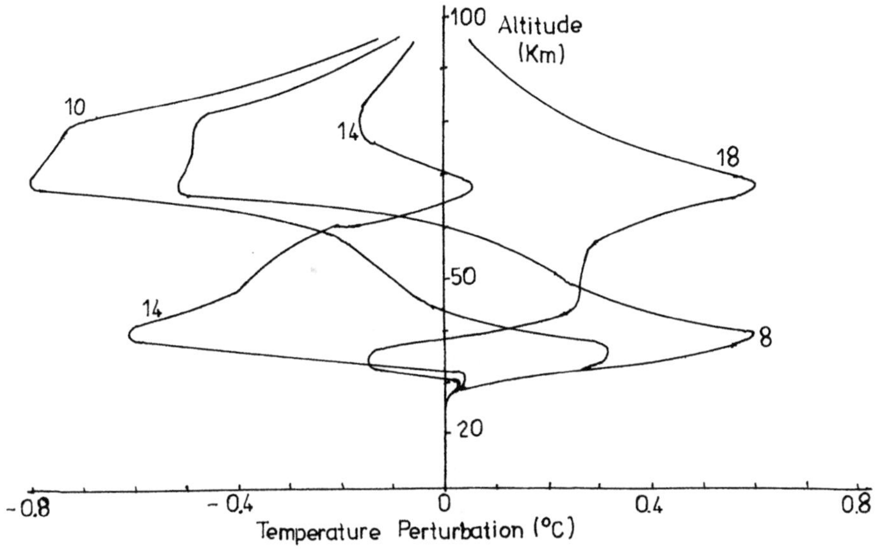

Figure 2. Temperature profiles for sinusoidal variation of UV radiation and solar cosmic rays.

By contrast, Figure 2 shows a family of profiles for the same variation of the UV flux but also including a variation in the solar cosmic ray flux. The profiles illustrated are for latitude 45°N, longitude 225°E, in April. In the stratosphere we essentially have the same variation as before, except that the maximum amplitude is reduced to ± 0.6°C. This is almost entirely due to the latitudinal variation of the flux of solar UV radiation. However, in the mesosphere, the results differ markedly from those of figure 1. The mesospheric temperature now oscillates out of phase with the stratospheric perturbation and with an amplitude ∼ ± 0.8°C. This effect is caused by the increased solar cosmic ray flux increasing [NO] in the mesosphere, and hence reducing the ozone density and temperature in that region. The two day time delay of the cosmic rays reaching the earth produces a phase difference. In the stratosphere variations of the concentrations of ozone and atomic oxygen are $\lesssim$ 1% whereas, in this instance, the mesospheric variation of $[O_3]$ has a maximum value of 10% at 80 km altitude.

We now consider a somewhat more realistic variation of the solar output with the forcing function given by,

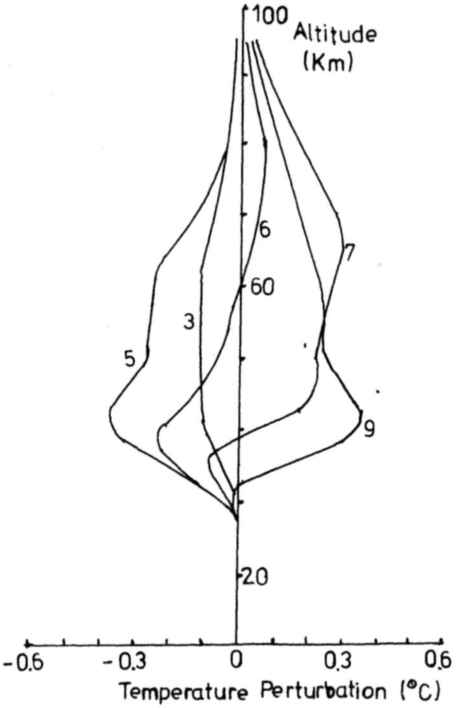

Figure 3. Temperature profiles for pulsed variation of solar output.

$$\begin{array}{lll} 0 & , & t = 4,\ t > 8 \text{ days} \\ -\exp\left[-\frac{(t-3)^2}{4} \right] & , & 0 \leqslant t \leqslant 3 \text{ days} \\ +\exp\left[-\frac{(t-5)^2}{4} \right] & , & 5 \leqslant t \leqslant 8 \text{ days} \end{array} \qquad (2)$$

Figure 3 shows the temperature response of the atmosphere to this stimulus at 45°N, 90°E for April. The stratosphere follows the variation of the solar constant, heating and cooling with an amplitude of $\lesssim 0.4°C$. The mesophere is again controlled by the cosmic ray variation. Figure 4 shows the corresponding family of profiles at 90°W where the effect of the cosmic rays is more noticeable. The stratosphere is little affected but the mesophere has a larger response $\sim \pm 0.8°C$.

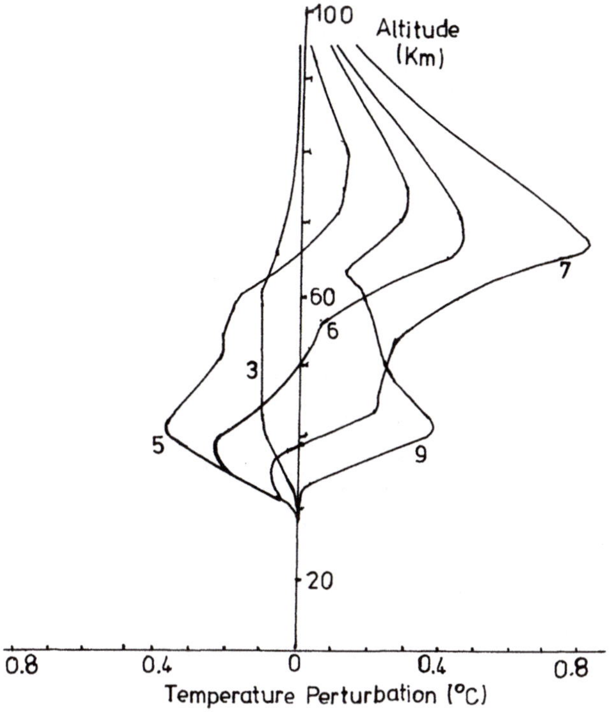

Figure 4. Temperature profiles for pulsed variation of solar output.

This longitudinal variation is caused by the cosmic ray flux being symmetric about the geomagnetic and not the geographic axis of the earth. Stratospheric variations of atomic oxygen and ozone concentration are again $\lesssim$ 1%. At an altitude of 80 km, the mesopheric variation of $[O_3]$ at 90°E, 45°N is ± 4%, whereas at 270°E it is ± 13%.

3. DISCUSSION

Frederick (1977) has used a more detailed chemical model of the atmosphere and has calculated the chemical response of the atmosphere to the maximum and minimum values of the solar constant reported by Heath (1973). We considered that until the temporal variation of the sun is better known that our simpler model gives reasonable estimates of the changes to be expected. However, there is clearly a need for far better knowledge of the solar variation.

Firstly Roberts and Olsen (1973), and then more extensively Wilcox et al. (1974), have related variations of the vorticity area index in the winter northern hemisphere to crossings of the interplanetary magnetic field sector boundaries by the earth. Hines (1974) has suggested that temperature variations in the stratosphere and mesosphere may affect the dynamics of the atmosphere sufficiently to produce the observed changes in the troposphere. It is thus necessary in future work to consider whether the calculated effects in the middle atmosphere are a physical link leading to the observed tropospheric effects or whether they are a separate effect derived from the same cause.

REFERENCES

Frederick, J.E. 1977, Planet.Space Sci., 25, 1.
Heath, D.F. 1973, J.G.R., 78, 2779.
Heath, D.F. et al. 1975, Solar Physics, 45, 79.
Hines, C.O. 1974, J.Atmos.Sci., 31, 589.
Roberts, W.O., and Olsen, R.H. 1973, Rev.Geophys. and Space Phys., 11, 731.
Ruderman, M.A., and Chamberlain, J.W. 1975, Planet.Space Sci., 23, 247.
Wilcox, J.W. et al. 1974, J. Atmos.Sci., 31, 581.

DISCUSSION

Cole: I have been considering similar scheme for sun-weather relations. The net heating rate in the mesosphere is the difference of two large numbers, one the heating rate of O_3 and the other the cooling rate of CO_2, and is equal to about 10% of either. It would follow that if the heating rate of O_3 varies by say 5% the net heating rate could vary by 50%. There is thus a multiplicative effect of a small perturbation of O_3 heating.

Rogers: A simple calculation can show the size of the effect suggested. A typical ozone heating rate is 10° K/day, and a typical infrared relaxation time is 2 days, allowing for photochemical acceleration. Thus a 10% change in ozone heating will lead to a change of about 2°K in the radiative equilibrium temperature. This is not significantly different from some of the values presented by Williams.

Williams: I agree with the comment by Rogers. The only additional comment I would make is that even if the infrared relaxation time was somewhat longer, then the model proposed only increases the ozone heating rate for a few days. Thus, we would again expect results of the order produced by the model.

LONG TERM VARIATIONS OF OH NIGHTGLOW EMISSION - RELATION TO STRATOSPHERIC HUMIDITY ?

G. Weill and J. Christophe

Institut d'Astrophysique, 98 bis Bd Arago, 75014 Paris
France.

1. INTRODUCTION

Hydrogen is provided to the middle atmosphere by diffusion and transport from the troposphere mainly in the form of methane and water vapour. After these compounds have been chemidissociated or photodissociated, hydrogen engages into active chemical reactions, whether in atomic or in compound forms, various types of effects on the ozone concentration have been identified, (Nicolet 1975 ; Booker et al. 1975). The effect of HO_x on O_3 concentration can be either positive or negative according to altitude, and the net global effect on middle atmosphere ozone cannot yet be assessed reliably. In the higher mesosphere atomic hydrogen reacts with ozone

$$H + O_3 \longrightarrow OH^* + O_2^* \text{ (exothermic 3.32 eV)}$$

and this reaction is the major cause of hydroxyl emissions which have been monitored in the nightairglow since their identification (Meinel 1950, Bates and Nicolet 1950 a, 1950 b). Because the middle atmosphere provides a link between the thermosphere, the variations of which with solar activity are well documented, and the troposphere where such effects are still debatable, we undertook a time analysis of two quantities closely related to the hydrogen contents of the middle atmosphere : OH emissions at the mesopause, water vapour concentration above the tropopause.

It is the purpose of this paper to present and discuss the experimental data.

B. Grandal and J. A. Holtet (eds.), Dynamical and Chemical Coupling, 85-90.

2. EXPERIMENTAL DATA

Since 1964, the vertical distribution of water vapour has been measured routinely by Mastenbrook (1971) near Washington, D.C., U.S.A., using a balloon-borne frost-point hygrometer. The air is sampled during the balloon descent from a 30 km altitude. The measurements cover the pressure range 15 to 300 mb. Within that range, data from the 70 mb level have been selected for the present work, in order to minimize the combined effect of experimental errors and of the annual variations which relate, at the lower heights, to the altitude variations of the tropopause (Mastenbrook 1976). The data available to us covered the period January 1964 through November 1975 on a quasi-monthly basis.

Systematic airglow observations were started at the Haute-Provence Observatory by D. Barbier in 1954. The equipment (Baillet 1955), the principles of operation and data reduction (Barbier 1956, 1961) have been carefully maintained until now providing a set of photometric data in eight wavelengths bands, every clear half hour for every moonless period through the 22 years period. We selected for the purpose of the present paper to use the wavelength band centered on 634 nm $\pm$ 19 nm, mostly comprising the OH (9-3) band with a slight contribution from the (5-0), (6-1) OH bands and from the airglow continuum.
The data, taken in the direction of the celestial pole are corrected for extraterrestrial background, and for contamination by the 630 - 636.4 nm OI triplet by means of a companion narrower filter centered on that radiation. The period covered is February 1954 through December 1975. A choice of 640 band over 670 band, also available with a same instrument, was based on a better signal to noise ratio, and also to guard against the possible effect of photocathode ageing and temperature effects at the red end of its sentivity domain. Because such an extended time series is not available elsewhere, it is given in table form below, in astronomical units for extended sources (Number of 10th / visual magnitude G_2 stars/degree square). The slant direction of observation is 44°05' zenith distance.

3. DISCUSSION

The long time variation and an auto-correlation analysis of the OH data confirm major periodicities at one and two cycles per year, as reported by Fukuyama (1977) who used 670 nm data from the same observatory.

We proceeded to correct the data for annual and semi-annual variations by normalizing each monthly mean intensity to the logarithmic mean over all the data in the same month. For coherence, the same procedure was applied to the H_2O data, although the annual variation at the 70 millibars level is not as marked (Figure 1). In the case of OH data, and in the absence of such a correction, the data from autumn.and early winter periods would outweigh the distribution . The time variation of OH intesity, H_2O concentrations, so corrected, are plotted on Figure 2 together with the solar flux at 2800 megahertz as an index of solar activity.

640 ± 19 nm OH S_{10}(vis) intensities
Mean of all half hourly observations in each New Moon period

1954	1955	1956	1957	1958	1959	1960	1961
	JAN 24 351	JAN 13 419	JAN 01 452	JAN 19 458	JAN 09 604	JAN 28 322	JAN 16 376
FEB 03 394	FEB 22 285	FEB 11 254	JAN 30 415	FEB 18 376	FEB 07 316	FEB 26 319	FEB 15 323
MAR 05 353	MAR 24 369	MAR 12 380	MAR 01 375	MAR 20 343	MAR 09 427	MAR 27 342	MAR 16 305
APR 03 273	APR 22 321	APR 11 358	MAR 31 391	APR 19 384	APR 08 302	APR 25 351	APR 15 339
MAY 02 336	MAY 21 294	MAY 10 329	APR 29 385	MAY 18 330	MAY 07 386	MAY 25 302	MAY 14 296
JUN 01 251	JUN 20 247	JUN 08 307	MAY 29 433	JUN 17 250	JUN 06 336	JUN 24 280	JUN 13 236
JUN 30 306	JUL 19 183	JUL 08 264	JUN 27 331	JUL 16 270	JUL 06 256	JUL 23 272	JUL 12 336
JUL 29 193	AUG 17 208	AUG 06 252	JUL 27 291	AUG 15 314	AUG 04 299	AUG 22 238	AUG 11 216
AUG 28 214	SEP 16 233	SEP 04 343	AUG 25 267	SEP 13 368	SEP 03 290	SEP 20 319	SEP 10 194
SEP 27 300	OCT 15 325	OCT 04 419	SEP 23 378	OCT 12 394	OCT 02 342	OCT 20 442	OCT 09 272
OCT 26 501	NOV 14 520	NOV 02 564	OCT 23 516	NOV 11 639	OCT 31 511	NOV 18 585	NOV 08 423
NOV 10 582	DEC 14 525	DEC 02 511	NOV 21 526	DEC 10 436	NOV 30 535	DEC 18 516	DEC 07 382
DEC 25 452			DEC 21 446		DEC 29 585		

1962	1963	1964	1965	1966	1967	1968	1969
JAN 06 346	JAN 25 342	JAN 14 291	JAN 02 373	JAN 21 331	JAN 10 466	JAN 29 320	JAN 18 368
FEB 05 312	FEB 24 267	FEB 13 246	FEB 01 338	FEB 20 265	FEB 09 290	FEB 28 374	FEB 16 351
MAR 06 285	MAR 25 295	MAR 14 229	MAR 03 273	MAR 22 221	MAR 11 313	MAR 28 305	MAR 18 293
APR 04 249	APR 23 229	APR 12 267	APR 02 339	APR 20 209	APR 09 307	APR 27 309	APR 16 313
MAY 04 209	MAY 23 237	MAY 11 209	MAY 01 248	MAY 20 214	MAY 09 294	MAY 27 355	MAY 16 335
JUN 02 233	JUN 21 219	JUN 10 175	MAY 30 228	JUN 18 189	JUN 08 296	JUN 25 282	JUN 14 290
JUL 01 203	JUL 20 169	JUL 09 184	JUN 29 160	JUL 18 199	JUL 07 212	JUL 25 240	JUL 14 201
JUL 31 215	AUG 19 181	AUG 07 160	JUL 28 183	AUG 16 166	AUG 06 193	AUG 23 262	AUG 13 190
AUG 30 174	SEP 17 255	SEP 06 163	AUG 26 199	SEP 14 248	SEP 04 299	SEP 22 324	SEP 11 282
SEP 28 291	OCT 17 325	OCT 05 316	SEP 25 279	OCT 14 333	OCT 03 400	OCT 21 465	OCT 11 333
OCT 28 573	NOV 16 470	NOV 04 596	OCT 24 493	NOV 12 480	NOV 02 612	NOV 20 410	NOV 09 498
NOV 27 475	DEC 16 318	DEC 04 340	NOV 23 477	DEC 12 435	DEC 01 458	DEC 19 574	DEC 09 437
DEC 26 385			DEC 22 369		DEC 31 485		

1970	1971	1972	1973	1974	1975
JAN 07 440	JAN 26 246	JAN 16 450	JAN 04 369	JAN 23 356	JAN 12 427
FEB 06 339	FEB 25 248	FEB 15 327	FEB 03 331	FEB 22 280	FEB 11 350
MAR 07 321	MAR 26 241	MAR 15 400	MAR 05 291	MAR 23 338	MAR 12 328
APR 06 282	APR 25 244	APR 13 322	APR 03 330	APR 22 278	APR 11 342
MAY 05 284	MAY 24 191	MAY 13 312	MAY 02 290	MAY 21 275	MAY 11 258
JUN 04 231	JUN 22 168	JUN 11 329	JUN 01 219	JUN 20 227	JUN 09 222
JUL 03 240	JUL 22 200	JUL 10 294	JUN 30 225	JUL 19 225	JUL 09 180
AUG 02 185	AUG 20 167	AUG 09 347	JUL 29 248	AUG 17 212	AUG 07 186
AUG 31 181	SEP 19 309	SEP 07 273	AUG 28 204	SEP 16 266	SEP 05 210
SEP 30 384	OCT 19 486	OCT 07 261	SEP 26 391	OCT 15 464	OCT 05 286
OCT 30 454	NOV 18 579	NOV 06 444	OCT 26 442	NOV 14 576	NOV 03 286
NOV 28 347	DEC 17 429	DEC 05 424	NOV 24 462	DEC 13 421	DEC 03 440
DEC 28 326			DEC 24 403		

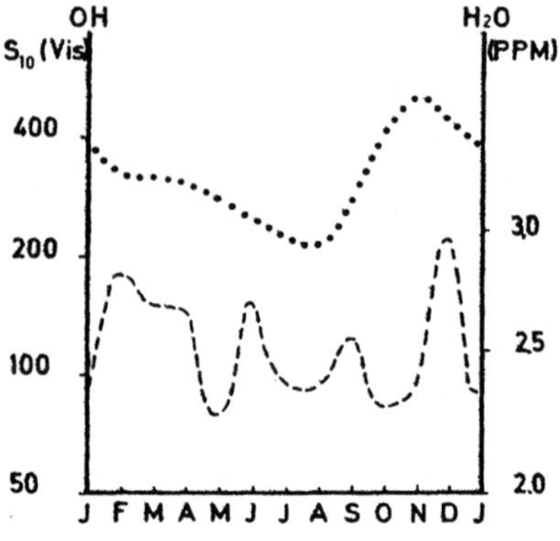

Fig. 1. OH (upper curve) and H_2O (lower curve) annual variations.

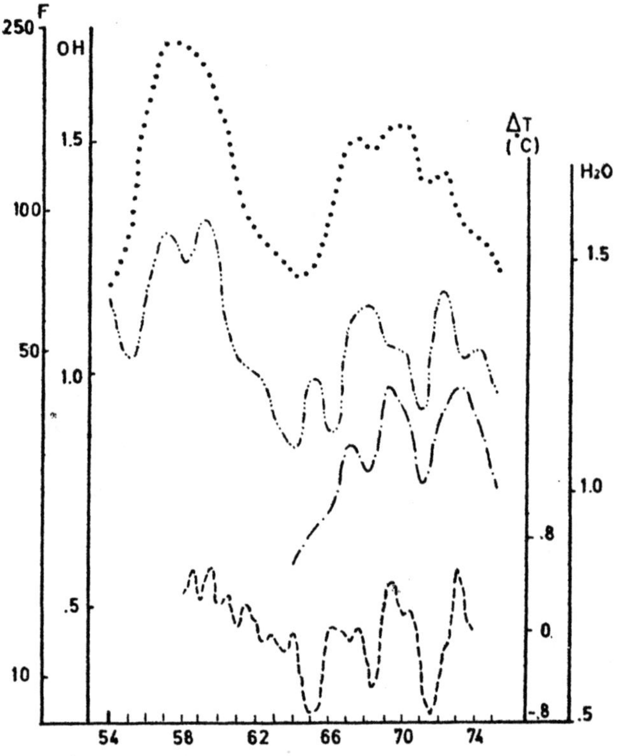

Fig. 2. From top to bottom : Solar flux, OH, H_2O and ΔT, long term variations.

The correlation between solar activity and OH intensity is, confirming the indication by Wiens and Weill (1973) based on a shorter time period and a different type of sampling. Other authors (Barbier 1965, Fukuyama 1977) also working on shorter time periods and on the more questionable 670 nm band of the same instrument, had not identified such an undecennal variation.

There is also some correlation between stratospheric water vapour and both the solar flux and OH intensities, with an indication that the water vapour variations may be lagging behind the other two parameters. The best correlation between H_2O and solar flux is .38 $\pm$.06 for a time lag of eight months. The quasi-biennal stratospheric oscillation seems to reflect in the two atmospheric curves. Because the H_2O data are yet available only for a comparatively limited time period, it would be worthwhile to obtain some additional evidence of a solar cycle control. Mastenbrook (1971) in his analysis suggested that stratospheric water vapour was connected to mean global temperatures at the tropopause, particularly the tropical tropopause. The time variations of ΔT (tropical tropospheric temperature deviation from seasonal averages) shown in Figure 2 after Angell and Korshover (1975) illustrates a moderate level of covariation.

Within the stratosphere and mesosphere, the concentration of hydrogen compounds and ozone are connected through chemistry. Recent reviews of this connection (Nicolet 1975 and Booker et al. 1975) recall that the impact of a water vapour increased concentration on ozone concentration may be either positive or negative, according to the height range within the middle atmosphere, and that the net effect on total ozone is difficult to assess. The long standing global network for total ozone measurements provides an opportunity for a more refined analysis of solar cycle variations, and there is an indication that total ozone variation is about 10% over a solar cycle with a minimum of ozone near the maximum of solar activity, except for the last solar cycle where the expected ozone variation may have been masked by man-made and or natural (volcanic) effects (Christie 1973).

Coming back to the mesopause region, it should be recalled that solar cycle effects are more significant and deeper than further down in the middle atmosphere : Hernandez (1976) analysis of his interferometric temperatures of the night airglow green line show that the temperature near the 100 km altitude varied by some 30° K and was minimum at solar maximum. Our own intensity data for the same green line show that the intensity varies in phase with the solar cycle with a maximum to minimum ratio of 2.4. On the basis of excitation of the line by triple collisions of oxygen atoms, and after correcting for the temperature variation of the relevant rate coefficient, it implies an increase in atomic oxygen maximum density of some 20%. The OH data in this paper indicate that around the 85 km level, the product $[O_3][H]$ is positively related to the solar cycle with a max to min ratio of the order of 1.6. The ratio would have to be increased by a further 50% to take into account the temperature variation of the rate coefficient, if the temperature at that level varies in parallel with that at the green line level.

The data in this paper provide a faint suggestion that the cause of events at stratospheric levels lags by half a year behind marked undecennal variations that are known to occur at the mesopause level. It would be interesting to confirm such a time lag by improved data and data analysis over a longer period of time, as it might provide a clue to possible solar cycle control of the lower atmosphere. Let us open for criticism the suggestion that the solar cycle modulates the physical characteristics of vertical transport at the mesopause and in the mesosphere, thereby inducing variations in the concentration of radiationally important constituants of the stratosphere such as H_2O and O_3, finally bringing about a minute variation of the radiation budget at the tropopause and below.

REFERENCES

Angell, J.K., and Korshover, J. 1973, Mon. Wea. Rev.101,426.
Angell, J.K., and Korshover, J. 1975, Mon Wea. Rev.103,1007 .
Baillet, A. 1955, Revue d'Optique 34, 361.
Barbier, D. 1956, The Airglow and the Aurorae, Ed. Pergamon Press, London and New York, 38
Barbier, D. 1961, Année Géophysique Internationale, Série IV, fasc. 1, C.N.R.S.
Barbier, D. 1965, Ann. Géophys. 21, 265.
Bates, D.R., and Nicolet, M. 1950 a, J. Geophys.Res. 55, 301
Bates, D.R., and Nicolet, M. 1950 b, Publ. Ast. Soc.Pacif.62.106
Booker,H.G. et al. Environmental Impact of Stratospheric Flight 1975, National Academy of Science, Washington D.C. ,Appendix A.
Christie, A.D. 1973, Pure Appl. Geophys. 106-108, 1000.
Fukuyama, K. 1977, J. Atmos. Terr. Phys. 39,1.
Hernandez, G. 1976, J. Geophys. Res. 28, 5165
Mastenbrook, H.J. 1971, J. Atm. Sc. 28, 1495.
Mastenbrook, H.J. 1976, Private Communication.
Meinel, A.B. 1950, Astrophys.J. 111,207.
Nicolet, M. 1975, Rev. of Geophys. 13, 593.
Wiens, R.H., and Weill, G. 1973, Planet. Space Sci. 21,1011.

PART II: THE IONOSPHERE

ION COMPOSITION AND ELECTRON-AND ION-LOSS PROCESSES IN THE EARTH'S ATMOSPHERE

F. Arnold and D. Krankowsky

Max-Planck-Institut für Kernphysik, Heidelberg

Our current understanding of the atmospheric plasma is well developed as far as the ionosphere located above 90 km is concerned. Regarding the height range below 90 km our knowledge seems to be far from having reached a satisfactory level, although considerable progress has been made in recent years. This lack of knowledge is mainly due to two reasons which both are associated with the high gas densities characterizing this part of the atmosphere: first, the complex interaction of neutral trace gases and charged species and second, the difficulties associated with in situ measurements in an environment of high ambient densities. The purpose of this article is to summarize the present knowledge of the atmospheric plasma in the altitude range 20 - 100 km and to point out those areas in which progress has been made recently. Important contributions came from in situ ion composition measurements by rocket-borne mass spectrometers which will be discussed in some more detail. The basic aspect of this article will be the attempt to identify those processes which are important in controlling the ionisation - recombination cycle and thereby determine the plasma densities.

1. THE ATMOSPHERIC PLASMA IN THE ALTITUDE RANGE 20 km - 100 km: AN OVERVIEW

Most of our knowledge about the bulk plasma parameters in this part of the atmosphere which covers also the lowest part of the ionospheric E- (>85 km) and D-regions (60-85 km), was obtained by rocket- and balloon - borne measurements of electron densities, total ion densities, conductivities, and mobilities by electrostatic probe and wave propagation methods, and by ground-based

B. Grandal and J. A. Holtet (eds.), Dynamical and Chemical Coupling, 93-127.

measurements of radio waves partially reflected by D-region electrons. A summary of these experimental methods is given e.g. by Rishbeth and Gariott (1969). The stratospheric conductivity and mobility measurements made by rocket- (Hale et al. 1968 , Hale et al. 1972, Pedersen et al. 1966, Widdel et al. 1971, Rose and Widdel 1972) and balloon-borne (Ishikawa et al. 1969) blunt probes and Gerdien-condensors are not included in the standard textbooks. From the in situ measurement the following picture of the atmospheric plasma- and electron densities emerged (Fig. 1): During daytime plasma densities are typically about $10^5 cm^{-3}$ around 100 km and decrease steeply at the bottom of the E-layer (95-85 km). D-region-densities decrease from about $10^3 cm^{-3}$ (85 km) to about $10^2 cm^{-3}$ (65 km). Below the D-region the plasma densities increase monotoniquely throughout the stratosphere reaching a maximum of about $10^4 cm^{-3}$ in the lower stratosphere. The concentration of free electrons decreases rapidly relative to positive ions somewhere between 70 and 60 km. At night plasma densities are considerably smaller and exhibit pronounced structures above 80 km. Below this height they increase rapidly and may be even somewhat higher than during daytime. Below about 60 km they are close to the daytime values. The concentration of free electrons decreases very steeply around 80-70 km. The plasma- and electron density profiles exhibit a considerable variability which may be a systematic (diurnal, seasonal, latitudinal, solar cycle) or an irregular (particle precipitation at high latitudes, excessive winter anomaly) variation of plasma densities. These variations are most pronounced in the D-region where enhancements by factors as high as ten to hundred are possible.

The most important sources of ionisation during quiet daytime are shown in Fig. 2. The lower E-region is mainly created by solar extreme ultraviolet radiation and solar X-rays. While X-rays ionize all neutral atmospheric gases, Lyman -ß (1026 Å), a major EUV-line, can just ionize molecular oxygen. All these sources are relatively well known, although the shorter wavelengths undergo substantial variations depending on solar activity. In the D-region photoionisation of the trace gas nitric oxide by solar Lyman -α radiation (1215.7 Å) is most important. Although the Lyman -α flux is rather well known and does only weakly depend on solar activity varying by about a factor of two from solar minimum to maximum (Hinteregger 1970), this ionisation source is quantitatively not well established. The reason is the poor knowledge of nitric oxide concentrations and the strong variability of this trace gas to be exspected on theoretical grounds. Other D-region sources are photoionisation of the metastable oxygen molecule in the $^1\Delta_g$ electronic state by radiation in the wavelength range 1118 - 1027 Å (Hunten and McElroy 1968, Huffman et al. 1971) and hard solar X-rays (<10 Å).

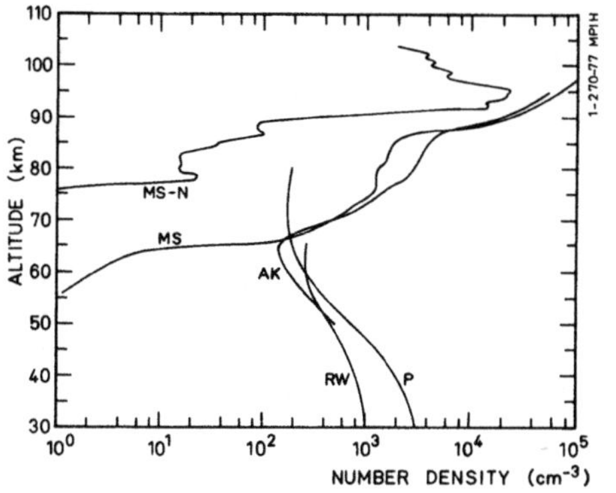

Fig. 1: Number densities of electrons during day (MS) and night (MS-N) from Mechtly and Smith 1970. Number densities of positive ions during day (AK, P, RW) from Arnold and Krankowsky; unpublished material (AK); Pedersen (P) 1966; Rose and Widdel 1972 (RW).

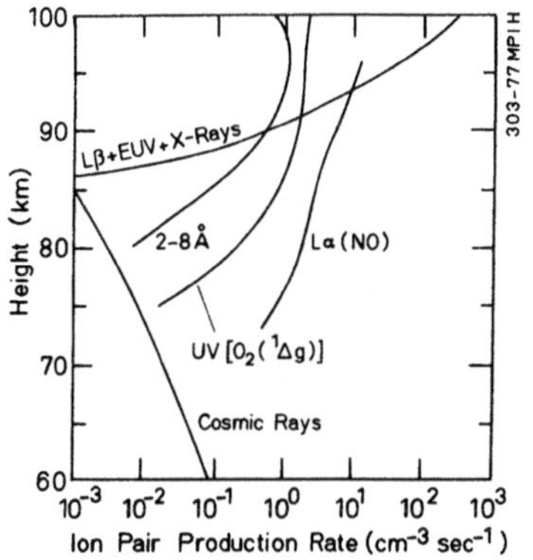

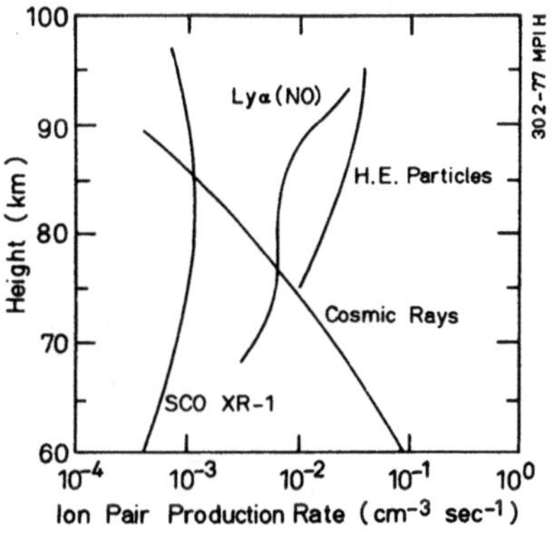

Fig. 2 (left) and 3 (right): Day - (2) and nighttime (3) ionisation rates. After Thomas (1974). The Lyman-α ionisation of NO is reduced by a factor of two due to a revision of the transition probabilities (see Poland and Broida (1971).

Strong enhancements of the ionisation by X-rays (solar flares) and electrons and protons (auroral activity) may eventually occur. Below about 60 km galactic cosmic rays (Webber 1962), which can deeply penetrate into the atmosphere, represent the only important source of ionisation. This source increases with geomagnetic latitude and decreases with solar activity. In addition solar cosmic rays associated with solar flares may eventually cause strong enhancements of the ionisation above about 35 km. The quiet nighttime ionisation sources (Fig. 3) are aside from galactic cosmic radiation much smaller and not well known. They are solar Lyman-α radiation scattered by the hydrogen geocorona which ionises nitric oxide (Ogawa and Tomatsu 1966), precipitating energetic electrons (Tunilov et al. 1966, Potemra and Zmuda 1970), and galactic X-ray sources (Poppoff and Whitten 1969).

The uncertainty of the most important ionisation source of the daytime D-region associated with our poor knowledge of nitric oxide, represents a major problem. Because of its importance the concentration of nitric oxide and its chemistry will be discussed in chapter 5.

The major ionisation products formed by the various sources discussed above are electrons and the positive ions N_2^+, O_2^+, N^+, O^+, and NO^+. The purpose of the following chapters will be to trace the fate of the primary ionisation products in order to identify those positively and negatively charged species which finally recombine with each other and which determine the bulk recombination coefficient of the ionospheric plasma. It is the bulk recombination coefficient which aside from the ion-electron production rate determines the plasma densities.

2. THE IONISATION-RECOMBINATION CYCLE

The most important loss process for atmospheric charged species in the altitude region of interest is mutual recombination in the gas phase. Anihilation of charges by interaction with particulate matter as aerosoles is negligible (Zigmunda und Mohnen 1972). The recombination coefficient strongly depends on the nature of charged species involved (Tab. 1).

No.	Process	Educts	Products	Typical α	Reference
1	Radiative R.	$A^+ + e$	$A + h\nu$	10^{-11}	a
2	Dissociative R.	$AB^+ + e$	$A + B$	$5 \times 10^{-7}(T/300)^{-0.5}$	b
3	Electron R. of Complex - Ion	$A^+(B)_n + e$		$\leq 10^{-5}(T/300)^{-0.5}$	c
4	Ion-Ion R. (2-Body)	$A^- + B^+$	$A + B$	$5 \times 10^{-8}(T/300)^{-0.5}$	d
5	Ion-Ion R. (3-Body)	$A^- + B^+ + M$	$A + B + M$	$2 \times 10^{-25}(T/300)^{-2.5}$	e

Tab. 1: Recombination processes for positive and negative ions. R. coefficients in cm^3s^{-1} (2-body) and cm^6s^{-1} (3-body). (a) Bates et al. (1962); (b,c) Biondi (1970); (d,e) Smith and Church (1976).

While atomic ions recombine only by radiative recombination which is characterized by a low $\alpha = 10^{-12} cm^3 s^{-1}$ the recombination of electrons with complex ions containing many atoms has a high $\alpha = 10^{-5} cm^3 s^{-1}$. Thus a variation of the bulk recombination coefficient α_B by 6 orders of magnitude is possible depending on which charged species dominate the plasma composition. By converting primary ions and electrons to positively and negatively charged products which recombine with a α_B different from that of the educts, ion chemistry may control atmospheric plasma densities. The recombination lifetime of a charge is given by

$$\tau_R = (\alpha_B \cdot [N^{\pm}])^{-1} \tag{1}$$

where $[N^{\pm}]$ is the number density of the positive or negative charge it recombines with. Considering α_B and minimum $N^{\pm}$ values as given in Tab. 1 and Fig. 1, we estimate $\tau_R \sim 10^4$ sec for dissociative electron recombination. This implies that τ_R is much smaller than time constants for transport, and at most altitudes also smaller than time constants for changes in the ionisation rates and the composition of the neutral atmosphere. Consequently, local ionisation-recombination equilibrium can be assumed. This does not apply to atomic positive ions recombining by radiative recombination (10^6s to 10^9s). Atomic positive ions will therefore be subject to transport (see chapter 3). At heights around 60 km where 2-body ion-ion recombination dominates and where relatively low plasma densities exist τ_R may be as large as one day. Again transport as well as significant deviations from local equilibrium have to be considered.

Aside from the exceptions discussed above local equilibrium may be assumed and the continuity equation

$$\frac{d[N^+]}{dt} = Q - \alpha_B [N^+][N^-] \tag{2}$$

relates the densities of positively $|N^+|$ and negatively $|N^-|$ charged species to the ionisation rate Q and to α_B. According to equation (2) a decrease of α_B by e.g. a factor 200, associated with a transition from electron recombination of complex ions to ion-ion recombination (Tab. 1), results in an increase of $|N^{\pm}|$ by a factor of 14. This example demonstrates the strong influence ion chemistry may have on plasma densities. Furthermore, a slight temperature influence on the plasma densities is introduced through temperature dependence of recombination coefficients. Only 3-body ion-ion recombination which becomes important below about 30 km introduces a significant temperature influence which may be as high as a factor of about 2.

3. POSITIVE IONS

3.1. COMPOSITION OF POSITIVE IONS

Our present knowledge of the atmospheric positive ion composition in the altitude range of interest is mainly based upon rocket-borne in situ ion mass spectrometer measurements. Since the first measurement of positive ions in the D-region (Narcisi and Bailey 1965) considerable progress has been made in experimental technique and our understanding of the ion population in the lowest parts of the ionosphere as demonstrated by the first measurements of negative ions (Narcisi et al. 1971 and Arnold et al. 1971) and stratospheric positive ions (Arnold et al. 1977a). Because of the high ambient pressures below 100 km the experimental method to be applied is much more difficult than at higher altitudes. The mass spectrometer has to be mounted inside a vacuum tank which is pumped by a fast cryo- or sublimation-pump. Atmospheric neutral and ionized gases enter the system through a small inlet orifice having a diameter of typically 1 mm. As the sensitivity of the mass spectrometer is approximately proportional to the particle flux through the inlet orifice the pumping system mainly determines the instrument's sensitivity. Due to pressure problems, until recently all measurements were limited to altitudes above 60-70 km. A detailed description of the experimental method of rocket-borne pumped mass spectrometers is for example given by Arnold et al. (1977 b). The first mass spectrometric measurements of positive ions extending to altitudes below 60 km were recently carried out by use of an improved experimental method which allowed to obtain data down to stratospheric altitudes as low as 24 km (Arnold et al. 1977a).

Since the pioneering measurements of Narcisi and co-workers of the AFGL in 1963, which as a great surprise, revealed the presence of cluster ions in the D-region, in situ mass spectrometer experiments were carried out under a wide variety of geophysical conditions. These measurements were made by four laboratories, the Air Force Geophysical Laboratory (AFGL), the Max-Planck-Institut für Kernphysik, Heidelberg (MPK), the Goddard Space Flight Center of NASA (GSFC), and the Universität Bern (UBE) and lead to the following picture of the D- and lower E-region positive ion composition:

a) The undisturbed daytime lower E-region is dominated by the molecular ions NO^+ and O_2^+. Atomic metal ions of meteoric origin (mainly Mg^+ and Fe^+) are also present forming a permanent layer around 85-100 km and superimposed narrow layers which are highly variable regarding height and number density. At heights of the narrow layers metallic ions may eventually dominate over the molecular ions.

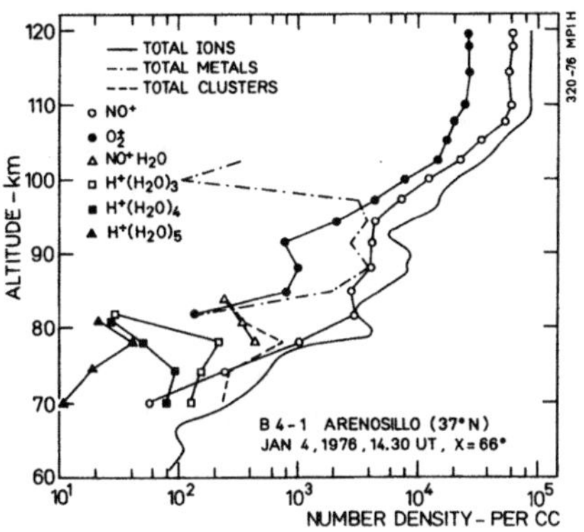

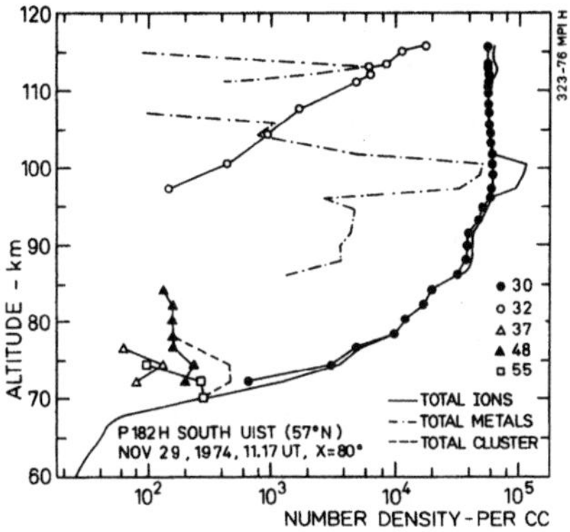

Fig. 4 (left): Daytime positive ion composition at middle latitudes during winter. From Arnold et al. (1976).

Fig. 5 (right): Daytime positive ion composition at high middle latitudes during winter. From Arnold et al. (1976).

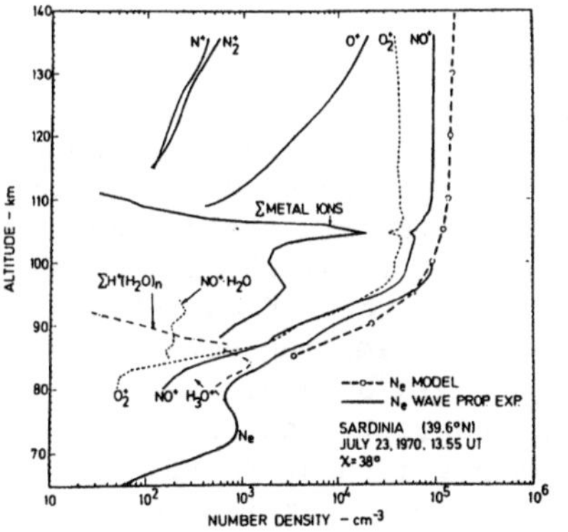

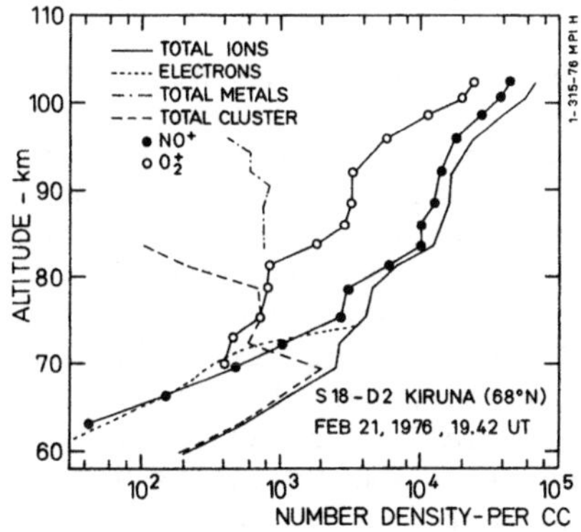

Fig. 6 (left): Daytime positive ion composition at middle latitudes during summer. From Johannessen and Krankowsky (1974).

Fig. 7 (right): Nighttime positive ion composition at high latitudes during winter and during an energetic electron precipitation event. From Arnold and Krankowsky, unpublished material.

b) The undisturbed nighttime lower E-region is dominated by NO^+ and atomic metal ions which dominate more often than during day.

c) Cluster ions, mainly proton hydrates ($H^+(H_2O)_n$) become dominant below an altitude varying between about 90 km and 70 km depending on the geophysical conditions. The transition from molecular to cluster ions is rather abrupt, typically occuring within few kilometers.

d) The degree of hydration n depends on altitude and conditions. Usually the distribution of proton hydrates peaks around $n = 2$ to 4. Under low temperature conditions around the mesopause (~ 85 km) proton hydrates with $n \leq 9$ have been observed.

e) Atomic metal ions are usually absent inside the cluster ion regime. However, they always disappear below about 80 km even when the cluster ion regime starts only at lower heights.

f) Collisional cluster-ion fragmentation occuring during sampling is evident in many experiments. It may be induced by high-shock temperatures or by electric fields applied for ion sampling from the plasma.

So far, the problem of cluster ion fragmentation in the detached shock wave forming upstream the blunt-shaped mass spectrometer probe was only tackled by the MPK-group (Arnold et al. 1974; Arnold et al. 1977 b). A sampling cone is utilized by which the shock wave becomes attached to the knife-shaped edges of the inlet orifice, thus allowing to extract ions from an undisturbed flow regime. Furthermore, the improved flow characteristic upstream the inlet orifice increases the sensitivity of the instrument. Based upon laboratory simulation experiments of electric field-induced cluster-ion fragmentation (Marien and Arnold 1973) the MPK-group used low ion-extraction potentials of 2 to 5 volt. From inflight diagnostics it was obvious that by use of a sampling cone and a low potential cluster ion fragmentation could be decreased substantially, and as it seems now, can be neglected at least above about 70 km. Comparing various measurements it has, therefore, to be kept in mind that most MPK-measurements were carried out by use of the improved experimental method whereas others were not.

Typical examples of positive ion-composition measurements carried out by the MPK-group are shown in Figures 4 to 9. A daytime situation at middle latitudes during winter (Fig. 4) exhibits a transition from molecular ions NO^+ and O_2^+ to cluster ions, mainly $H^+(H_2O)_3$ and $H^+(H_2O)_4$, around 75 km. The permanent metal ion layer around 85-90 km gives rise to a pronounced relative maximum in the plasma density. Another quite different daytime-winter situation at high middle latitudes (Fig. 5) shows NO^+ to be the only important ion at altitudes above 72 km.

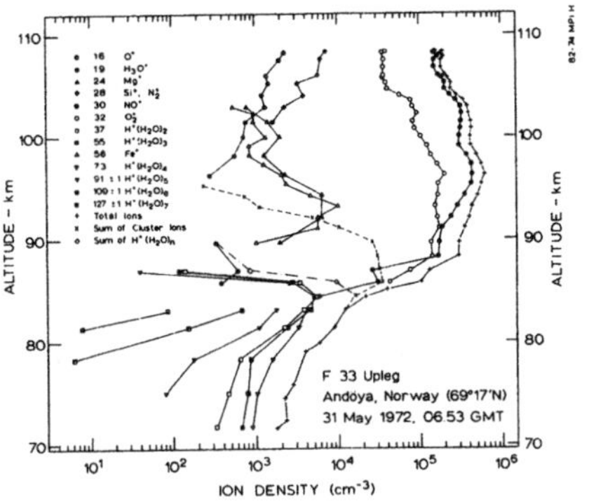

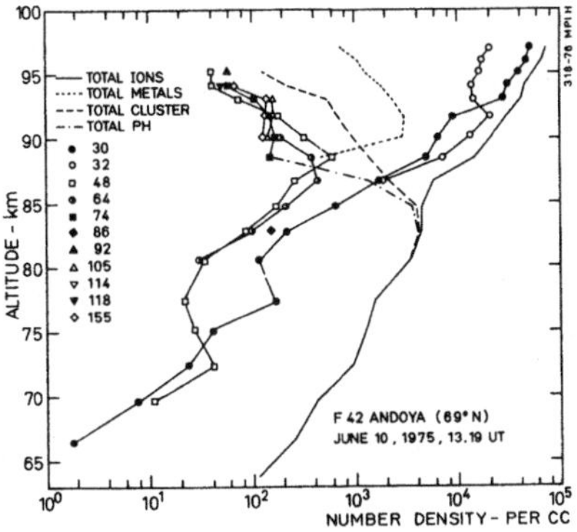

Fig. 8 (left) and 9 (right): Daytime positive ion composition at high latitudes during spring and during an energetic particle precipitation event (8) and during summer (9). From Arnold et al. (1974).

Metal ions are important only in a narrow layer around 100 km. A daytime-summer situation at middle latitudes (Fig. 6) exhibits cluster ions to become dominant already below about 86 km. A comparison of a winter- (Fig. 7) and a summer situation (Fig. 8), both at high latitudes during enhanced energetic electron ionisation and a quiet summer day (Fig. 9), again reveals an upper boundary of the cluster-ion regime at a low altitude (74 km) during winter and at high altitudes (87 km) during summer. In addition to proton hydrates a great number of other cluster ion species, mainly populating the transition region between about 87 km to 95 km, was observed (Fig. 8). These cluster ions include species such as NO^+H_2O, NO^+CO_2, O_4^+, $O_2^+H_2O$, NO^+N_2 and others containing N_2, O_2, CO_2 or H_2O.

After this phenomenological excursion the relevance of the observational features to the ionisation-recombination cycle will be discussed. Obviously we are dealing with three types of positive ions, molecular-, cluster- and atomic ions. Both groups of ions recombine with electrons by considerably different recombination coefficients (Tab. 1). Excluding the highly variable atomic metal ions from the discussion for a moment the two regimes of molecular ions and cluster ions, which are separated from each other by a rather sharp boundary, are associated with two regimes of different recombination coefficients (Reid 1970, Haug and Landmark 1970) of about $5\cdot10^{-7}(T/300)^{-1/2}$ cm^3s^{-1} and $0.5\text{-}1\cdot10^{-5}(T/300)^{-1/2}$ cm^3s^{-1}, respectively. The implication is a drop in plasma densities by a factor of about 4 below the boundary. Considering the observed variability of the boundary from about 70 to 90 km a corresponding variability of the plasma densities as a result of α_B-variations has to be expected. Realizing the importance of the total gas density, the gas temperature, and the atomic oxygen density for the ion composition (see chapter 3.2. and 4.2.) the MPK-group developed a mass spectrometer probe which measures the composition of positive and negative ions as well as the densities of atomic oxygen, N_2, O_2, Ar and CO_2 almost simultaneously. Furthermore, the total gas density is measured by an ionisation gauge integrated in the mass spectrometer probe (Arnold et al. 1975). Temperatures can be obtained from the measured scale heights of the total gas density, N_2, O_2, A, and CO_2. Examples of densities of N_2, O_2 and O measured by this instrument are shown in Fig. 10 and Fig. 11.

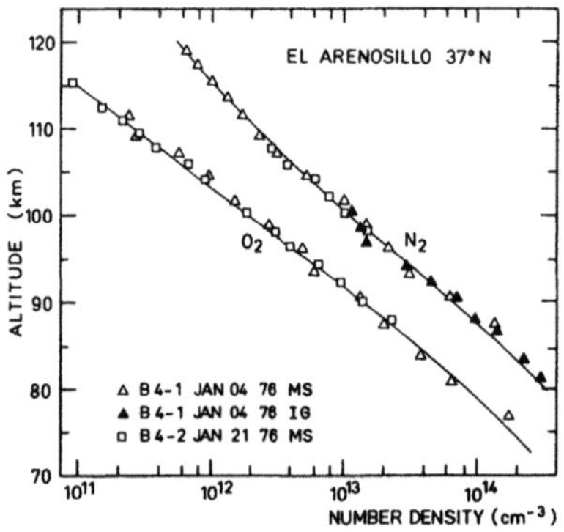

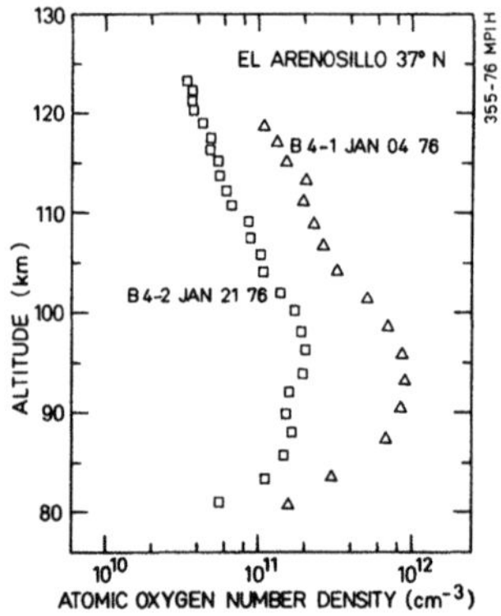

Fig. 10 (left): Number densities of N_2 and O_2 as measured by mass spectrometers. From Arnold et al. (1976).

Fig. 11 (right): Number densities of atomic oxygen as measured by mass spectrometers. From Arnold et al.(1976) .

We will now leave the D-region and discuss the positive ions recently observed below 70 - 60 km (Arnold et al. 1977a). These measurements revealed that proton hydrates (PH) remain the dominant positive ions to altitudes as low as about 40 km. Around this height an abrupt change in the positive ion population takes place and new ion species, non-proton-hydrates (NPH), not observed in the higher atmosphere become dominant (Fig. 12). Their atomic mass numbers and tentative interpretations are given in Tab. 2. It should be stressed that these ions probably represent only fragments of higher order cluster ions which became decomposed during sampling. This process, however, is expected to strip off just one or more molecules attached to the fragment ion itself.

Observed mass numbers and tentative identifications of ions

Mass (AMU)	Tentative identification of ions
19±2	H_3O^+
29±2	H^+CH_2O
37±2	$H^+(H_2O)_2$
42±2	$C_2H_2O^+$
55±2	$H^+(H_2O)_3$
60±2	$H^+(CH_2O)_2$; $C_2H_2O^+H_2O$
80±2	$H^+(CH_2O)_2H_2O$; $C_2H_2O^+$ $(H_2O)_2$

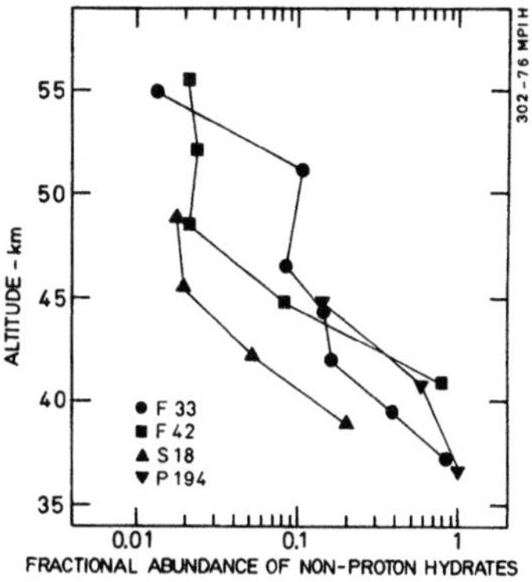

Tab. 2: Mass numbers and tentative identifications of positive ions measured in the stratosphere. From Arnold et al. (1977a).

Fig. 12: Fractional abundance of stratospheric positive ions being no proton hydrates. From Arnold et al.(1977a) .

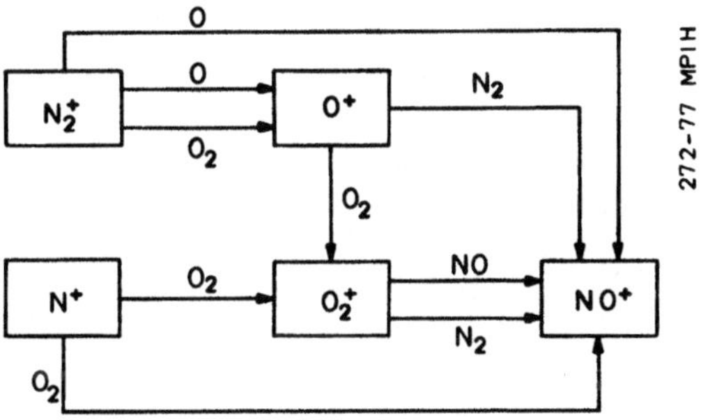

Fig. 13: Reaction scheme for the chemistry of atomic and molecular ions in the lower E-region and D-region.

3.2. CHEMISTRY OF POSITIVE IONS

In the following the ion chemical processes which are responsible for the conversion of primary ions (N_2^+, O_2^+, O^+, N^+, and NO^+) to the most abundant ions NO^+, O_2^+, and $H^+(H_2O)_n$ determining α_B will be discussed. The molecular ions NO^+ and O_2^+ usually dominating the lower E-region are formed from primary ions by fast twobody ion-molecule reactions summarized in the following reaction scheme (Fig. 13). The corresponding reactions are summarized in Tab. 3.

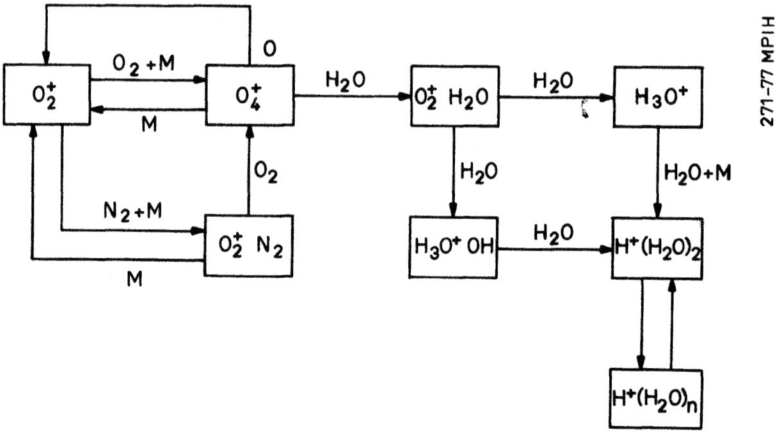

Fig. 14: Reaction scheme for the conversion of O_2^+ to cluster ions in the D-region.

All reactions are faster than one second except the NO charge transfer of O_2^+ which may take up to 500 seconds depending on the NO density. As the time constant for dissociative recombination of molecular ions typically exceeds 25 seconds in the lower E-region, NO^+ and O_2^+ have to be the dominant ions. The ratio $[NO^+]$ / $[O_2^+]$ mainly depends on the densities of NO and electrons. As the recombination coefficients of NO^+ $(4.1 \cdot 10^{-7}(T/300)^{-1/2}cm^3 s^{-1})$ and O_2^+ $(2.1 \cdot 10^{-7}(T/300)^{-1/2}cm^3 s^{-1})$ differ only by a factor of 2 from each other, the ratio $[NO^+]$ / $[O_2^+]$ is not very important for the plasma density.

Atomic metal ions have long lifetimes (10^6s to 10^9 s, see chapter 2) and, therefore, may be subject to transport. They are mainly formed by charge transfer with NO^+ and O_2^+ of neutral metal atoms originating from meteor ablation and also by photoionisation. As the time constant for charge transfer ionisation of a neutral metal atom (~ 10^4 seconds) is much smaller than the recombination lifetime of the atomic metal ion, the metal atoms should all be ionized in the E-region. Oxidation of ionized and probably also of neutral metal atoms is prevented by the action of atomic oxygen (e.g. Brown 1973). Transport of atomic metal ions proceeds in the following way: at heights above about 140 km where the collisions frequency with neutral gas is small the atomic metal ions are coupled to the magnetic field and gyrate around the magnetic field lines. At heights where the ion-neutral gas collision frequency is much higher than the gyrofrequency they follow the motion of the neutral gas. In the intermediate height region these ions are pushed by horizontal winds and thus experience a Lorentz force e(v x B) which introduces a vertical velocity component. Due to wind shears the atomic metal ions may become concentrated into narrow layers located at the nodes of the shears which are known to propagate downwards (Rosenberg and Justus 1966). A layer of atomic metal ions follows the downward propagation of the wind node until collisions become too frequent (~ 90 km) (Chimonas and Axford 1968). From subsequent observations of metal ion layers vertical downward transport velocities of the order of 1 m s^{-1} are inferred at heights above about 100 km (Smith 1970). Thus ionized metal atoms are continuously swept from the region below about 150 km into the permanent layer around 85 - 100 km. As ablation of meteors and micrometeorites mainly occurs between 80 and 140 km (Lebedinets and Shoshkova 1970), depending on mass and velocity, practically all metal atoms become deposited in the permanent layer. The origin of the atomic metal ions was clearly identified as being meteoric due to the close agreement in elemental abundance of metal ion layers and chondrites (Narcisi 1973; Aikin and Goldberg 1973; Krankowsky et al. 1972; Zbinden et al. 1975). From gound-based optical observations of neutral sodium (Gibson and Sandford 1971) it is known that there exists a layer of Na-atoms between about 80 and 100 km which closely resembles in shape the permanent layer of atomic metal ions. In particular, a sharp lower boundary is also found. The concentration of neutral Na-atoms is about 10 times higher than the concentration of Na^+ which implies that neutralisation of Na^+ seems to be efective at these heights. Most likely, the process responsible involves the formation of a metal oxide ion or a clustered atomic metal ion which recombine by the fast dissociative recombination.

We will now discuss the formation of cluster ions from the major molecular ions NO^+ and O_2^+. NO^+ is the most important primary ion in the D-region and should therefore be the major precursor of the D-region cluster ions. The molecular ion O_2^+ eventually becomes an important precursor of cluster ions only under disturbed conditions when the production of O_2^+ at D-region levels is high. While the conversion of NO^+ to cluster ion is still not well established, the appropriate reaction chain for the O_2^+ conversion (Fig. 14) has been identified (Fehsenfeld and Ferguson 1969, Good et al. 1970a).The corresponding reaction coefficients are given in Tab.4. For the built-up of the second proton hydrate $H^+(H_2O)_2$ this scheme involves only one 3-body reaction representing the first step, and subsequent binary reactions which are relatively fast (~1s) in the D-region. The time constant of the 3-body reaction forming O_4^+ steeply increases with height due to the decreasing gas density (Fig. 15). Because of the strong inverse temperature dependence of the 3-body reaction coefficient (Payzant et al. 1973) a significant influence of D-region temperatures has to be expected. Taking a

No.	Educts	Products	Rate Constant *
1	$N_2^+ + O_2$	$O_2^+ + N_2$	$(6.0 \pm 1.0) \times 10^{-11}$
2	$N_2^+ + O$	$NO^+ + N$	$(1.4 \pm 0.7) \times 10^{-10}$
3	$N_2^+ + O$	$O^+ + N_2$	$(1.1 \pm 0.5) \times 10^{-11}$
4	$O_2^+ + N_2$	$NO^+ + NO$	$< 1 \times 10^{-15}$
5	$O_2^+ + NO$	$NO^+ + O_2$	$(6.0 \pm 1.5) \times 10^{-10}$
6	$O^+ + N_2$	$NO^+ + N$	$(1.2 \pm 0.1) \times 10^{-12}$
7	$O^+ + O_2$	$O_2^+ + O$	$(2.0 \pm 0.1) \times 10^{-12}$
8	$N^+ + O_2$	$NO^+ + O$	$(6.0 \pm 1.0) \times 10^{-10}$
9	$N^+ + O_2$	$O_2^+ + N$	$6 \times 10^{-10} - 2 \times 10^{-9}$

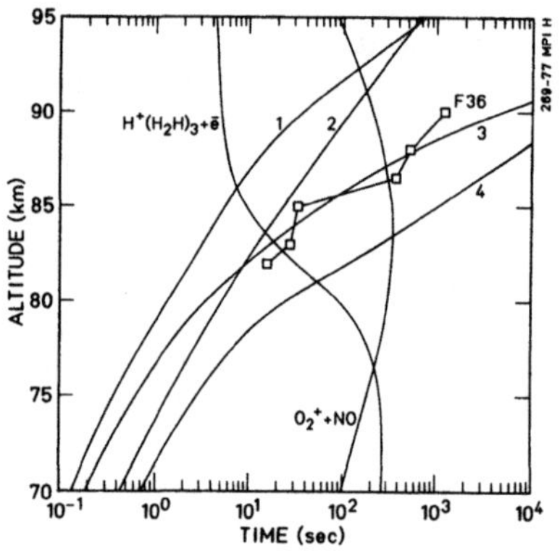

Tab. 3: Reactions involved in the formation of O_2^+ and NO^+ in the D- and lower E-region. Rate constants in cm^3s^{-1}.

Fig. 15 (right): Time constants for the O_2^+-chemistry. The numbers denote: Time constant for $O_2^+ + O_2^+ + M = O_4^+ + M$ summer (1) and winter (2). Effective time constant for $O_2^+ \longrightarrow O_2^+ \cdot H_2O$ also considering the process $O_4^+ + O \longrightarrow O_2^+ + O_2$; summer (3) and winter (4). [M] and temperature were taken from CIRA 1972. For [O] see text. The data points (squares) denote effective time constants for O_2^+-clustering as derived from a D-region positive ion composition measurement at high latitudes during summer (see text).

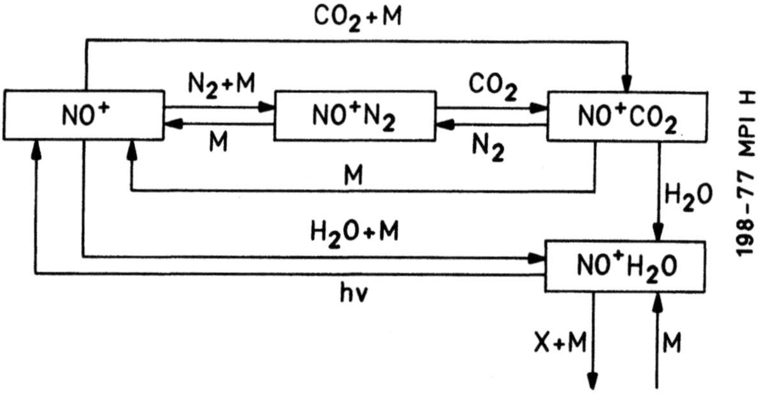

Fig. 16: Reaction scheme for the conversion of NO^+ to cluster ions in the D-region. From Arnold and Krankowsky (1977 d).

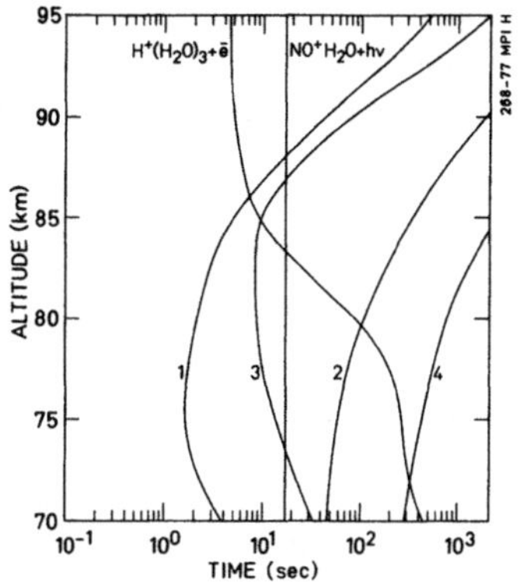

Fig. 17 Time constants for the NO^+-cluster ion chemistry in the D-region. Numbers denote: effective time constants for the conversion of NO^+ to $NO^+ \cdot H_2O$, summer (1), winter (2). Temperature enhancement by 20 degrees at all heights summer (3) and winter (4).

typical electron density of $2 \cdot 10^3 cm^{-3}$ and an effective recombination coefficient of $4.5 \cdot 10^{-6}(T/300)^{-0.5}$ corresponding to $H^+(H_2O)_2$ (Biondi 1969) a recombination lifetime τ_R of cluster ions (Fig. 15) may be estimated. As seen from Fig. 11 the time constants for cluster ion formation and loss are equal around 86 km which defines about the upper boundary of the regime of cluster ions produced from O_2^+. Due to the steep increase of O_2^+ a variation of the electron density even by a factor of 10 has only little influence on the boundary. However, atomic oxygen which converts O_4^+ back to O_2^+ may lower the boundary. From a steady state-treatment of the continuity equations

$$[O_2^+]\,\tau^{-1}(O_2^+) = [O_4^+]\,(K_1[H_2O] + K_2[O]) \qquad (3)$$

$$[O_4^+]\,(K_1[H_2O]) = [O_2^+ H_2O]\,\tau_L^{-1} \qquad (4)$$

an effective time constant $\tau_{eff}(O_2^+)$ for the conversion of O_2^+ to cluster ions

$$\tau_{eff}(O_2^+) = [O_2^+]\,(1 + \frac{K_2[O]}{K_1[H_2O]}) \qquad (5)$$

may be obtained. Taking total gas densities and temperatures from the CIRA 1972 reference atmosphere, a measured H_2O-mixing ratio by volume of $5 \cdot 10^{-6}$ (Arnold and Krankowsky 1977 a) and O densities as measured by neutral gas mass spectrometers (Arnold et al. 1976 the $\tau_{eff}(O_2^+)$ values shown in Fig. 15 are obtained. The strong seasonal variation is mainly due to different temperatures and O densities. Now the boundary is located around 80 km which is in excellent agreement with the in situ observations. Also shown for comparison are experimental time constants for the conversion of O_2^+ to cluster ions derived from a positive ion composition measurement which took place during summer (Arnold and Krankowsky, unpublished material). These data compare well with the $\tau_{eff}(O_2^+)$ curve for summer which is based upon measurements of atomic oxygen and H_2O carried out at the same time as the ion composition measurement. The agreement strongly supports the concept of O_2^+ conversion via O_2^+ followed by H_2O exchange which was outlined above.

Time constants for NO charge transfer (Fig. 15) of O_2^+ which are based upon the NO profile of Meira (1971) are smaller than $\tau_{eff}(O_2^+)$ at heights above about 81 km. It may be noted, however, that particularly under disturbed conditions when the O_2^+ chain becomes important NO may be enhanced by factors up to about 100 around 80 - 85 km (see chapter 5). Consequently, O_2^+ ions might become converted to NO^+ rather than form cluster ions. In conclusion we may state that the upper boundary of the regime of cluster ions formed from O_2^+ is mainly controlled by atomic oxygen and by the electron concentration. Because of reaction (11) any additional process converting O_4^+ to more stable cluster ions would be of great interest. In particular, a CO_2 exchange of O_4^+ leading to O_2^+ CO_2 may be considered. Unfortunately, there exists no accurate experimental information on the bond-energy of $O_2^+ \cdot CO_2$ and, therefore, the possible role on this reaction remains uncertain. On the other hand, an ion interpreted as $O_2^+CO_2$ has been observed in the D-region (Arnold and Krankowsky 1974) which, provided the interpretation is right, would imply at least that an O_2 exchange of $O_2^+CO_2$ should be extremely slow. It is conceivable that the CO_2 exchange of O_4^+ has only a low exothermicity and, therefore, in the atmosphere, where $[CO_2] / [O_2] = 1.5 \cdot 10^{-3}$, only becomes important at sufficiently low temperatures. In that case a strong temperature dependence of the O_2^+ conversion to cluster ions would be introduced and the importance of atomic oxygen would be reduced.

Now we will discuss the conversion of NO^+ to cluster ions. According to current D-region reaction schemes the first step is clustering of N_2, CO_2 or H_2O to NO^+ (Heimerl and Vanderhoff 1971, Heimerl et al. 1972). Due to the high abundance of N_2 the formation of cluster ions by N_2 association is most effective followed by CO_2 and H_2O association (Fig. 16). Reaction coefficients are summarized in Tab. 5 . The occurrence of N_2 clustering in the D-region was inferred from in situ measurements of the ion $NO^+ \cdot N_2$ and from the high abundances of NO^+ cluster ions relative to NO^+ (Arnold and Krankowsky 1974) at a time when there existed only conflicting room temperature laboratory measurements of the corresponding reaction coefficient. From the D-region in situ measurements the reaction coefficient was determined as a function of temperature

$$k = 2\cdot 10^{-31}(T/300)^{-5}\,cm^6 s^{-1} \qquad (6)$$

(Arnold et al. 1974; Arnold and Krankowsky, 1977 b).

No.	Educts	Products	Rate Constant	Reference	
10	$O_2^+ + 2\,O_2$	$O_4^+ + O_2$	$2.6 \times 10^{-30}(T/300)^{3.2}$	Payzant et al.	1973
11	$O_4^+ + O$	$O_2^+ + O_3$	$(3 \pm 2) \times 10^{-10}$	Fehsenfeld and Ferguson	1972
12	$O_4^+ + H_2O$	$O_2^+(H_2O)$	1.5×10^{-9}	Howard et al.	1972
13	$O_2^+(H_2O)+H_2O$	$H_3^+O(OH)+O_2$	1.0×10^{-9}	Howard et al.	1972
14	$O_2^+ H_2O + H_2O$	$H_3O + OH + O_2$	2×10^{-10}	Howard et al.	1972
15	$H_3^+O(OH)+H_2O$	$H^+(H_2O)_2+OH$	1.4×10^{-9}	Howard et al.	1972
16	$H_3O^+ + H_2O + M$	$H^+(H_2O)_2+M$	$1.2 \pm 0.4) \times 10^{-27}$	Howard et al.	1972
17	$O_2^+(H_2O) + h\nu$	products	$0.6\ s^{-1}$	Vanderhoff and Niles	1976

Tab. 4: Reactions involved in the conversion of O_2^+ to proton hydrates in the D-region. Rate constants in cm^3s^{-1} (2-body) and cm^6s^{-1} (3-body).

No.	Educts	Products	Rate Constant	Reference	
18	$NO^+ + 2N_2$	$NO^+(N_2) + N_2$	$2 \times 10^{-31}(300/T)^{4.4}$	Johnsen et al.	1975
19	$NO^+ + N_2 + M$	$NO^+(N_2) + M$	$2 \times 10^{-31}(300/T)^5$ (125-180 K)	Arnold and Krankowsky	1977
20	$NO^+ + N_2 + NO$	$NO^+(N_2) + NO$	$(2.0 \pm 0.6) \times 10^{-31}$	Heimerl and Vanderhoff	1974
21	$NO^+(N_2) + CO_2$	$NO^+(CO_2) + N_2$	$\sim 10^{-9}$	Heimerl and Vanderhoff	1974
22	$NO^+(N_2) + M$	$NO^+ + N_2 + M$	3×10^{-13} (210-220 K)	Arnold and Krankowsky	1977
23	$NO^+ + CO_2 + N_2$	$NO^+(CO_2) + N_2$	$(2.5 \pm 1.5) \times 10^{-29}$ (200 K)	Dunkin et al.	1971
24	$NO^+(CO_2) + H_2O$	$NO^+(H_2O) + CO_2$	$\sim 10^{-9}$	Dunkin et al.	1971
25	$NO^+ + H_2O + N_2$	$NO^+(H_2O) + N_2$	1.6×10^{-28}	French et al.	1973
26	$NO^+(H_2O)+H_2O+N_2$	$NO^+(H_2O)_2+N_2$	1.1×10^{-27}	French et al.	1973
27	$NO^+(H_2O) + h\nu$	products	$0.06\ s^{-1}$	Vanderhoff and Niles	1976
28	$NO^+(H_2O)_2+H_2O+N_2$	$NO^+(H_2O)_3+N_2$	1.9×10^{-27}	French et al.	1973
29	$NO^+(H_2O)_3+H_2O$	$H^+(H_2O)_3+HNO_2$	7×10^{-11}	Fehsenfeld et al.	1971

Tab. 5: Reactions involved in the conversion of NO^+ to proton hydrates in the D-region. Rate constants in cm^3s^{-1} (2-body) and cm^6s^{-1} (3-body).

Furthermore, a NO^+N_2 bond-energy of about 0.2 eV was inferred. Laboratory measurements at low temperature carried out recently are in reasonable agreement with the above values (Johnsen et al. 1975, Turner and Conway 1976). Due to the low bond-energy of NO^+N_2, thermal decomposition of this ion starts to compete with CO_2 exchange (Tab. 5) at temperatures above 180 K in the atmosphere. Thus the effective time constant for the NO^+ conversion to cluster ions is increased to an extent which critically depends on the NO^+N_2 bond-energy, the temperature, and the room temperature reaction coefficient for thermal decomposition of NO^+N_2. A theoretical effective time constant based upon a bond-energy of 0.219 eV (Turner and Conway 1976) and a decomposition rate constant k(300 K) = $10^{-11} cm^3 s^{-1}$ which is half the upper limit inferred from Heimerl and Vanderhoff (1974) is shown in Fig. 18 (Curve 4).

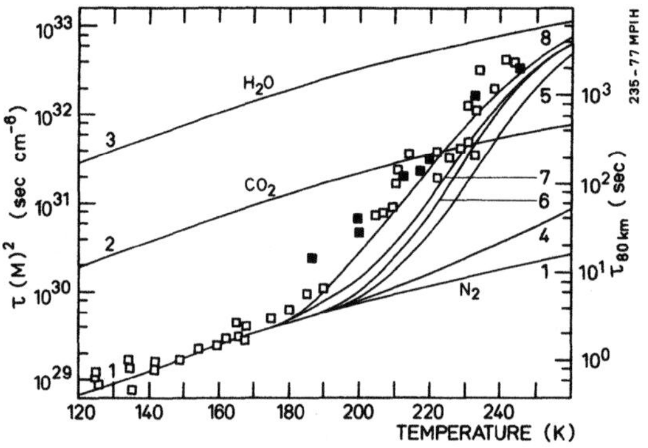

Fig. 18: The product of the effective time constant for the conversion of NO^+ to cluster ions and the square of the gas density as a function of temperature. Data points are derived from D-region positive ion composition measurements. Curves give theoretical values (see text). From Arnold and Krankowsky (1977d).

By comparing curves 4 and 2 it is seen that even if NO^+N_2 decomposition is considered N_2 clustering still remains the dominant process for the NO^+ conversion. After further hydration of NO^+H_2O proton hydrates may be formed from $NO^+(H_2O)_3$ in a binary reaction with H_2O (Good et al. 1970 b, Fehsenfeld and Ferguson 1969). Realizing that this chain is by far to slow in order to account for the rapid transition from NO^+ to proton hydrates observed in the D-region it has been suggested that the formation of $NO^+(H_2O)_2$ and $NO^+(H_2O)_3$ also involves N_2 or CO_2 clustering (Heimerl and Vanderhoff 1971, Heimerl et al. 1972). However, as no laboratory data exist on N_2 and CO_2 clustering of $NO^+(H_2O)$, and $NO^+(H_2O)_2$, there is no way at present to judge on the validity of such a concept. Considering the recent finding of NO^+H_2O - photolysis by visible sunlight with a photodissociation coefficient J = 0.06 s^{-1} (Vanderhoff and Niles 1976), there is an even greater need for a rapid conversion of

NO^+H_2O to higher order cluster ions. Based upon D-region positive ion composition measurements it has been proposed recently that NO^+CO_2 undergoes thermal decomposition and N_2 exchange at temperatures above 180 K (Arnold and Krankowsky 1977 d) causing a steep increase of the effective time constant for the NO^+ conversion to NO^+H_2O. The experimental data points, on which the above conclusion is based, are shown in Fig. 12 along with theoretical curves. Curve 8 considers thermal decomposition and N_2 exchange of $NO^+ \cdot CO_2$ assuming a bond-energy of 0.39 eV which is consistent with the lower limit given by Dunkin et al. (1971). Although being endothermic by about 0.17 eV, the N_2 exchange of NO^+CO_2 may become important at higher temperatures because of the high ratio $|N_2| / |CO_2| = 2.7 \cdot 10^3$. The importance of this reaction depends strongly on the accurate values for the bond-energies of NO^+CO_2 and NO^+N_2. Regarding the further reactions leading from NO^+H_2O to proton hydrates we propose, in contrast to current schemes, that not $H^+(H_2O)_3$ but $H^+(H_2O)_2$ is the first proton hydrate formed by the NO^+ chain. This conclusion is based upon the observation of $H^+(H_2O)_2$ in the D-region under conditions when NO^+ was the only precursor of importance (see for example Fig. 4). The reaction responsible probably involves $NO^+(H_2O)_2$ and a neutral hydrogen compound. A scheme of the NO^+ cluster ion chemistry including thermal decomposition and N_2 exchange of NO^+CO_2 as well as the link between $NO^+(H_2O)_2$ and $H^+(H_2O)_2$ is shown in Fig. 13. The effective time constants for the conversion of NO^+ to NO^+H_2O are given as a function of altitude for winter and summer at middle latitudes (Fig. 17). Also shown for comparison are the time constants for cluster ion recombination and for NO^+H_2O photolysis. Mainly due to seasonal temperature variations a strong variability of τ_{eff} (NO^+) is introduced. The upper boundary of the cluster ion regime is located around 86 km (summer) and around 80 km (winter) which agrees well with the D-region observations. Aside from boundary variations introduced by seasonal and latitudinal temperature variations also irregular temporal temperature variations which are characteristic for the winter D-region (c.f. Schmidlin 1976) have to be considered. As seen from Fig. 14 an increase by 20 degrees has little effect on the boundary during summer, but a drastic effect during winter. In winter the boundary may be shifted downwards by 8 km depending on the ionisation rate which is mainly determined by the NO-concentration and the solar zenith angle. Regarding the further built-up of cluster ions from NO^+H_2O a very fast reaction is required in order to compete with NO^+H_2O photolysis. This reaction needs to be much faster than the conversion of NO^+ to NO^+H_2O. Consequently, there are only few possibilities either a 2-body reaction leading to H_3O^+ which according to laboratory studies, however, does not seem to occur (Fehsenfeld and Ferguson 1972; Fehsenfeld et al. 1975) or a very rapid 3-body reaction having a strong inverse temperature dependence.

The implications of the above scheme for D-region plasma densities are important. The strong temperature dependence of $\tau_{eff}(NO^+)$ introduces a pronounced temperature control of D-region plasma densities and thereby a sensitive response to seasonal, latitudinal, and irregular temporal fluctuations. The latter seem to be much more pronounced during winter when the average D-region temperatures (210 K) are close to the steepest portion of the $\tau_{eff}(NO^+)$ curve. Another factor emphasizing wintertime variations is the small gradient of the τ_R curve below 80 km. This temperature control may be at least one explanation for the well known winter variability of D-region plasma densities and radio wave absorption, the most dramatic manifestation of which is the winter anomaly. Another well known feature of the D-region plasma density profile namely the steep ledge occuring around 85 km during summer is explainded by the location of the boundary around 85 km (Fig. 17). This leads to a coincidence of an abrupt decrease of α_B with the steep increase of the ionisation by EUV radiation and X-rays which due to the lower solar zenith angle penetrate to about this height during summer. The winter situation is completely different, here the EUV and X-radiations do not penetrate into the mesosphere and the ionisation rate does not change as drastically with height below about 90 km. On the other hand, the change of α_B associated with the boundary occurs somewhere around or below 80 km. This explains why usually no steep ledge is observed during winter. The interesting question of proton hydrate growth will not be touched here because little influence on α_B should be expected. The reason is that the electron recombination coefficients of the third and fourth proton hydrate which usually are dominant are lower than the upper limit of $10^{-5}cm^3s^{-1}$ (Leu et al. 1973) only by a factor of about two.

At heights below the D-region the formation of proton hydrates from primary ions, mainly N_2^+ and O_2^+, proceeds the faster the higher the gas density becomes. At 40 km the time constant should be less than 0.01 seconds. Also the equilibrium distribution of proton hydrates which depends on the H_2O density and the temperature is rapidly established within less than seconds. For typical densities and temperatures the distribution of proton hydrates should peak around $H^+(H_2O)_5$. The non-proton-hydrates (NPH) are probably formed by reactions of proton hydrates with a neutral molecule. These molecules should have higher proton affinity or higher permanent dipole moment than H_2O, thus allowing proton transfer or exchange reactions to occur. Furthermore, the NPH of lowest mass (29±2 amu involves a molecule of mass 28±2 attached to a proton. This molecule may be identified as formaldehyde (H_2CO) which is expected in considerable concentrations as a product of the stratospheric CH_4 oxydation (e.g. Stewart and Hoffert 1975). The proton affinity of H_2CO slightly exceeds that of H_2O (Beauchamp and Butrill 1968, Kebarle et al. 1976). This has been directly proven by the laboratory measurement of the reaction.

$$H_3O^+ + H_2CO = H^+H_2CO + H_2O \qquad (7)$$

(Beauchamp and Butrill 1968). Regarding H_2CO reactions of higher order proton hydrates no measurements are given in the literature. Recent observations at our laboratory (Arnold et al. 1976) infer that $H^+(H_2O)_n$ with n = 1 to 4 reacts with H_2CO. Assuming equilibrium conditions the continuity equation

$$[H^+(H_2O)_n] \; K \; [X] = [NPH^+] \; \alpha_{ii} \; [N^-] \qquad (8)$$

relates the concentration of the unknown stratospheric molecule reacting with proton hydrates to the measured ratio of ion concentrations and to the density of negative ions which at these heights equals the plasma density. The reaction coefficient k may be assumed to be $10^{-10} cm^3 s^{-1}$ and the ion-ion recombinations coefficient α_{ii} = $5 \cdot 10^{-8} cm^3 s^{-1}$ (Tab. 1). The [X] profile is shown in Fig. 19 along with a theoretical H_2O profile based upon a photochemical model including vertical turbulent diffusion (Stewart and Hoffert 1975) and with a profile of the total gas density [M]. A comparison of [X] and [M] shows that [X] decreases much steeper than [M] below the stratopause and thus identifies X as a typical stratospheric molecule. The rough agreement of [X] and [H_2CO] regarding both relative as well as absolute numbers may provide another argument in favour of H_2CO or at least a hydrocarbon related to the CH_4 chemistry. It should be noted, however, that many stratospheric trace gases have similar profiles and that the argument may not be as conclusive as it seems from Fig. 19.

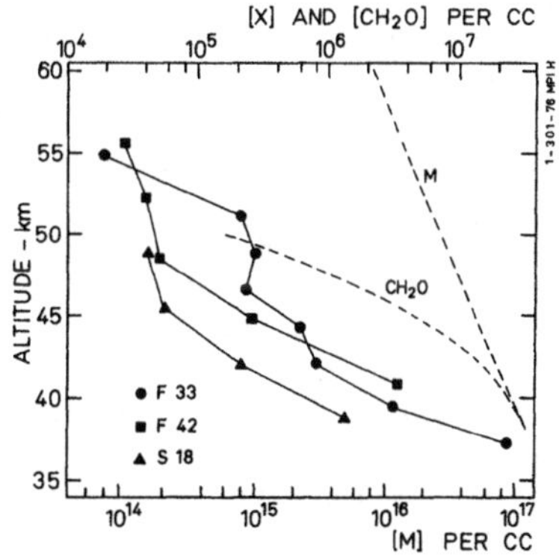

Fig. 19: Number densities of the unknown reaction partner X of stratospheric proton hydrates as derived from stratospheric positive ion composition measurements (F33, F42, S18). A theoretical CH_2O-profile (Stewart and Hoffert 1976) and a profile of the total gas density [M] are shown for comparison (Arnold et al. 1977a).

Coming back to our basic line of argumentation we are confronted with the question whether the conversion of proton hydrates to NPH has any impact on the effective recombination coefficients. A conclusive answer cannot be given at the present as there exist no laboratory measurements of α_{ii} values for these ions. However, according to the limited data existing for binary ion-ion recombination coefficients of complex ions (Smith and Church 1976) no strong influence should be expected. Regarding 3-body recombination the answer is even less certain. Considering the very strong inverse temperature dependence of this process an influence of ion complexity is conceibable. In that case one question is whether NPH contain more atoms than proton hydrates at a given height. The other question is whether a possible difference in complexity would affect the effective recombination coefficient.

In summarizing the discussion of positive ions we may conclude that we seem to approach a satisfactory understanding of the important ion chemical processes controlling the D-region positive ion composition while our quantitative knowledge of the ionisation rates is still somewhat uncertain due to the nitric oxide concentrations which are not well established (see chapter 5).

4. NEGATIVE IONS

4.1. COMPOSITION OF NEGATIVE IONS

There exist only few mass spectrometric in situ measurements of atmospheric negative ions which is in contrast to the great number of positive ion composition measurements. The major reason is that negative ions probably are present in concentrations comparable to those of positive ions only at altitudes below about 80 - 70 km which is close to the lower height limit of previous rocket-borne mass spectrometer measurements. Furthermore, it seems as if the negative ion population contains more species of comparable abundance than the positive ion population. This implies lower partial densities of negative ion species which are more difficult to detect.

All measurements existing to date were carried out by the AFGL and the MPK groups. First results published in 1971 (Narcisi et al. 1971, Arnold et al. 1971) were rather conflicting both ion species and upper boundaries of the negative ion regime. While the nighttime measurement of the MPK group revealed a sharp upper boundary around 78 km (Fig. 20) the AFGL daytime data (Fig. 21) infer a boundary around 90 - 95 km. The major ion species found by MPK are CO_3^-, HCO_3^-, and Cl^- while the AFGL data mainly show $NO_3^-(H_2O)_n$. Subsequent measurements by the AFGL group have only deepened the mystery showing high negative ion count rates even at E-region altitudes (Narcisi et al. 1972 a,b). In contrast to the AFGL results recent MPK measurements (Arnold and Krankowsky, unpublished

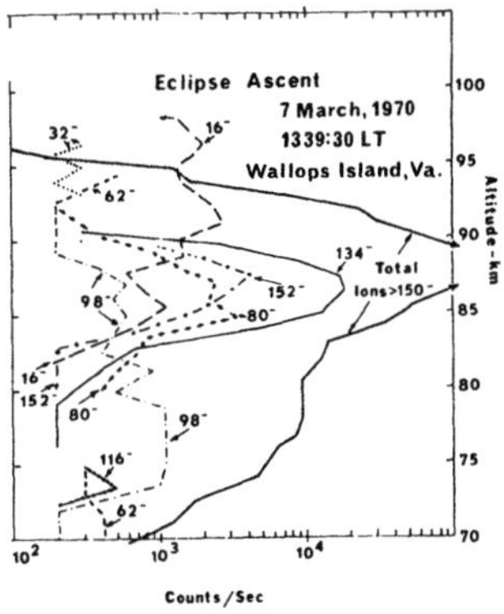

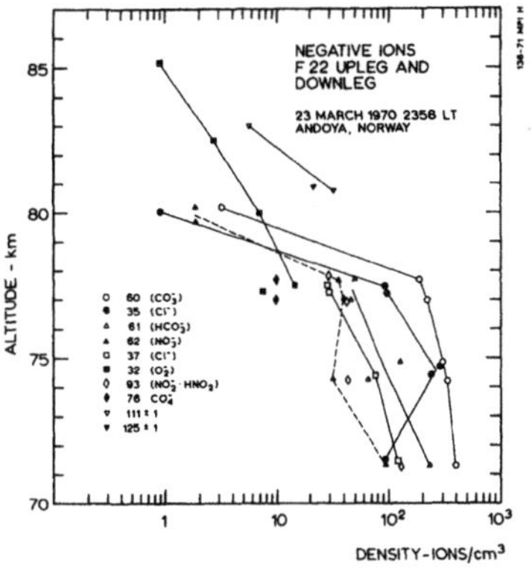

Fig. 20 (left): Daytime negative ion composition at middle latitudes during spring. From Narcisi 1972 .

Fig. 21 (right): Nighttime negative ion composition at high latitudes during spring. From Arnold and Krankowsky (1971).

material) again exhibit sharp boundaries of the negative ion regime located at altitudes of 74 km and 70 - 78 km (night) and 68 km (day) (Fig. 22).

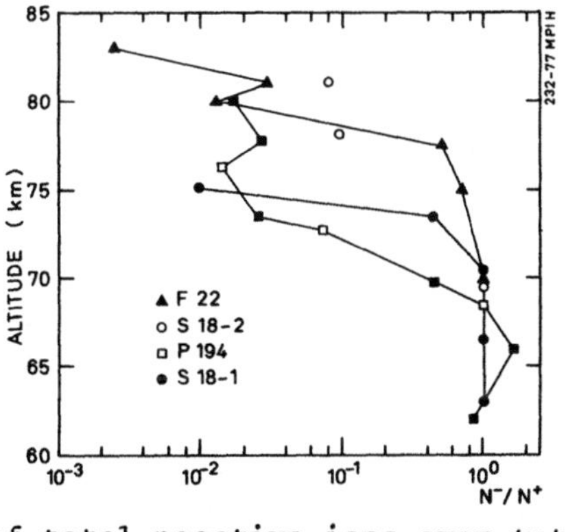

Fig. 22: Ratios of total negative ions over total positive ions as measured by ion mass spectrometer measurements (Arnold and Krankowsky, unpublished material). (S18-1, S-18,2) winter night at high latitudes during strong energetic particle precipitation. (F22) spring night at high latitudes. (P194) winter day middle latitudes.

Also included in Figure 22 is the first MPK measurement of negative ions (F22). It seems interesting to note that the MPK nighttime measurements consistently show the boundary around 70 - 78 km while the daytime boundary is located at a somewhat lower altitude. An example of simultaneous measurements of the total negative ion density (MPK mass spectrometer), the total positive ion density and the total positive cluster ion density (both MPK mass spectrometer) as well as the electron density (Faraday rotation, Bjorn 1977 private communication) is shown in Figure 23.

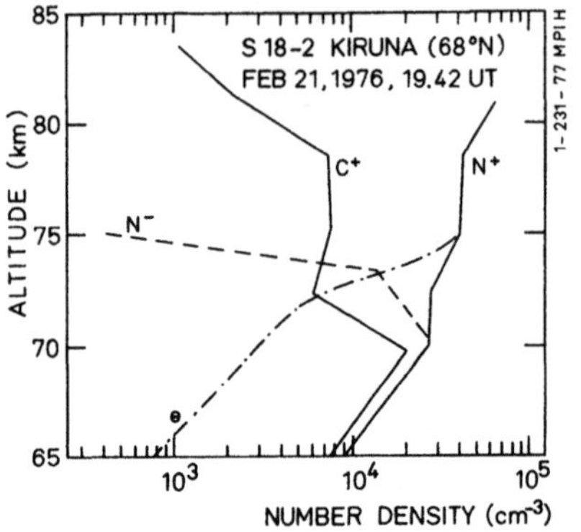

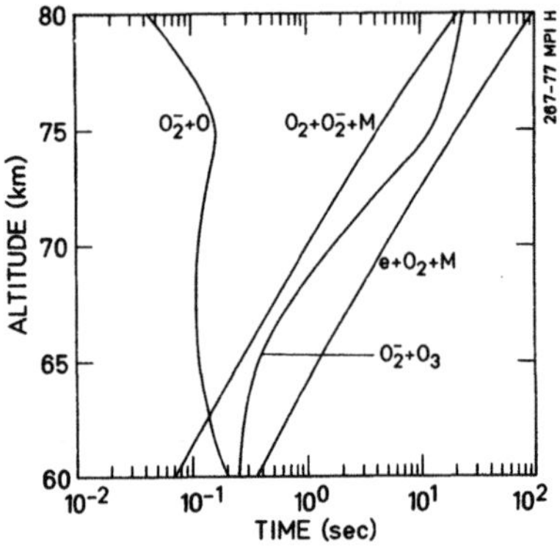

Fig. 23 (left): Simultaneous mass spectrometric measurements of densities of total positive ions (N^+), total negative ions (N^-) and total positive cluster ions (C^+) (Arnold and Krankowsky unpublished material). Electron densities (e) also measured simultaneously by the Faraday rotation method (Björn unpublished material) are also shown.

Fig. 24 (right): Time constants for the chemistry of negative ions in the D-region.

The experiment took place at high latitudes during strongly enhanced ionsation at night. The nice complemantarity of total negative ions and electrons observed clearly demonstrated the existence of a sharp boundary of the negative ion regime. It may be noted that the positive ion composition significantly responds to the abrupt change in the nature of negatively charged species around 74 km. Obviously, due to the abrupt decrease of α_B below 74 km, the recombination lifetime of positive cluster ions increases causing an abrupt increase in their relative abundance. This may be regarded as another independent support for the location of the negative ion boundary. Simultaneous measurements of electron densities carried out during the other MPK mass spectrometer measurements also supported the locations of the negative ion boundaries shown in Fig. 22. Additional support is provided in the experiment F22 by the observation of an abrupt change of the negative current measured by an electrostatic probe (Folkestad 1970, private commun.). This behaviour was attributed to an abrupt change of the dominant negatively charged species. Considering the conflicting results obtained by the AFGL and MPK groups, it seems unlikely that these can be explained by natural variations of atmospheric negative ions alone. Therefore the major differences of the experimental methods applied by the two groups will be briefly described.

The AFGL instruments exclusively measured negative ions and the mass scale was based upon laboratory calibrations. As the sensitivity strongly depends on altitude (see part 2) in a way which may considerably vary from experiment to experiment depending on rocket velocity and altitude and on pump performance no direct determination of number densities was possible in previous measurements. In the case of positive ions by a normalisation of the sum of all partial count rates to the total positve ion density measured independently by probes, absolute number densities can be obtained. For negative ions such a possibility does not exist at altitudes where the total negative ion density markedly differs from the total positive ion density. Because of this problem the AFGL data give only count rates and no number densities for negative ions. The MPK-instrument measures negative and positive ions and recently also neutrals almost simultaneously (Arnold et al. 1977b) Therefore, an inflight calibration of the mass scale and the altitude dependence of the sensitivity is possible allowing the determination of negative ion number densities. Other differences are the sampling cone, lower extraction potentials and higher pumping speed of the MPK-instruments except for experiment F22.

In concluding the discussion of negative ion composition measurements another striking difference of the AFGL and MPK data should be noted. While the latter reveal the presence of high relative abundances of Cl^- in the D-region (Arnold et al. 1971) no such indication is shown by the former results. As discussed by Arnold et al.(1971)and by Arnold and Krankowsky(1971)an identification as contamination which was favored by the AFGL workers (Narcisi et al. 1972a)seems to be unlikely. Instead, it was proposed that Cl^- results from reactions of other negative ions with neutral atmospheric Cl compounds which were estimated to have a minimum concentration of $10^5 cm^{-3}$ around 75 km corresponding to a mixing ratio by volume of about 10^{-10}. This conclusion is strongly supported by recent measurements of stratospheric HCl indicating a mixing ratio of about 10^{-9} (Farmer et al. 1976; Lazarus et al. 1976). As no loss process for Cl compounds is known above the tropopause and as HCl is expected to be the dominant Cl compound in the stratosphere a mixing ratio for Cl compounds of the same order should be expected at D-region altitudes. This is in a strikingly good agreement with the rough estimate based upon Cl^- measurements in the D-region. Therefore, it seems likely that Cl^- is an atmospheric ion which consequently should be included in the discussion of the atmospheric negative ion chemistry.

4.2. CHEMISTRY OF NEGATIVE IONS

As suggested by the variety of negative ion species observed in the D-region the underlying chemistry seems to be more complex when compared with that of positive ions. A reaction scheme for negative ions, mainly based upon the one proposed by Ferguson (1974), but in addition including Cl^- is shown in Fig. 25 . The corresponding reaction coefficients are compiled in Tab. 6. Clustering of negative ions is not included except for reaction (32). The most important formation process for atmospheric ions currently considered is 3-body electron attachment to molecular oxygen. After formation O_2^- is rapidly converted to CO_3^- and CO_4^- which more slowly react to form NO_3^- and Cl^-. All ions except NO_3^- are subject to attack by fast reactions involving atomic oxygen, atomic hydrogen, or photons. These reactions either cause direct electron detachment or lead to less stable negative ions. The nighttime upper boundary of the negative ion regime is rather sharply defined by the steep increase of atomic oxygen above 75 - 80 km which inhibits the built-up of stable negative ions. A gradual increase of the height of the boundary within the 70 - 80 km region associated with the gradual recombination of atomic oxygen after sunset should be expected (Fig. 24).

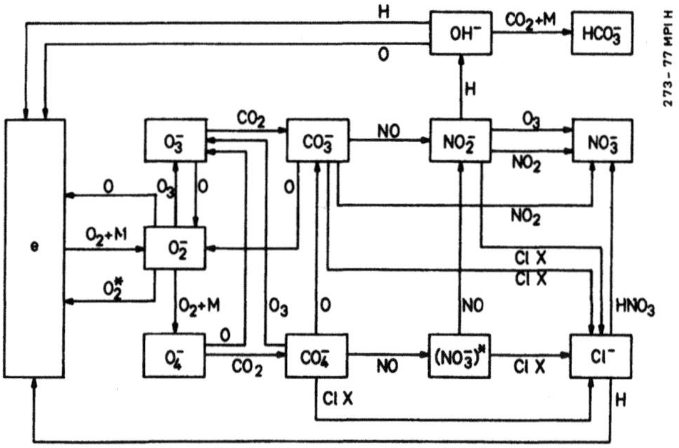

Fig. 25: Reaction scheme for negative ions in the D-region based upon the scheme of Ferguson (1974) but in addition including reactions involving chlorine compounds and HNO_3.

During daytime atomic oxygen is abundant throughout the D-region and in addition photolyses and photodetachment complicate the chemistry. As reaction (32) is faster than (33), at heights where we expect the boundary, O_2^- converts mainly to CO_3^- which in turn is predominantly lost by reaction (44). Consequently, the ratio $[CO_3^-]/[e^-]$ should be low for $(K_{31}[O_3]+K_{32}[O_2][M]) < K_{44}[O]$ corresponding to an altitude around 63 km. However, a major uncertainty involved is the unknown temperature dependence of reaction (32). If we assume a similar temperature dependance as found for production of the analogue ion O_4^+ the boundary would be located at a somewhat higher altitude around 66 km for typical D-region temperatures. Because of the stability of NO_3^- even a relatively low production of this ion allows a high ratio $[NO_3^-]/[e^-]$.

No.	Educts	Products	Rate Constant$^+$	Reference	
30	$e + O_2 + M$	$O_2^- + M$	1.5 - 30	Chanin et al.	1959
31	$O_2^- + O_3$	$O_3^- + O_2$	3.0 - 10	a	
32	$O_2^- + O_2 + M$	$O_4^- + M$	4.0 - 31	a	
33	$O_2^- + O$	$O_3 + e$	33 - 10	a	
34	$O_2^- + O_2(^1\Delta g)$	$2\ O_2 + e$	2 - 10	a	
35	$O_4^- + O$	$O_3^- + O_2$	4.0 - 10	a	
36	$O_4^- + CO_2$	$CO_4^- + O_2$	4.3 - 10	a	
37	$O_4^- + NO$	$(NO_3^-)^* + O_2$	2.5 - 10	a	
38	$O_3^- + CO_2$	$CO_3^- + O_2$	5.5 - 10	a	
39	$O_3^- + O$	$O_2^- + O_2$	1 - 10	Le Levier Branscomb	1968
40	$O_3^- + NO$	products	1.0 - 11	a	
41	$CO_4^- + O$	$CO_3^- + O_2$ $O_3^- + CO_2$	1.5 - 10	a	
42	$CO_4^- + NO$	$(NO_3^-)^* + CO_2$	4.8 - 11	a	
43	$CO_4^- + O_3$	$O_3^- + O_2 + CO_2$	1.3 - 10	a	
44	$CO_3^- + O$	$O_2^- + CO_2$	8.0 - 11	a	
45	$CO_3^- + NO$	$NO_2^- + CO_2$	9.0 - 12	a	
46	$CO_3^- + NO_2$	$NO_3^- + CO_2$	8.0 - 11	a	
47	$(NO_3^-)^* + NO$	$NO_2^- + NO_2$	1.5 - 11	a	
48	$NO_2^- + H$	$OH^- + NO$	4.0 - 10	a	
49	$NO_2^- + O_3$	$NO_3^- + O_2$	1.8 - 11	a	
50	$NO_2^- + NO_2$	$NO_3^- + NO$	4 - 12	a	
51	$OH^- + O$	$HO_2 + e$	2.0 - 10	a	
52	$OH^- + H$	$H_2O + e$	1.0 - 9	a	
53	$OH^- + CO_2 + O_2$	$HCO_3^- + M$	7.6 - 28	Fehsenfeld and Ferguson	1974
54	$Cl^- + H$	$HCl + e$	1.0 - 9	a	
55	$Cl^- + HNO_3$	$NO_3^- + HCl$	1.6 - 9	Fehsenfeld et al.	1975
56	$CO_3^- + h\nu$	$O^- + CO_2$	0.18	Cosby et al.	1976
57	$CO_3^- \cdot H_2O + h\nu$	$CO_3^- + H_2O$	1.2	Cosby et al.	1976
58	$NO_3^- + HCl$	$Cl^- + HNO_3$	<2 - 12	Ferguson et al.	1972
59	$NO_2^- + HCl$	$Cl^- + HNO_2$	1.4 - 9	Ferguson et al.	1972

Tab. 6: Negative ion-processes in the D-region. Rate constants in cm^3s^{-1} (2-body) and cm^6s^{-1} (3-body). Photodissociation coefficients in sec^{-1}. For example 1.5 - 30 denotes 1.5×10^{-30}. (a) atomic and nuclear data tables (Ferguson 1973).

Assuming equilibrium of NO_3^- production and loss from the continuity equation

$$[CO_3^-]\ K\ [NO] = [NO_3^-]\ \alpha_{ii} [N^+] \tag{9}$$

we obtain approximately

$$[NO_3^-] / [CO_3^-] = 1.8 \cdot 10^{-4}\ [NO] / [N^+] \tag{10}$$

For typical values of [NO] and $[N^+]$ the ratio $[NO_3^-]$ / $[CO_3^-]$ becomes 240 at 70 km yielding a value of $\lambda \sim 2$ at this height. This implies a boundary located around 70 km. If there exist NO_3^- loss processes much faster than ion-ion recombination, such as for example photodetachment, NO_3^- becomes less important and consequently the daytime boundary should be defined by $[CO_3^-] / [e^-]$. The ion Cl^- should not be important during daytime as it is rapidly destroyed by atomic hydrogen. At night Cl^- should be important below about 80 - 75 km where atomic hydrogen is virtually absent. As a precursor of Cl^- the abundant ions CO_3^- and CO_4^- seem to be most favorable while NO_3^- due to its high electron affinity (3.9 eV, Ferguson et al. 1972) should not react with Cl (3.76 eV, Bailey, 1958). As CO_3^- and CO_4^- also represent the most important precursors of NO_3^- we may estimate a maximum ratio

$$[Cl^-] / [NO_3^-]\ \alpha\ 10^{-7} \frac{[ClX]}{[NO]} \tag{11}$$

if CO_3^- is the most important precursor and as long as $|NO| >> |NO_2|$. At altitudes below about 70 km where NO_2 becomes important the ratio $[Cl^-]$ / $[NO_3]$ should become lower. Finally, Cl^- should decrease rapidly at heights where HNO_3 becomes abundant. According to phtochemical models including vertical transport the density of HNO_3 increases steeply below about 50 km (e.g. Shimazaki and Whitten 1976).

As already mentioned clustering of negative ions which is with one exception not considered here may alter the negative ion chemistry considerably, particularly when clustering reactions become sufficiently fast to compete with reactions of the molecular ion scheme shown in Fig. 24. Thus, for example, at night ions like CO_3^- or CO_4^- might undergo clustering before reacting with NO and the clusters, in turn, might react in a way different from the corresponding molecular ions.

Regarding the influence of negative ions on atmospheric plasma densities the following conclusions can be drawn:

a) According to the limited laboratory data of 2-body ion-ion recombination coefficients for complex negative and positive ions no strong influence of the ion nature is expected. The values are close to $5 \cdot 10^{-8} cm^3 s^{-1}$. Consequently, the negative ion chemistry will have no significant influence on the bulk recombination coefficient aside from controlling the upper boundary of the negative ion regime.

b) The upper boundary of the negative ion regime at night is sharply defined by the bottom of the atomic oxygen layer located around 75 - 80 km. During daytime the boundary will again be rather sharp and will be located at a height not lower than about 66 - 63 km depending on atomic oxygen, ozone, and possibly also on temperature. Depending on the value of the unknown photodetachment coefficient of NO_3^- and on the concentration of nitric oxide the boundary can be located as high as about 70 km.

c) Free electrons will be virtually absent below about 80 - 75 km at night. During daytime they will decrease more gradually relative to positive ions below about 70 - 65 km. Below approximately 55 km they will be virtually absent.

d) Associated with the sharp boundaries of the negative ion regime rather abrupt changes of the bulk recombination coefficient are expected. Consequently, plasma densities will be relatively high below 80 - 75 km at night and below 70 - 60 km during day in spite of the low ionisation rates at these altitudes.

5. DIAGNOSTIC ASPECT OF ION COMPOSITION MEASUREMENTS IN THE ATMOSPHERE

The importance of the diagnostic aspect of atmospheric ion composition measurements may be realized by consideration of the following arguments: From the preceeding discussion of the atmospheric ion chemistry it was obvious that various parameters determine the ion composition. Besides others they include neutral atmospheric constituents and rate coefficients of ion chemical processes. Consequently, it should be possible to determine such parameters from in situ ion composition measurements, provided the underlying processes are quantitatively known. In favorable cases a density ratio of two ion species alone must be known. Such an example we have already seen in chapter 3.1. where the density of a stratospheric molecule possibly formaldehyde (H_2CO) was determined. Other examples include nitric oxide (Donahue et al. 1970; Zipf et al. 1970; Arnold et al. 1975; Swider and Narcisi 1977), H_2O_2 (Arnold and Krankowsky 1974), H_2O, H2O (Arnold and Krankowsky 1977b), metals (Zbinden et al. 1975; Krankowsky et al. 1972), ClX (Arnold et al. 1971). Also, as discussed in chapter 3.2., a reaction coefficient of a process which is extremely difficult to measure in laboratory, namely the N_2 association to NO^+, has been determined from atmospheric ion composition measurements (Arnold and Krankowsky 1977b). Another conceivable application is the determination atmospheric temperatures. This might be done by exploiting the strong temperature dependence of the NO^+ conversion to cluster ions (Fig. 18) or by making use of the temperature sensitivity of the distribution of proton hydrates. The advantage of the method would be a high accuracy and altitude resolution for the measurement of relative temperatures.

The reason for the powerful diagnostic possibilities of ion composition measurements regarding the determination of neutral trace gases is the very high sensitivity. As seen from Fig. 19, a detection limit of 10^4 molecules cm^{-3} was possible which can be lowered by a factor of 10 without great difficulty.

Coming back to the line of argumentation of this article we will discuss only those out of the above results which are relevant to atmospheric plasma densities. They are water vapor, nitric oxide, and temperature.

Water vapor densities were determined by Arnold and Krankowsky (1977a) in the following way: at low temperatures NO^+H_2O is predominantly formed by H_2O exchange of NO^+CO_2 (see chapter 3.2.). At heights above 85 km NO^+H_2O and under conditions of enhanced ionisation also NO^+CO_2 is mainly destroyed by electron recombination. Therefore, the H_2O density may be determined from the measured densities of NO^+H_2O, NO^+CO_2, and electrons from the continuity equation:

$$[NO^+CO_2]\ K\ [H_2O] = [NO^+H_2O]\ \alpha\ [e^-] \qquad (12)$$

Another reaction path providing an independent H_2O density also used by the above authors is:

$$[O_2^+H_2O] + [H_2O] \qquad [H_3O^+OH] + [O_2] \qquad (13)$$

$$[H_3O^+OH] + [H_2O] = [H^+(H_2O)_2] + [OH] \qquad (14)$$

The results of 2 experiments (Fig. 26) infer H_2O mixing ratios by volume between 4 and $6 \cdot 10^{-6}$. The most important implications of this result concerning plasma densities relates to the degree of hydration of proton hydrates and to the transformation of NO^+ to cluster ions (see chapter 3.2.).

Another example for deriving neutral trace constituents from ion composition is nitric oxide. Its concentration is poorly known at D-region altitudes because the experimental method mostly used, the measurement of the NO γ-band emission (Barth 1966, Pearce 1969, Meira 1971), is rather uncertain at these heights. Another reason is the limited number of direct measurements which is insufficient to cover the range of variability to be expected for the long lived nitric oxide. Consequently, an indirect determination of nitric oxide is important. Various NO density profiles have been derived from positive ion composition measurements during auroral conditions while little information exists about middle latitudes. Two middle latitude profiles derived from positive ion composition measurements carried out during winter anomalous conditions (Arnold and Krankowsky 1978) are shown in Fig. 27.

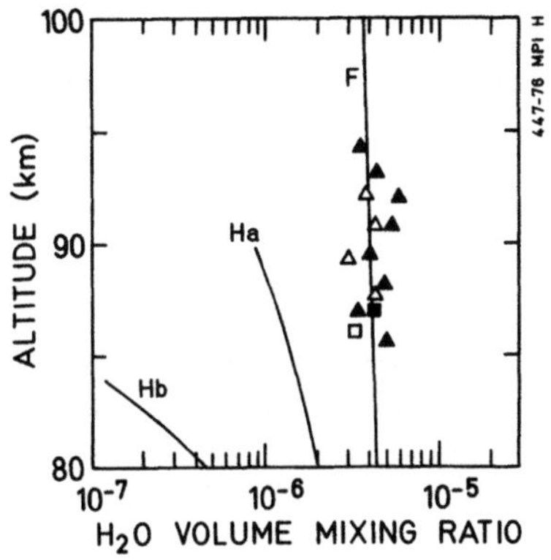

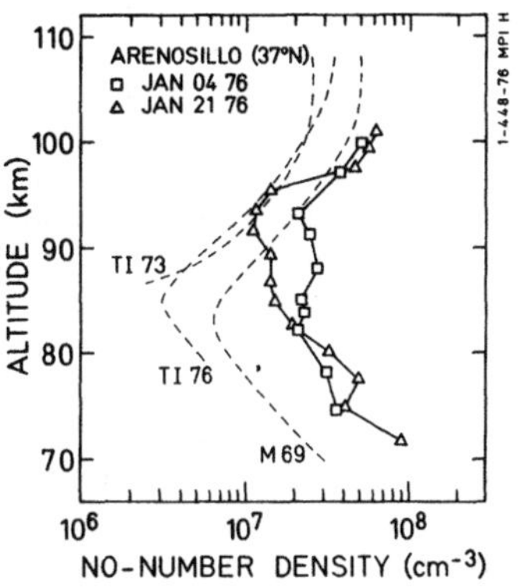

Fig. 26 (left): Water vapor mixing ratios as derived from D-region positive ion composition measurements (from Arnold and Krankowsky 1977a). Theoretical values of Fukuyama 1974 (F) and Hesstvedt 1970 (Ha strong turbulent transport; Hb weak turbulent transport) are shown for comparison.

Fig. 27 (right): Nitric oxide densities as derived from D-region positive ion composition measurements under winter anomalous conditions (from Arnold and Krankowsky 1977c). Optical measurements of Meira(1971)(M69) and Tomatsu and Iwagami 1969 (M69) and(1973)(M73) are shown for comparison.

Along with optical measurements (Meira 1971; Tomatsu and Iwagami 1973, 1976). While the NO densities derived from ions compare reasonably well with Meira's values above about 90 km they are higher by a factor of about 3 to 4 at heights below. This enhancement may have been related to the winter anomalous conditions prevailing during the mass spectrometric measurements. From a comparison of NO densities measured at D-region altitudes a substantial variability is inferred. Consequently, a corresponding variability of D-region ionisation rates should be expected (Fig. 2). Another possible implication concerning plasma densities is related to the location of the daytime upper boundary of the negative ion regime which may be controlled by the production of NO_3^- which is determined also by the NO density (see chapter 4.2.).

6. CONCLUDING REMARKS

In summarizing the discussion of the atmospheric plasma at heights between 20 and 100 km it may be concluded that our understanding has made considerable progress during recent years. Although many details of the ion composition and ion chemistry are still uncertain it seems, as if we were approaching an understanding of the major features of the ion composition and of the underlying processes relevant to the bulk plasma parameters. In particular, the D-region plasma (60-90 km) which undergoes considerable variations seems to be much better understood now due to an improved insight into the transformations of positive molecular ions to cluster ions and of electrons to negative ions which control the bulk recombination coefficient. A major finding seems to be the strong temperature control of the D-region plasma introduced by the transformation of NO^+ and O_2^+ to cluster ions and possibly also by the transformation of O_2^- to O_4^- determining the built-up of negative ions. A major problem of the D-region remains the poor knowledge of nitric oxide which determines the ionisation rates and possibly also the upper daytime boundary of the negative ion regime.

Based upon improved ion mass spectrometric in situ measurements and an improved understanding of the ion chemistry the diagnostic possibilities of atmospheric ion composition measurements can be exploited. Striking examples are the determination of mesospheric water vapor and a stratospheric trace gas as well as the determination of the rate coefficient of a complicated ion molecule reaction. Until a full understanding of the major plasma processes can be achieved, future work will be needed. In particular, significant contributions may be expected from further atmospheric in situ measurements such as mass spectrometric measurements of positive and negative ions as well as neutrals coupled with measurements of temperature and of nitric oxide. Regarding laboratory work measurements of various temperature dependant ion clustering and decomposition reactions are needed.

REFERENCES

Arnold, F., Kissel, J., Krankowsky, D., Wieder, H., and Zähringer, J. 1971, J. Atmosph. Terr. Phys., 33, 1963

Arnold, F., Joos, W., Hettmannsperger, E., Krankowsky, D., Marien, K.H., and Grochulska, J. 1974, Yearly Report of the Max-Planck-Institute for Nuclear Physics, Heidelberg, Germany

Arnold, F., Böhringer, H., Hettmannsperger, E., Krankowsky, D., Stanisławska, J., and Zettwitz, E. 1975, Yearly Report of the Max-Planck-Institute for Nuclear Physics, Heidelberg , Germany

Arnold, F., Böhringer, H., Harke, U., Hettmannsperger, E., Krankowsky, D., Marien, K.H., Stanisławska, J., and Zettwitz, E. 1976, Yearly Report of the Max-Planck-Institute for Nuclear Physics, Heidelberg, Germany

Arnold, F., Krankowsky, D., and Marien, K.H. 1977a, Nature, 267, 30

Arnold, F., Krankowsky, D., Marien, K.H., and Joos, W. 1977b, submitted to J. Geophys.

Arnold, F., and Krankowsky, D. 1971, J. Atmosph. Terr. Phys., 33, 1693

Arnold, F., and Krankowsky, D. 1974 in Proceedings of the International Symposium on Solar-Terrestrial Physics, COSPAR, Sao Paulo

Arnold, F., and Krankowsky, D. 1974, Geophys. Res. Lett., 1, 243

Arnold, F., and Krankowsky, D. 1977a, Nature, in press

Arnold, F., and Krankowsky, D. 1977b, J. Atmosph. Terr. Phys., 39, 625

Arnold, F., and Krankowsky, D. 1977c, J. Atmosph. Terr. Phys., in press

Arnold, F., and Krankowsky, D. 1977d, to be published

Arnold, F., and Krankowsky, D. 1978, J. Atmosph. Terr. Phys. to be published

Arnold, F., and Krankowsky, D. 1977c, submitted to J. Geophys.

Barth, C.A. 1966, Ann. Geophys., 22, 198

Bailey, T.L. 1958, J. Chem. Phys., 28, 792

Bates, D.R., Kingston, A.E., and McWhirter, R.W.P. 1962, Proc. Roy. Soc. London, Ser. A 267, 297

Beauchamp, J.L., and Butrill, S.E. 1968, J. Chem. Phys., 48, 1783

Biondi, M.A. 1969, Can. J. Chem. , 47, 1711

Biondi, M.A. 1970, Adv. Atom. Phys., 6, 85

Brown, T.L. 1973, Chemical Review, 73, 645

Chimonas, G., and Axford, W.I. 1968, J. Geophys. Res., 73, 111

Donahue, T.M., Zipf, E.C., and Parkinson, T.D. 1970, Planet. Space Sci., 18, 171

Dunkin, D.B., Fehsenfeld, F.C., Schmeltekopf, A.L., and Ferguson, E.E. 1971, J. Chem. Phys., 54, 3817

Farmer, C.B., Raper, O.F., and Norton, R.H. 1976, Geophys. Res. Lett., 3, 13

Fehsenfeld, F.C., and Ferguson, E.E. 1969, J. Geophys. Res., 74, 221

Fehsenfeld, F.C., and Ferguson, E.E. 1972, Radio Sci., 1, 113

Fehsenfeld, F.C., Mosesman, M., and Ferguson, E.E. 1971, J. Chem. Phys., 55, 2120

Fehsenfeld, F.C., Carleton, J.H., Harrop, W.J., and Ferguson, E.E. 1975, J. Geophys. Res., 80, 2229

Ferguson, E.E., Dunkin, D.B., and Fehsenfeld, F.C. 1972, J. Chem. Phys.,57, 4, 1459
Ferguson, E.E., 1973, Atomic and Nuclear Data Tables, Vol. 12,2, 159
Ferguson, E.E. 1974, Rev. Geophys. and Space Phys. , 12, 703
Fisk, G.A., Mahan, B.H., and Parks, E.K. 1967, J. Chem. Phys., 47, 2649
French, M.A., Hills, L.P., and Kebarle, P. 1973, Can. J. Chem., 51, 456
Fukuyama, K. 1974, J. Atm. Terr. Phys., 36, 1297
Gibson, A.J., and Sanford, M.C.W. 1971, J. Atm. Terr. Phys., 33, 1675
Good, A., Durden, D.A., and Kebarle, P. 1970a, J. Chem. Phys., 52,222
Good, A., Durden, D.A., and Kebarle, P. 1970b, J. Chem. Phys., 52,212
Hale, L.C. 1967, Space Research VII, Vol. 1, p. 140, North-Holland, Amsterdam
Hale, L.C., Nickell, L.C., Kennedy, B., and Powell, T.A. 1972, Radio Sci., 7, 89
Hale, L.C., Hoult, D.P., and Baker, D.C. 1968, Space Res., 8, 321
Haug, A., and Landmark, B. 1970, J. Atmosp. Terr. Phys., 32, 405
Heimerl, J.M., and Vanderhoff, J.A. 1974, J. Chem. Phys., 60, 4362
Heimerl, J.M., and Vanderhoff, J.V. 1971, Trans. Am. Geophys. Union, 52, 870
Heimerl, J.M., Vanderhoff, J.V., Puckett, L.D., and Niles, F.E. 1972, U.S. Army Ballistic Research Laboratories Report No. 7570
Hesstvedt, E. 1970, Report of the Institute of Geophysics, University of Oslo
Hinteregger, H.E. 1970, Ann. Geophys., 26, 547
Howard, C.J., Bierbaum, V.M., Rundle, H.W., and Kaufmann, F. 1972, J. Chem. Phys., 57, 3491
Huffman, R.E., Paulsen, D.E., Larrabee, J.C., and Cairns, R.B. 1971, J. Geophys. Res., 76, 1028
Hunten, D.M., and McElroy, M.C. 1968, J. Geophys. Res., 73, 2421
Ishikawa, M., Morita, Y., and Takagi, M. 1969, Planetary Electrodynamics, Vol. 2, Gordon and Breach Science Publisher, New York, London, Paris
Johnsen, R., Huang, Ch. M., and Biondi, M.A. 1975, J. Chem. Phys., 63, 3374
Johannessen, A., and Krankowsky, D. 1974, J. Atmosph. Terr. Phys., 36, 1233
Krankowsky, D., Arnold, F., Wieder, M., and Kissel, J. 1972, Int. J. Mass. Spectrom. Ion Phys., 8, 379
Kebarle, P., Yamdagi, R., Miracka, K., and McMahon, T.B. 1976, Int. J. Mass. Spectrom. Ion Phys., 19, 71
Lazarus, A.L., Gandrud, B.W., Woodward, R.N., and Sedlacek, W.A. 1976, J. Geophys. Res., 81, 1067
Lebedinets, V.N., and Shuskova, V.B. 1970, Planet. Space Sci., 18, 1659
Leu, M.T., Biondi, M.A., and Johnsen, R. 1973, Phys. Rev., 7A, 292
Marien, K.H., and Arnold, F. 1973, Yearly Report of the Max-Planck-Institute for Nuclear Physics, Heidelberg, Germany
Mechtly, E.A., and Smith, L.G. 1970, Radio Sci., 5, 1407

Meira, L.G. 1971, J. Geophys. Res., 76, 202
Narcisi, R.S., and Bailey, A.D. 1965, J. Geophys. Res., 70, 3687
Narcisi, R.S., Bailey, A.D., Della Luca, L., Sherman, C., and Thomas, D.M. 1971, J. Atmosph. Terr. Phys., 33, 1147
Narcisi, R.S., Bailey, A.D., Wlodyka, L.E., and Philbrick, C.R. 1972a, J. Atmosph. Phys., 34, 647
Narcisi, R.S., Sherman, S., Philbrick, C.R., Thomas, D.M., Bailey, A.D. Wlodyka, L.E., Wlodyka, R.A., Baker, D., and Federico, G. 1972b, in Proceedings of COSPAR Symposium on "Solar Partile Event of November 1969", AFGRL-72-0474 Special Report No., 1141 Boston, Mass.441-420
Narcisi, R.S. 1973, Space Research, VIII, 360
Ogawa, T., and Tomatsu, T. 1966, Rep. Ionosph. Space Res. Japan, 20, 395
Payzant, J.D., Cunningham, A.J., and Kebarle, P. 1973, J. Chem. Phys., 59, 5615
Pearce, J.B. 1969, J. Geophys. Res., 74, 853
Pedersen, A. 1966, Uppsala Ionospheric Observatory Report No. 15, 1966, 9
Poland, P.H., and Broida, H.P. 1971, J. Quant. Spectroscop. Radiat. Transfer, 11, 1863
Popoff, F.G., and Whitten, R.C. 1969, Nature, 224, 1187
Potemra, T.A., and Zmuda, A.J. 1970, J. Geophys. Res., 75, 7161
Reid, G.C. 1970, J. Geophys. Res., 75, 2551
Rishbeth, H., and Garriott, O.K. 1969 in Introduction to Ionospheric Physics, Academic Press, New York and London
Rose, G., and Widdel, H.V. 1972, Radio Sci., 7, 81
Rosenberg, N.W., and Justus, C.G. 1966, Radio Sci., 1, 149
Schmidlin, F.J. 1976, Geophys. Res. Lett., 3, 173
Shimazaki, T., and Whitten, R.C. 1976, Rev. Geophys. Space Phys., 14, 1
Smith, D., and Church, M.J. 1976, Int. J. Mass. Spectrom. Ion Phys., 19 185
Stewart, R.W., and Hoffert, F. 1975, J. Atmosph. Sci., 32, 195
Swider, W., and Narcisi, R.S. 1977, Planet. Space Sci., 25, 103
Thomas, L. 1974, Radio Sci., 9, 121
Tohmatsu, T., and Iwagami, N. 1975, Space Res., 15, 241
Tohmatsu, T., and Iwagami, N. 1976, J. Geomag. Geoelectr. , in press
Tunilov, V.F., Shibaeva, L.V., and Jakovlev, S.G. 1969, Space Res., 9, 231
Turner, D.L., and Conway, D.C. 1976, J. Chem. Phys., 65, 3944
Vanderhoff, J.A., and Niles, F.E. 1976, EOS, Trans. Am. Geophys. Union, 57, 303
Webber, W. 1962, J. Geophys. Res., 67, 5061
Widdel, H.U., Rose, G., and Borchers, R. 1971, Pure and Applied Geophys., 84, 154
Zbinden, P.A., Hidalgo, M.A., Eberhardt, P., and Geiss, J. 1975, Planet. Space Sci., 23, 1621
Zikmunda, J., and Mohnen, V.A. 1972, Meteorologische Rundschau, 25. Jahrgang, 1. Heft
Zipf, E.C., Borst, W.L., and Donahue, T.M. 1970, J. Geophys. Res., 75, 6371

MID- AND HIGH LATITUDE REFERENCE IONOSPHERE

K. Rawer

Institut für physikalische Weltraumforschung,
Fraunhofer-Gesellschaft, Freiburg (F.R.G.)

1. GENERALITIES

In the International Reference Ionosphere (I.R.I.), a joint effort by URSI and COSPAR, representative profiles of the main parameters of the ionosphere are to be assembled for reference purposes. Agreement has been obtained that significant experimental evidence and not aeronomic theory should be the basis of IRI. Theoretical explanation should be left to individual scientists while assembling and reviewing experimental results is an adequate task for international scientific Unions.

The original request from the Unions is for coherent sets of vertical profiles at a few suitably chosen locations. Extension towards a continuous global description could be a second step when enough measured data will be available to this end.

Work on IRI is promoted by a Steering Committee, the membership of which stands at present as follows: S.A. Bowhill, USA, K.S.W. Champion, USA, A.D. Danilov, USSR, E.A.Lauter (J.Taubenheim, substitute), GDR, K. Rawer (Chm.), FRG, M.J. Rycroft, UK.
Since 1974 two ad-hoc parties are engaged with particular problems:
(a) Description of conditions above the thermopause, in particular in the plasmashere (M.J. Rycroft, Chm.),
(b) Ion composition and negative ions below the 100 km level (D. Krankowsky, Chm.).
A large number of correspondents have taken interest in the work of IRI, either by furnishing measured data or be assisting at discussion meetings, which took place at international assemblies of IRI or COSPAR since 1969. A particular symposium on the lower ionosphere was held in 1973 at Konstanz (Rawer 1974).

B. Grandal and J. A. Holtet (eds.), Dynamical and Chemical Coupling, 129-143.

The unanimously accepted conclusions of this meeting decided which data sources are to be taken as primary inputs. Status Reports on IRI were given by Rawer (1970, 1975) and by Ramakrishnan and Rawer (1972, 1974a, b). A preliminary set of profiles was published by Rawer et al. (1975) and was distributed at the 1975 General Assembly of URSI at Lima (Peru).

2. SUMMARY DESCRIPTION

It was soon found out that nowadays the majority of the potential users will apply computing facilities. Therefore, instead of profile tables, the first Reference Ionosphere to be edited by URSI in early summer of 1977 contains computer programs (in ALGOL 60 and FORTRAN IV). These programs derive from mathematical expressions describing continuous vertical profiles in suitable chosen height sections. The parameter values of these functions must be fed-in and are specified either in tables or by mathematical relations with the independent parameters hour, season, solar activity. In this presentation the meaning of most parameters is straight-forward physical; for example the peak electron densities of the different layers are directly fed in.

Following a proposal made by H.G. Booker, another kind of descriptive functions will also be given in the preliminary edition of IRI. These are whole-profile analytical functions obtained from a "skeleton function", H(z), by an exponential:

$$\text{const.}\exp((z_o - z)/H(z)).$$

H is a kind of equivalent scale height and is given as a smoothed-out analytical "staircase function" with continuous transitions following the well-known "EPSTEIN-step-function". H takes positive values where the profile is essentially decreasing with altitude and negative ones where it is increasing. In fact, the steps of the staircase function are merging and the transitions at the chosen characteristic altitudes, z_n, are not independent from each other. Therefore, straight-forward introduction of physical parameters is not possible in this kind of description. The parameters of the skeleton staircase and the width of the different transitions had to be found by a trial-and-error procedure, independently from each profile. In spite of this deficiency such whole profile analytical functions might be useful for some applications, e.g. for full-wave analysis of propagation problems.

Profiles are given in the IRI for the following physical parameters of the ionospheric plasma:

- electron density (in m^{-3})
- electron temperature (in K)
- ion temperature (in K)
- ion composition (in %).

Redundant descriptions of the characteristic ratios :

- electron density ratio N/Nmax
- temperature ratio Te/Ti

might be useful for quite a few applications.
The density ratio is in fact a primary result since it gives the shape of the profile which is more consistently deduced from the observations than electron densities at different altitudes. Since for the peak density there exists a CCIR compilation in program form (see following Section) the absolute electron density profile was obtained by multiplying the shape function (with peak value 1) by the peak density.

3. BASIC DATA SOURCES

Most of the data used in establishing the IRI-profiles stem from one of the following four techniques of measurement :

(1) ground-based upward ionospheric sounding (ionosondes)
(2) ground-based incoherent scatter sounding
(3) downward sounding from satellites (top-side sounders)
(4) in situ measurements near rockets and satellites.

Results from ionosondes (1), available on a day-to-day, hour-to-hour basis for a large number of stations and more than three solar cycles, were -after reduction to 'true height'- used for most of the bottomside electron density profiles and, indirectly, for peak electron densities and the relevant altitudes. For the latter two parameters the computer program compilation of the "Comité Consultatif International des Radiocommunications"(C.C.I.R.) was a great help. This program (Report 340) summarizes an enormeous number of individual measurements as monthly medians by a suitable combination of Fourier development (at each station, after the hour) and a particular kind of spherical development in which geographic latitude is replaced by a suitably chosen combination with magnetic inclination. The magnetic control of ionospheric parameters is so better taken account of. The final presentation for the user is in conventional geographic coordinates.

Results from incoherent scatter sounders (2), though available from very few stations on a two-days-per-month basis only, is the most important input since these techniques allow at least three (sometimes even four) of our physical parameters to be simultaneously determined. Such results are still our main source for topside profiles of electron density and almost whole profiles of electron and ion temperature. Since these measurements end up directly with profiles, they can readily be introduced. Unfortunately, the small number of observing days is a serious draw-back. It is well known that the ionospheric parameters admit large day-to-day fluctuations such that data from idividual days are not necessarily representative.

Top-side sounding (3) is a very valuable technique for obtaining electron density profiles of the top-side ionosphere. A very large number of such measurements has meanwhile been gathered in the ALOUETTE and ISIS programs many of which were converted into profiles. Unfortunately, these cannot readily be introduced into a station-oriented presentation like that of C.C.I.R. Also most of them are not full profiles down to the peak altitude so that it is difficult to merge them with the bottomside. Using a large compilation on magnetic tape (Bent and Llewellyn 1973) a new effort was just started to resolve this problem.

Results from in-situ measurements (4) are available from a limited but increasing number of rocket launches and, with very large numbers of measurements now from near-Earth satellites. As for the latter, the problem of description in a fixed-coordinate system is not yet completely resolved. Since satellites cannot orbit at altitudes below about 200 km rockets are the only in-situ sources at the lower altitudes. Data from rocket measurements were in fact our main source for electron density and ion composition in the lower ionosphere, in a few cases for temperatures too. Satellite in-situ measurements on the other hand are our only source for ion composition near and above the peak altitude of the ionosphere. Efforts are under way to describe electron density and electron and ion temperature data from satellites in a form suitable for producing vertical profiles.

4. ELECTRON DENSITY PROFILES IN THE LOWER IONOSPHERE

Ground-based sounding (technique (1) of Section 3) ends up with an 'ionogram' giving apparent reflection heights as a function of the sounding radio-frequency. An example is shown in Fig. 1. The principal ("ordinary") echo trace is due to a reflection at the level where $X \equiv (f_N/f)^2 = 1$. f is the instantaneous sounding radio-frequency and f_N the plasma frequency at the reflecting level. f_N^2 is directly proportional with the electron density N at the same level. In spite of this simple reflection condition the recorded echo trace cannot be taken as a direct analogon of the electron density vs. height profile because the signal velocity is not constant on the sounding path. In the ionospheric plasma the sounding pulse travels with a group velocity which is less than the propagation velocity in free space, c_o. In particular, near the reflecting level there occurs an important "delay" which is the greater, the smaller is the vertical gradient of electron density near the level where X = 1. Such delays may produce peaks of the "virtual height", h', which can be delayed by more than 100 km.

Therefore, the virtual height trace is a very distorted picture of the true profile. The latter can be deduced from the former by resolving an integral equation which would be of Abelian type but is further compilated by the influence of the terrestrial magnetic field in the plasma dispersion equation. Numerical resolution procedures are now well established so that the true profile can readily be determined from the ionogram. Fig. 1 also shows the true profile below the virtual one. The altitude of the peak of the ionosphere (the F2-peak) cannot be directly found from these procedures but must be determined by extrapolation.

The bottomside profile of the F2-layer may have different shapes at low latitudes but is nearly parabolic at temperate and high latitudes. However, in summer there appears a "kink" at a well determined plasma density which produces a peak of the virtual height (see Fig. 1). This is the so-called F1-layer. Its characteristic frequency shows a more regular behaviour than that of the F2-layer. Mohler (1937) was the first to suppose that both layers might be due to the same ionization process the F1 kink occurring at the level of maximum ionization production and F2 being due to a rather sensitive balance between ionization and deionization. In fact Massey (1937) announced a process of 'dissociative recombination' by which the observed deionization rates can be explained. This is a two-step process including a charge-transfer from O^+ ions to diatomic molecules. In the F2-layer both, production and deionization decrease with increasing height so that the balance is rather sensitive and its time-scale becomes large. Therefore, the F2-layer is very sensitive to transport processes and thus largely variable. Quite different for the lower layers F1 and E the maximum electron density of which obeys rather simple laws as function of the zenith angle under which the ionizing radiation arrives.

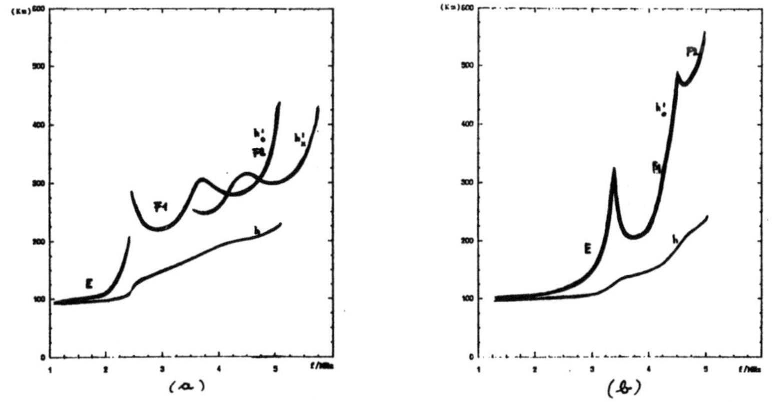

Fig.1. Main ionogram trace (bold) and computed profile (thin). Dourbes 18 June 1964, 06 h(a) and 12 h(b). (L.Zandarin 1975).

In fact, these functions summarize the characteristic electron density values as were deduced with quite high accuracy from ionograms. E.g. the dispersion of such data for the E-region is only ± 10% near noon. Note that F1 uses to appear only in summer.

The reduction procedure from virtual to true height becomes inadequate where the electron density profile is not monotonuously increasing. Between F1 and E (or between F2 and E, when F1 is absent) a 'valley' is quite often found. At medium latitude it is rather shallow or even disappears near noon, but it is quite important at other hours. All efforts to deduce a valley profile for ionograms were in vain. This is not only due to the limited measuring accuracy but mainly to natural distorsions of the ionospheric layers, e.g. by gravity waves (Paul 1972). Fortunately, there are some incoherent scatter observations now available (technique no. (2) of Section 3) from which valley shapes could be determined (examples are presented in Figs. 2a and b).

It is worthwhile to note that a daytime valley may be an indication of vertical transport processes in the range above the E-region. Aeronomic computations of the ionizing effect of solar e.u.v. radiation showed that below F1 the production should be a monotonuously increasing function of height so that no valley should appear in the production profile itself (Rawer 1974).

Below the valley region we have the very regular E-region peak. The shape of the electron density profile below the peak cannot accurately enough be deduced from routine ionograms. In-situ measurements (technique no. (4) of Section 3) were used instead. We have one compilation of published rocket data by Maeda (1969, 1970, 1971, 1972) and another one by Soboleva (1972) with partially different sources. For daytime an agreeable shape could be found. At night, however, Soboleva had more data and these were used for the bottom part of Fig. 2c where a very deep valley occurs above the still persisting E-layer.

Meanwhile these profiles were used by H.G. Booker for a full-wave computation of u.l.f. resonance in the cavity formed by ground and ionosphere which were first announced by Schumann (1951). Booker (communication by letter 1976) found that in order to agree with the observations the daytime shape should show a characteristic inflection point at low electron density. This appeared to be agreeable with Maeda's data points though not with the curve originally drawn through these. For the nighttime profile a very deep valley with reduction down to 7% as given in Soboleva's model was not acceptable. So we went back to Maeda's night valley which achieves a reduction to only 23% of the E-region peak density.

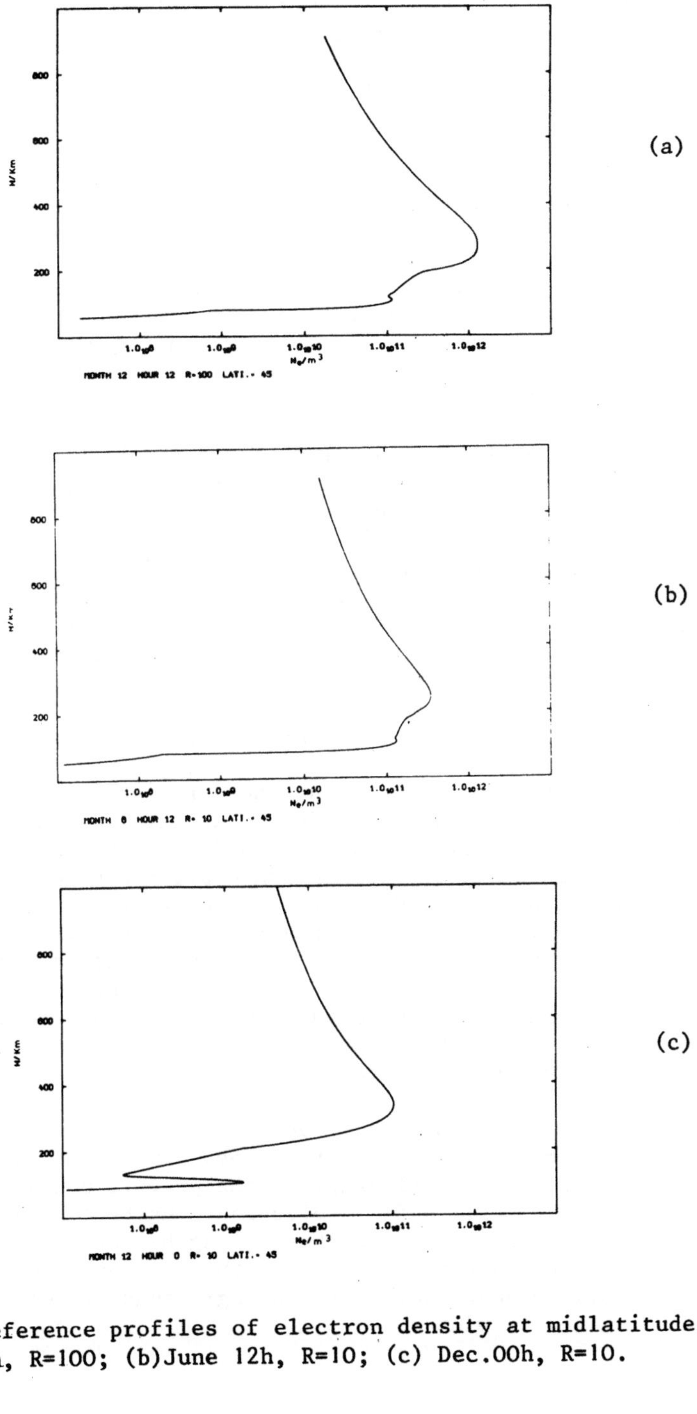

Fig. 2. Reference profiles of electron density at midlatitude(45°). (a)Dec.12h, R=100; (b)June 12h, R=10; (c) Dec.00h, R=10.

After sunset, due to the small density of molecules in the upper thermosphere the F2-layer continues to exist though its electron density uses to go down considerably during night. In the lower layers, however, there is a rapid decrease. Below 200 km the electron density shows a rather important gradient its value going down from between 10^9 and 10^{10} to about 10^8 m^{-3}. However, rocket measurements (technique no. (4) from Section 3) have shown that an E-region with peak electron density of 10^9 m^{-3} and more still exists at night, see Fig. 2c. Such 'persisting E-region' was also inferred from sounding experiments with technique no.(1) (Piggott and Rawer 1961). As to the origin of this layer different speculations are under discussion. It may be that metallic ions play some role as they certainly do for sporadic layers. Further, since a weak scattered Lyman-alpha radiation of solar origin is also present at night, this might be the cause of nighttime E-ionization.

The lowest ionospheric region of interest is the D-region at 65 to 90 km of altitude. Its ionization is always small as compared with that of the upper layers. Nevertheless it has some importance for radio propagation, in particular at medium and low frequencies. The (electron-neutral) collision frequency is rather high at this level such that even at higher radio frequencies absorption is important. Also the D-region is rather complex and thus interesting from the aeronomist's viewpoint. Technique no. (1) is not applicable to it since at the appropriate sounding frequencies absorption is too large. Also techniques no.(2) and (3) cannot be applied such that in-situ measurements, no. (4), provide the only useful method. There is, however, still another technique giving information about the D-region, namely radio wave propagation experiments. After a detailed discussion at the Konstanz Symposium (Rawer 1974) it could be stated that field-strength data obtained in propagation experiments between ground stations cannot be inverted accurately enough so as to be applicable as a primary source of information about the profile in the D-region. In-situ probes, on the other hand, suffer from calibration difficulties particularly in a height range where collisional effects cannot be neglected. Therefore, in-situ probing checked by a simultaneous propagation experiment between ground and rocket was found to be the best choice.

Following these criteria Mechtly and Bilitza (1974) assembled a rather large number of rocket data. They found that in most of these a characteristic inflection in the profile could be detected at a rather well defined electron density called NmD which is under solar control. The corresponding height hmD is between 78 and 85 km. At temperate latitudes the seasonal variation of NmD is not simply under solar zenith angle control but shows increased values in winter.

This "regular winter anomaly" is taken account of in I.R.I. but not of the "excessive winter anomaly" which gives much larger electron densities in the D-region during certain groups of days. It was not possible to deduce a latitudinal variation in the undisturbed profile from the variable data sources but, apparently, this variation is not too large in the lowest ionosphere. Our original guess admitting solar control comparable with that of the E-layer was an overestimate as Gnanalingam has shown by comparing with his absorption measurements made in Ceylon(communication by letter 1976).

The D-region disappears rapidly after sunset. During the night the electron density is extremely small at altitudes below 90 km. This is at least true for undisturbed conditions. When corpuscular radiation arrives the electron density in the D-region can go up. This occurs very often in the auroral zone but is rather seldom at temperate latitudes. For the time being the Reference Ionosphere ignores such disturbances.

5. ELECTRON AND ION TEMPERATURES

In the ionosphere thermal conditions are quite different from what is usually found in the terrestrial systems. They are in fact very far from those in a "closed system" since during sunlight hours an important energy input arrives via solar radiation. Also above about 100 km, thermal equilibrium does not exist between the different constituents. Since the ionospheric plasma is thin, i.e. the ionized species are very rare compared with neutral atoms and molecules, the bulk of the thermal energy resides in the neutrals. However, the main heat source is photo-ionization and the heat input goes into the electron component only. In each photo-ionization process the excess energy (quantum minus ionization potential, of the order of 1 eV) is distributed between ion and electron after the reciprocal mass ratio; thus it goes essentially to the electron. As another consequence of the mass ratio collisional heat transfer to the heavier particles remains small such that the photo electrons first heat up the electron population as such. Now, the heat contact of electrons is much better with ions than with neutrals (because Coulomb collisions between charged particles have a much larger cross-section). Therefore, the second step in heat transfer is from the electrons to the ions, i.e. between the rare populations. The third step only is heating of the neutrals by collisions with ions. These heat flow conditions are explained in Fig. 3 by a comparison with water flow through several sections of pipes in series connection with the current held constant by a pump. The pressure difference over one section of given length is larger for a narrow pipe than for a wide one. Since the pipes are in series (i.e. heat flows as explained above) we have the highest pressure on top (i.e.the highest temperature

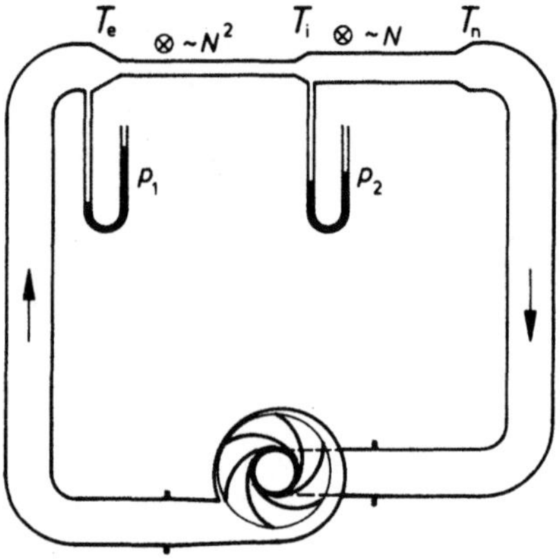

Fig. 3. Water flow simile explaining heat flow between constituents.

for the electrons) and find an intermediate value after the first narrowest pipe (i.e. for the ion temperature). The differences may become quite important if the pipes are very narrow. In our example water flow is held constant (i.e. constant radiation input). Therefore the pressure difference depends on the cross-sections only. This means that the electron population reaches high temperature if its thermal conductivity towards the ions is small, i.e. when collisions are rare. The probability of these, however, depends on the number density of electrons and ions, such that it goes with the square of the plasma density. On the other hand the collision probability of ions with neutrals is directly prportional with the plasma density. Therefore, the heat flow equilibrium is controlled by the plasma density, electron temperature and density are inversely related - a fact which in-situ plasma measurements have shown at different occasions (e.g. Spenner 1974).

This describes local heat transfer between different constituents which is most important in the lower thermosphere. There is, of course, also heat propagation, namely by heat conductivity in the electron gas. But this term prevails only in the upper thermosphere, above about 400 km. Finally, below about 90 km collisions between electrons and neutrals are so frequent that direct heat transfer to the neutrals prevails so that the comparison given in Fig. 3 is no more applicable. At this level the temperature difference between charged particles and neutrals becomes negligible: temperature equilibrium is practically reached.

Temperature measurements are only provided by methods (2) and (4) of Section 3. In-situ measurements with probes (4) had many difficulties to overcome in the past. Even now not all rocket and satellite data are reliable. Unfortunately, all deficiencies of probes produce errors of the same sign, viz. they provoke a too high temperature indication. Errors of a few 100 K are not rare. Therefore, our principal data source is still incoherent scatter (2) which technique when carefully reduced yields reliable electron and ion temperatures. It has also the advantage against satellites that vertical profiles are directly obtained.

Let us first consider the ion temperature, Ti. Figs 4a and b show daytime and nighttime measurements obtained near Boston,USA (courtesy: J. Evans). The most striking feature is a break near the 500 km level.

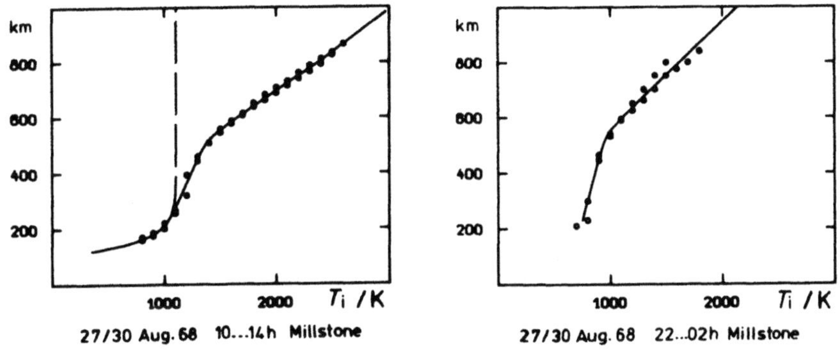

Fig. 4. Typical ion temperature profiles day (a) and night (b).

It has a physical reason: Above this level the heat contact between neutrals and ions becomes poorer and poorer because the neutral density decreases rapidly with height. Speaking in terms of our water-flow simile (Fig. 3) the second pipe becomes narrower at greater heights so that the intermediate pressure, p_2, rises; it remains, of course, below the value of p_1, which rises too. Thus the ion temperature goes toward the electron temperature without reaching it. Below the break the ion temperature profile follows a different law. In fact towards lower levels the heat contact with the neutrals becomes excellent (pipe 2 becomes wide) and the ion temperature practically equals the neutral one. This is the situation in the lower thermosphere. There, I.R.I. takes the neutral temperature profile of CIRA (1972) as approximation for Ti.

Coming now to the electron temperature, Te, the situation is different. In the lower thermosphere there is a clear difference between Te and Ti, which almost equals T_n. The vertical gradient of Te is important; in daytime, it may be as high as 5 K/km, at night about half of this. Figs. 5a and b show day and nighttime profiles also measured near Boston (courtesy: J. Evans).

At great heights the profile rarely reaches a fixed asymptotic value but continues to increase, though more slowly. The temperature differs between day and night by about a factor of 2. As explained above nighttime electron temperatures are much nearer to Ti and T_n than during daytime.

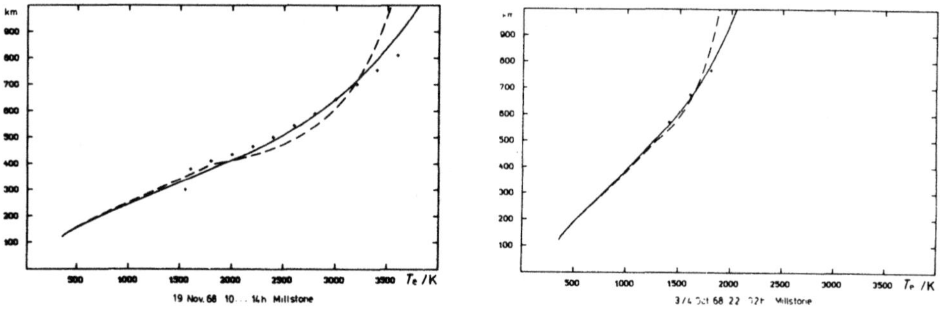

Fig. 5.Typical electron temperature profiles day(a) and night(b).

Figs. 4 and 5 show examples taken from typical "monthly" incoherent scatter observations with measurements executed during a few consecutive days only. These are not necessarily representative for the whole month. On the other hand, up to the F2 peak, the electron density data stem from daily measurements so that their median is representative for a month. In order to take account of the variability the temperature data were corrected by comparing the electron density profile measured during the same period by incoherent scatter with the monthly median profile obtained from ionosonde observations. The difference was translated into a temperature correction by using a suitable relation between Te and Ne. Such relation was originally established using an approximate heat flow theory (Lejeune and Waldteufel 1970). Subsequent comparison with measurements showed that more empirical laws gave a better representation (Bilitza 1974). These are now used.

The rule that electron density and temperature have opposite variations is rather general. It has sometimes astonishing consequences. For example,in winter when at temperate latitudes, the highest electron density values occur the electron temperature can be lower during sunspot maximum (when the electron density is highest) than during minimum. This is, of course, only true for daytime. At night Te is nearer to T_n which takes higher values for sunspot maximum. On the other hand there are now data from low latitude proving that the Te-profile can be pushed down around the F2 peak as a consequence of high thermal conductivity at this level. See Fig. 6 for an example.

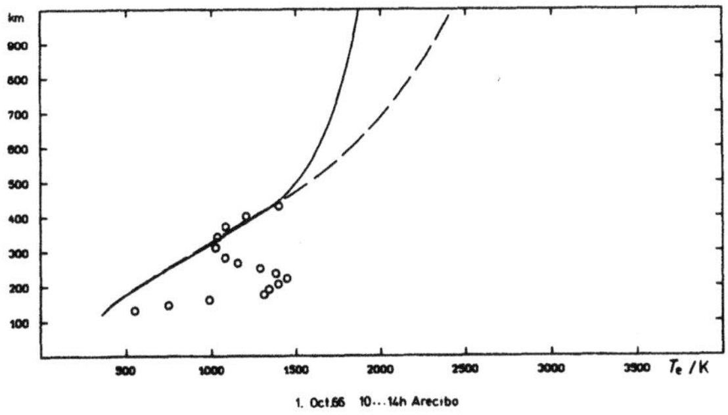

Fig. 6. Abnormal profile of electron temperature with a "valley" near the F2-peak.

6. ION COMPOSITION

The primary ion production depends on the density of the relevant neutrals, their ionization cross-sections and on the intensity of the ionizing radiation. There are, however, secondary processes provoking major changes. In the upper thermosphere where collisions are extremely rare ion diffusion is the main cause of change. In the lower thermosphere, however, collisional charge transfer is prevailing. The lower is the altitude the more complex is the ion chemistry.

Available data above 200 km stem from satellites, in particular from the OGO's and from AEROS. For these latter data are available in a readily usable form, namely as percentage values of the whole ion population. Below 200 km we obtain results from an increasing number of rockets most of which were launched at temperate latitude. The main basis is a compilation due to A.D.Danilov (communication by letter). Some more data were contributed by D. Krankowsky (communication by letter). These data are sufficient to produce one distribution chart for daylight hours, see Fig.7., covering the height range from 100 to 220 km. At the upper margin of this chart the primary O^+ ions prevail. However, below 180 km NO^+ ions take over. N_2^+ ions are not observed in percentage quantity. These are quickly transformed by secondary reactions into NO^+. Note that the ionization potential of NO^+ is smaller so that there is no way back to N_2^+ if NO^+ is once formed. Amongst the O_2^+ ions most should be primary ones.

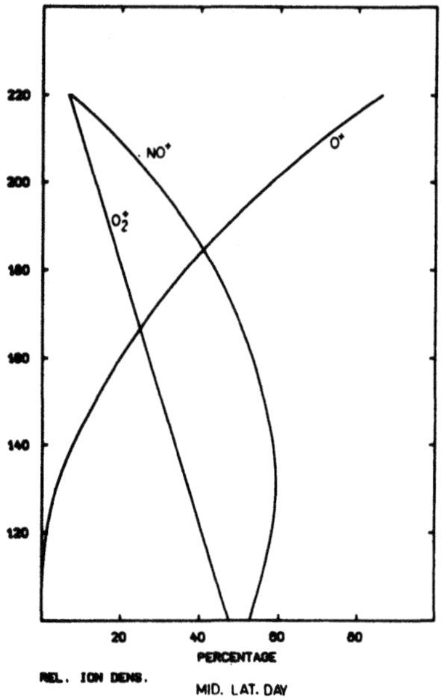

Fig. 7. Percentage distribution chart of the main ions in the 100....220 km range.

Unfortunately, nighttime data were so scarce and their results so scattered that we did not dare to establish a similar chart for night. The situation with low latitude data is similar. A.D. Danilov hopes to produce soon a new compilation including more recent data so that the deficiencies might be removed.

As for heights below 100 km the situation is worse. Below about 83 km negative ions appear. Also rather heavy cluster ions are formed. It cannot be excluded that some measuring instruments destruct clusters, or help building up them. D. Krankowsky feels that with the at present available data even the ratio of negative ions to electrons though it is measurable under given circumstances cannot be indicated as a reliable reference profile.

REFERENCES

Bent, R.B.,and Llewellyn, S.K. 1973, Report AFCRL-TR-73-0657,AFGL, Hanscom AFB, Bedford, Mass.

Bilitza, D. 1974, Diploma Thesis. Univ. of Freiburg, Freiburg.

CIRA 1972, COSPAR International Reference Atmosphere 1972. COSPAR Secretariat, Paris.

Lejeune, G.,and Waldteufel, P. 1970, Annls.Geophys., 26, 223.

Maeda, K.-I. 1969, J.Geomag.Geoelect., 21, 557.

Maeda, K.-I. 1970, J.Geomag.Geoelect., 22, 551.

Maeda, K.-I. 1971, J.Geomag.Geoelect., 23, 133.

Maeda, K.-I. 1972, Space Res., XII, 1229.

Massey, H.S.W. 1937, Proc.Roy.Soc., A 163, 542.

Mechtly, E.A.,and Bilitza, D. 1974, Report W.B.1, Institut für physikalische Weltraumforschung, Freiburg.

Mohler, F.L. 1937, Bur.Stand.J.Res., 19, 447 and 559.

Paul, A.K. 1972, Paper presented at Gen.Assembly U.R.S.I.,Warszawa

Piggott, W.R.,and Rawer, K. 1961, URSI Handbook of Ionogram Interpretation and Reduction. Elsevier, Amsterdam.

Ramakrishnan, S.,and Rawer, K. 1972, Space Res., XII, 1253.

Ramakrishnan, S.,and Rawer, K. 1974a, Annls.Geophys., 30, 347.

Ramakrishnan, S.,and Rawer, K. 1974b, p.35 in Methods of Measurements and Results of Lower Ionosphere Structure(ed.K. Rawer). Akademie-Verlag, Berlin.

Rawer, K. 1970, COSPAR Information Bulletin. COSPAR Paris (ed.).

Rawer, K. 1974, Methods of Measurements and Results of Lower Ionosphere Structure. Akademie-Verlag, Berlin.

Rawer, K. 1975, Space Res., XV, 295.

Rawer, K., Ramakrishnan, S.,and Bilitza, D. 1975, Report W.B.2, Institut für physikalische Weltraumforschung, Freiburg.

Schumann, W.O. 1951, Zs.Naturforschg., 79, 149.

Soboleva, T.N. 1972, Model'nye profil sutocnogo raspredelenija electronnoj Koncentracii spokojnoj ionosfery na srednih sirotah. IZMIRAN, Moskva.

Spenner, K. 1974, Space Res., XIV, 259.

Zandarin, L. 1975, Doctor Thesis, Univ.Catholique de Louvain, Louvain-la-Neuve.

IONIZATION PROCESSES

J. B. Reagan

Lockheed Palo Alto Research Laboratory
Palo Alto, California 94304 USA

ABSTRACT. In the middle and high latitude atmosphere below 100 km altitude the principal sources of ionization are 1) solar radiation in the x-ray and extreme ultraviolet wavelengths, 2) galactic cosmic rays and other galactic radiations, and 3) precipitating energetic particles of solar origin and from the earth's radiation belts. In this paper the relative contribution of each of these source terms to the maintenance of the dynamic ionosphere is reviewed. The significant role of electron precipitation and the associated bremsstrahlung to the ionization and modification of the neutral atmosphere is presented. The important role of solar particle events in both ionizing the atmosphere and in producing and modifying key minor neutral constituents such as ozone, atomic oxygen, and the nitric oxides is illustrated.

1. INTRODUCTION

The sources of ionization in the middle and high-latitude quiescent atmosphere below 100 km have been reviewed by several authors (Aiken 1971; Thomas 1971; Bauer 1973; Banks and Kockarts 1973; Potemra 1974; Paulikas 1975). Under daytime conditions solar radiations at wavelengths below ~ 3000Å, i.e., at ultraviolet and x-ray wavelengths, have traditionally been thought to be the principal ionizing sources of the neutral atmospheric constituents above 60 km altitude. The quantitative determination of the ionization rates at a given location is made difficult by the uncertainties in 1) the intensity of the source terms, 2) in the complex chemistry of the region, and 3) in the concentrations of minor but important neutral species such as nitric oxide, atomic oxygen, ozone and water vapor which vary with season, latitude, and time-of-day (Isaksen 1974). At night the ionization has traditionally been thought to result from excitation of nitric oxide

B. Grandal and J. A. Holtet (eds.), Dynamical and Chemical Coupling, 145-160.

by solar Lyman-α (HLy-α) radiation scattered from the opposite hemisphere. Other steady sources of ionization such as galactic x-rays, diffuse x-rays, and discrete cosmic x-ray sources have been shown to produce ionization significantly below that of scattered HLy-α (Potemra 1974) and hence will not be considered further.

Below 60 km altitude the principal ionization source under quiescent conditions is galactic cosmic rays whose intensity at the top of the atmosphere varies both with geomagnetic latitude and solar cycle. The maximum ionization from cosmic rays is experienced at polar cap latitudes during solar minimum conditions.

The ionization rates resulting from these sources are illustrated in Figure 1. The rates are quite low throughout the region, particularly at night. Hence, it is not surprising that several authors have investigated the possible significance of precipitating particles from the sun and the earth's radiation belts to the maintenance of the ionosphere (Zmuda 1966; Lauter and Knuth 1967; Maehlum 1967; Belrose and Thomas 1968; Rees 1969; Bailey 1970; Potemra and Zmuda 1970; Manson 1972; Manson and Merry 1971; Potemra 1973; Potemra and Rosenberg 1973).

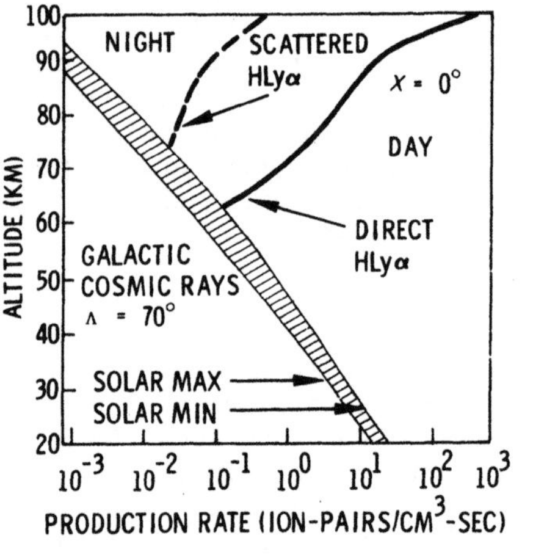

FIG. 1. Ionization rates in the quiescent day and nighttime D-region.

Most of the experimental evidence for the role of particulate radiation has been indirectly inferred from studies of abnormal electromagnetic wave propagation at VLF, LF, and HF (see references in Potemra and Rosenberg 1973). Association of ionospheric phenomena such as the D-region 'winter anomaly', 'post-storm' effects, and short-lived disturbances with electron precipitation and ionization have been made by several authors but quantitative data have been minimal. Larsen et al. (1976a) provided the first detailed experimental evidence that precipitating electrons from the radiation belt were the main cause of D-region post-storm conditions at mid-latitudes. Satellite data on the intensity, spectral, and spatial distribution of precipitating electrons obtained just above the atmosphere at Ottawa, Canada were shown to positively correlate with the enhanced ionization measured simultaneously by a ground-based sounder. The calculated ionization rates and measured electron densities from the precipitation exceeded those due to solar radiation and galactic cosmic rays at altitudes between 55 and 85-89 km for a period of several days following the storm of 17 December 1971.

The experimental results of Larsen et al. (1976) confirmed the theoretical predictions of Spjeldvik and Thorne (1975) that precipitating electrons could be the dominant D-region ionization source above 55 km during both daytime and nighttime at middle latitudes during the few-to-several-day period following strong magnetic storms. Those authors predicted that the resonant interaction of the trapped electron population with the naturally-occurring and post-storm enhanced whistler mode wave environment at ELF (3-3000 Hz) would diffuse electrons into the atmospheric loss cone and produce significant ionization enhancements.

It has recently been suggested by Thorne (1977) that energetic electron precipitation and the associated bremsstrahlung ionization, if it occurs more than a few percent of the time, can on an annual basis provide a local source of upper stratospheric nitric oxide molecules comparable to that from either galactic cosmic rays or energetic solar proton events. Since nitric oxide plays a major role in the removal of stratospheric ozone, the effects of electron precipitation events may have to be included in future modeling of the terrestrial ozone layer.

At the times of major disturbances on the sun, the earth's polar cap ionosphere above magnetic latitudes of $\sim 60°$ is greatly enhanced by solar energetic protons, electrons, alpha particles and x-rays. The occurrence, intensity and duration of these events are highly variable. Events occur much more frequently during solar maximum conditions and typically last for a few to several days. Notable among the events of the past solar maximum were those that occurred on 2 November 1969 and 4 August 1972. Extensive results on the ionospheric effects of these events are reported in Ulwick (1972) and Coffey (1973), respectively. The significant differences between electron precipitation events and solar particle events (SPE) are that the former occur more frequently, are confined to middle and auroral magnetic latitudes ($\sim 45°$ to $72°$) and enhance the ionization above ~ 60 km altitude, while the latter are confined to high magnetic latitudes ($> 60°$) and often produce their maximum ionization below ~ 60 km altitude.

Because the resulting ionization is often deep in the atmosphere, large SPE's have been shown to influence the neutral chemistry of the stratosphere in addition to enhancing the ionized component. During both the 2 November 1969 (Weeks et al. 1972; Swider and Keneshea 1973) and the 4 August 1972 (Crutzen et al 1975;

Reagan and Imhof 1976) events, stratospheric ozone was shown to be depleted as a result of increases in nitric oxide and odd-hydrogen species produced directly by the solar protons.

2. ELECTRON AND BREMSSTRAHLUNG IONIZATION

At the time of major magnetic storms, the trapped electron population in the outer radiation belt greatly increases as evidenced by the Explorer 45 satellite data taken during the 17 December 1971 storm and shown in the upper panel of Figure 2 (Lyons and Williams 1975). Note the significantly larger intensity increase in the more energetic electrons. The trapped electron population decays in a near exponential manner over the time period of ~ ten days. At low altitudes on these same field lines the energetic electron flux directly precipitating into the atmosphere over Ottawa, as observed by the 1971-089A satellite (Larsen et al. 1976a), is shown in the middle panel. The temporal features of both populations are similar, suggesting that some mechanism altered the equatorial pitch-angle distribution of the electrons such that enhanced fluxes were observed within the loss cone just above the atmosphere.

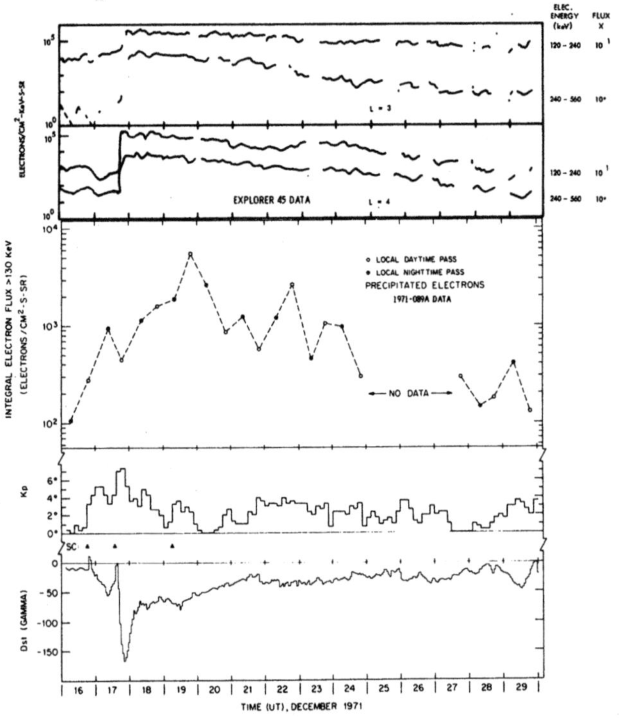

FIG. 2. Simultaneous observations of the equatorial trapped electron population (upper panel) and the low-altitude precipitating electron flux > 130 keV over Ottawa (middle panel) during the 17 December 1971 magnetic storm.

In the lower panels the storm-time indices K_p and D_{st} are shown. The precipitating flux increases at the time of increasing K_p but maximizes approximately two days later than the peak K_p. Note that at the time of peak precipitation the K_p index was quite low, indicating that the latter is not a reliable indicator of ionospheric effects due to electron precipitation.

The behavior exhibited in Figure 2 is consistent with the postulations of Spjeldvik and Thorne (1975) that wave-particle interactions occurring along the field lines, but principally in the equatorial region, are responsible for the enhanced precipitation rate. The peak precipitation rate occurring a few days after the storm time is also consistent with the experimental evidence (Thorne 1973) that the wave intensities are enhanced in the recovery phases of storms.

The latitudinal distribution of the electron precipitation is illustrated in Figure 3, taken from Larsen et al. (1976b). The locally trapped electron flux > 130 keV identified by the REES detector is shown on 16 December prior to the magnetic storm,

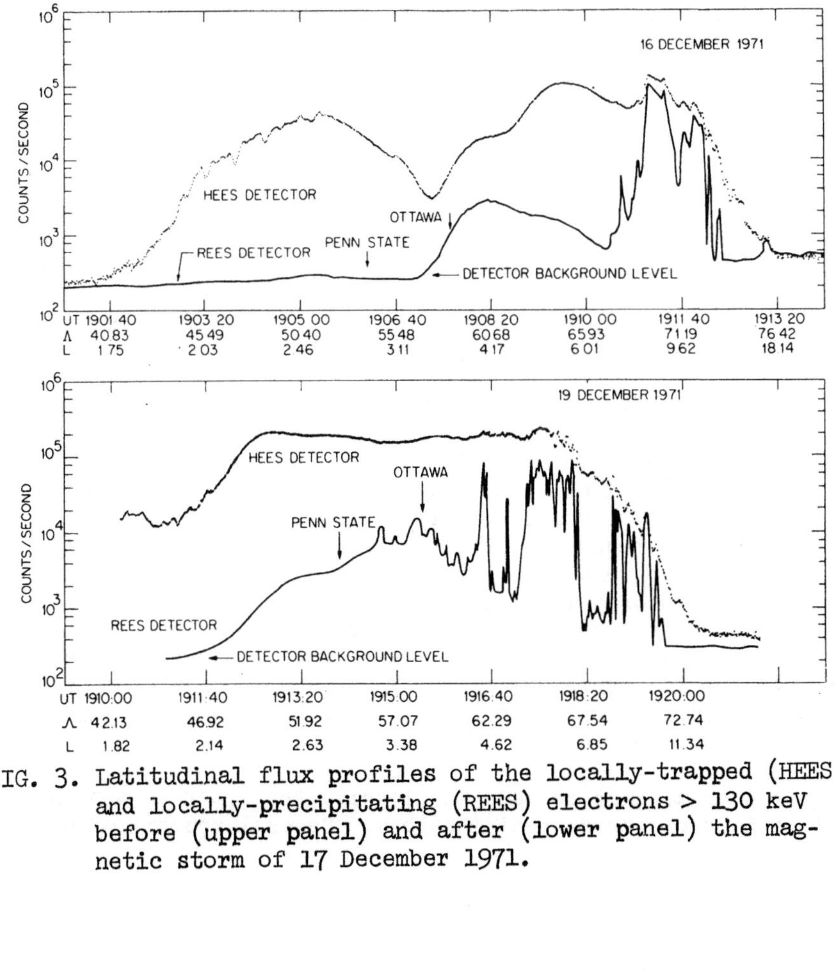

FIG. 3. Latitudinal flux profiles of the locally-trapped (HEES) and locally-precipitating (REES) electrons > 130 keV before (upper panel) and after (lower panel) the magnetic storm of 17 December 1971.

and on 19 December two days after the storm peak. In the upper panel significant precipitation can be observed only above a magnetic latitude of $\sim 57°$. Note the large difference between the precipitation occurring at the two ionospheric stations Ottawa and Pennsylvania State University which are approximately at the same longitude. The fact that Ottawa experiences greater electron precipitation, even on quiet days, might explain a number of previously unresolved dilemmas between the two sites. On 19 December the precipitation had increased by approximately an order of magnitude at each station. Precipitation now extends down to $\sim 45°$ magnetic latitude (L=2) which is also consistent with the predictions of Spjeldvik and Thorne (1975). At high latitudes the precipitation is intense between $\sim 66°$ to $73°$, even before the storm.

Attempts have been made to identify the longitudinal spatial/temporal features of the energetic electron precipitation. Ground-based optical techniques and balloon-borne x-ray measurements are capable of observing only a small portion of the precipitation oval. Single satellite measurements can likewise identify only two local time features of the electron population on a given pass. Dual or multiple satellite crossings of the precipitation region at the same time can begin to distinguish spatial/temporal features. Satellite observations of the x-ray bremsstrahlung emission from the atmosphere due to the precipitating electrons can add significantly to resolving spatial/temporal features. Imhof et al. (1975) utilized the technique of combined direct electron measurements on two polar-orbiting satellites and x-ray observations on one of the satellites to conclude that the major feature changes in the flux profiles over an extended longitude interval were primarily spatial in nature. In the future combined optical and x-ray imaging of the atmosphere from polar-orbiting satellites along with direct electron measurements will be required to obtain the worldwide picture of electron precipitation.

To obtain quantitative information on the effects of electron precipitation on the ionosphere the energy spectra and the pitch-angle distribution of the incoming electrons must be measured. There is an amazing lack of detailed electron spectra in the literature with which to work. Figure 4 shows the spectra of precipitating electrons measured with the 1972-076B satellite at mid-latitudes during several storm periods in 1972-74 (Imhof et al. 1976). The low-energy fluxes shown are comparable to those measured in moderate auroral events. Note the large intensity variations encountered in the energetic electrons > 10 keV that penetrate into the ionosphere below 100 km. Because of these variations, detailed spectral measurements such as shown in Figure 4 are necessary to obtain quantitative effects.

In Figure 5 an attempt is made to classify the spectrum and intensity of various electron precipitations. Several spectra of storm-

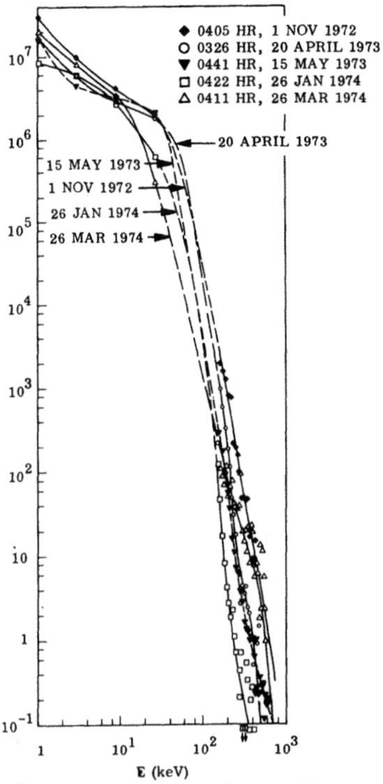

FIG. 4. Energy spectra of precipitating electrons measured at midlatitudes during several magnetic storm periods in 1972-74.

time electron precipitation occurring at local midnight at midlatitudes (magnetic latitudes, Λ, between 57° to 61°) are shown. The data were taken with the spinning polar-orbiting satellite 1972-076B (Imhof et al. 1976) which also measured the pitch-angle distribution of the particles. Note the especially energetic precipitation measured at the trapping boundary, $\Lambda = 61.3°$ in this case (Imhof et al. 1977). Also shown is a typical spectrum measured in the post-storm recovery period following the 17 December 1971 magnetic storm. The spectra were measured at local day at Ottawa (Larsen et al. 1976a). Finally the solar electron spectrum measured over Chatanika, Alaska in the intense event of 4 August 1972 (Reagan and Watt 1976) is shown for comparison.

The spectra shown in Figure 5 were input to a computer program called AURORA (Walt et al. 1968) to obtain the ion-pair production in the lower ionosphere. This program utilizes finite difference techniques to numerically solve the Fokker-

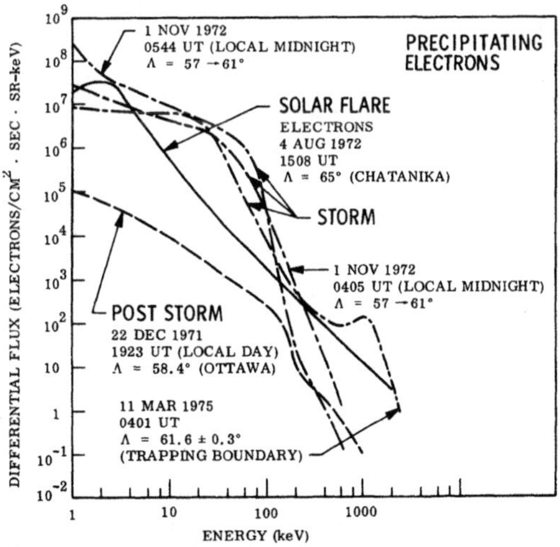

FIG. 5. Summary of the energy spectra measured in several classes of electron precipitation events.

Planck steady-state equation for the electrons diffusing through the atmosphere. The program takes into account the pitch-angle scattering of the electrons down to energies of ~ 1 keV. The ion-pair production resulting are shown in Figure 6, assuming that 35 eV are required to create each ion-pair. Also shown for reference is the ionization due to cosmic rays at a magnetic latitude of 65° in the epoch 1972, as determined by the technique of Heaps (1976). The contribution of solar radiation at a solar zenith angle of 70° is also shown for reference.

As is clearly evident in Figure 6, electron precipitation at the time of major magnetic storms totally dominates the ambient ionization sources between altitudes of 55 to 100 km, even in the daytime. The dominance of the electron ionization above 70 km is even greater at nighttime when the ambient ionization is as low as shown in Figure 1. Even in the post-storm recovery periods the electron-produced ionization dominates over the solar radiation source at altitudes between 55-85 km. On one of the quietest days measured, 9 January 1972, the electron ionization was still important in the 65 km altitude range of significance to VLF communications.

Note especially the ionization role of the electron precipitation at the trapping boundary. The intense energetic spectrum shown in Figure 5 results in ionization rates of 5×10^3 ion-pairs/cm^3 sec at altitudes near 60 km, which are higher than the solar electron ionization rates in a major flare and some 50,000 times greater than the ambient cosmic ray ionization. Such high ionization rates may have significant importance to the alteration of the neutral chemistry of the upper stratosphere, particularly through the generation of nitric oxides and odd-hydrogen molecules. The

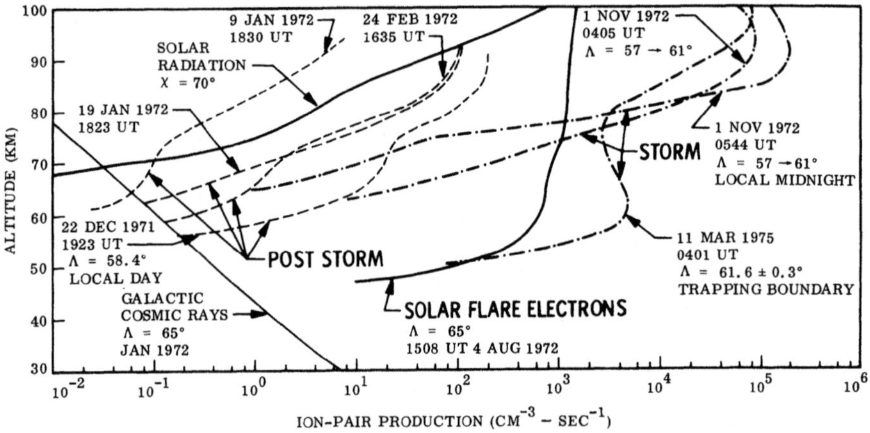

FIG. 6. Ion-pair production rates in the lower atmosphere due to the several classes of electron precipitation shown in Figure 5. Also shown for comparison are the solar radiation and cosmic ray sources.

narrow latitudinal extent of the region, ~ 0.6°, however must be taken into account when assessing the worldwide atmospheric impact.

In addition to the direct ionization created by the electrons, their slowing down in the atmosphere results in the generation of bremsstrahlung which can further penetrate and ionize. Figure 7 shows the significance of this secondary ionization as compared to the ambient sources. The bremsstrahlung produced by moderate electron auroras having integrated energy deposition of 10 ergs/cm^2 sec and exponential energy spectra with typical e-folding values of 6 keV and 12 keV is shown. The ionization due to the bremsstrahlung was calculated using the computer program BREMGAT (Newkirk et al. 1974). This ionization dominates over the ambient solar and cosmic radiation between ~ 45 to ~ 75 km altitudes. The important point illustrated is that during moderate activity the ionosphere below 75 km altitude in the auroral zones results principally from the secondary bremsstrahlung ionization created by the electrons that stop above 100 km altitude. Since the auroral electron source is highly variable in time and location around the oval, one can expect the lower ionosphere below this source to be equally variable.

The role of bremsstrahlung at storm-times is also illustrated. While the direct electron ionization in the 1 November 1972 storm was dominant down to altitudes of ~ 60 km, the bremsstrahlung ionization associated with these electrons extends the dominance over the cosmic ray source to altitudes near 35 km. Thus, the bremsstrahlung creates significant ionization in the lower atmosphere that exceeds the only known steady source, cosmic rays. As sug-

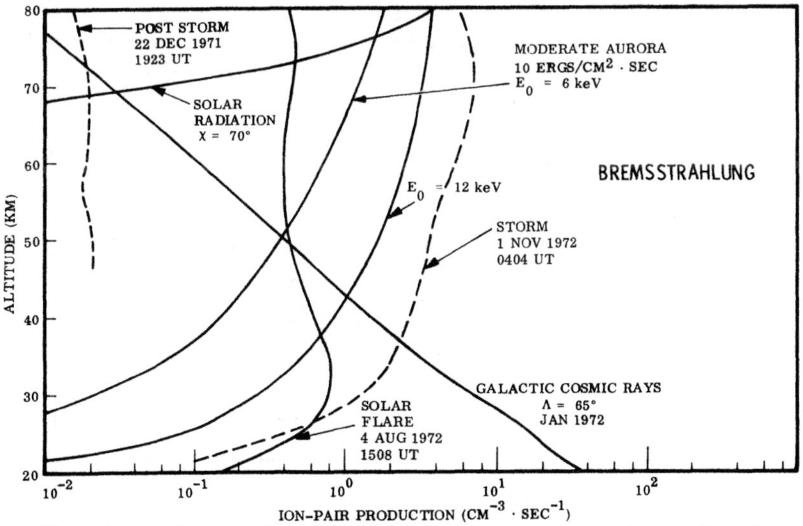

FIG. 7. Ion-pair production in the lower atmosphere due to bremsstrahlung created by precipitating electrons.

gested by Thorne (1977), if these storms occur frequently enough the bremsstrahlung source may be important in modifying the neutral chemistry of the upper stratosphere. For example, the storm-produced bremsstrahlung ionization rate at 50 km altitude is approximately ten times that due to cosmic rays. If this source is to be significant at this altitude on an annual basis, storms of this magnitude must occur greater than ten percent of the time. The auroral and storm-produced bremsstrahlung sources are not dominant over cosmic rays in the lower stratosphere, as earlier shown by Johnson and Imhof (1973).

The ionization shown in Figure 6 leads to significantly enhanced electron densities in the D-region ionosphere. Utilizing nighttime values of the effective electron loss rate, ψ, measured by Reagan and Watt (1976) under disturbed conditions, the production rates have been converted into corresponding electron densities from the relation $N_e = (Q/\psi)^{\frac{1}{2}}$. The resulting storm-time electron densities are shown in Figure 8. The peak electron densities of $4 \times 10^5/cm^3$ at 90 km altitude are over 1000 times higher than the typical ambient nighttime ionosphere. The total dominance of the precipitation-induced electron densities at all altitudes in the D-region nighttime ionosphere is clearly evident.

3. SOLAR PARTICLE EVENTS

In contrast to electron precipitation events, solar particle events (SPE) occur less frequently but cover the entire polar cap regions down to $\sim 60°$ magnetic latitude for a period of a few to several days per event. The region irradiated by the solar protons, alpha particles and sometimes electrons covers about 17 percent of the earth's surface area and hence is significant in

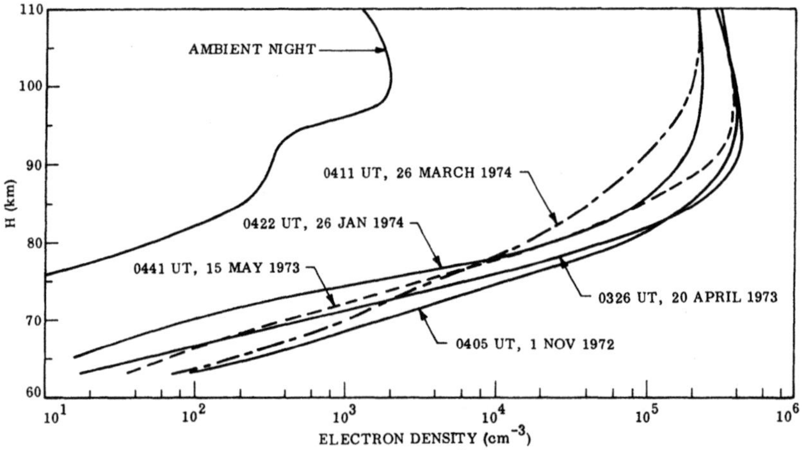

FIG. 8. Electron densities in the nighttime D-region ionosphere resulting from the storm-time precipitation shown in Figure 4.

terms of worldwide effects. Also in contrast to electron events the incident solar particle source tends to be more uniform and slowly varying over the irradiated region. The energy input from moderate and large SPE's can not only greatly increase the ionization but can alter the neutral chemistry of the middle and lower stratosphere. The neutral chemistry changes can have profound effects on both the ionization and physical/chemistry characteristics of the region long after the SPE is over.

An example of the ion-pair production resulting from the large SPE of 4 August 1972 is given in Figure 9, taken from Reagan and Watt (1976). The solar protons extending to several hundred MeV in energy produced the main ionization in the atmosphere below 60 km altitude in this event. The peak ionization occurred near 40-45 km altitude and exceeded the cosmic ray production by 30,000 times. The solar electrons, shown earlier, were important to the total ionization above 70 km and became the dominant source above 90 km altitude. The solar alpha particles were only a minor contributor to the total ionization. Simultaneous measurements of the electron density in the D-region were performed with the incoherent scatter radar at Chatanika, Alaska (Reagan and Watt 1976). Electron densities in excess of $10^5/cm^3$ were measured near 90 km and above $10^4/cm^3$ down to altitudes of 50 km near the peak of the event.

A detailed analysis of the neutral and ion chemistry occurring during the August 1972 event has been performed by Gunton et al.

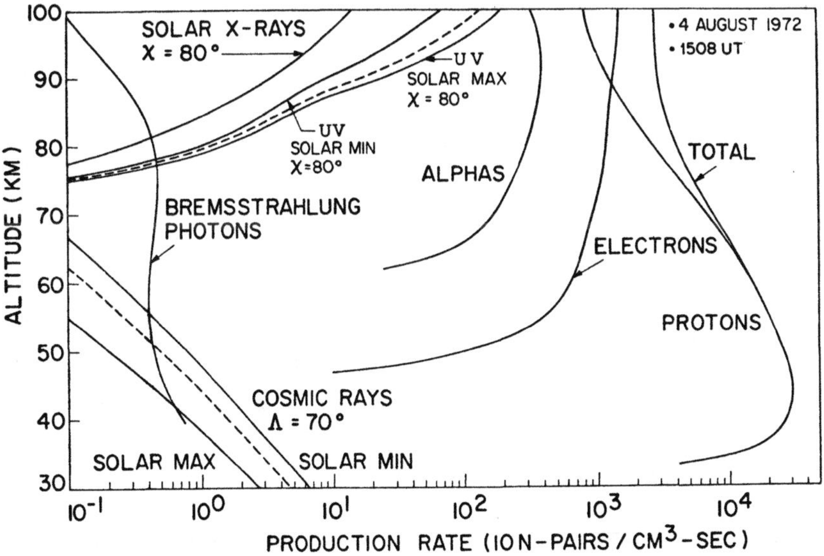

FIG. 9. The contribution of the various solar particle, solar radiations, bremsstrahlung and cosmic rays to the total ion-pair production in the August 1972 solar particle event.

(1977). An example of the results obtained at Chatanika at altitudes of 80, 70, and 60 km is shown in Figure 10. In the top panel the effective electron loss rate, ψ, determined from the chemistry analysis (solid line) is compared to experimental data

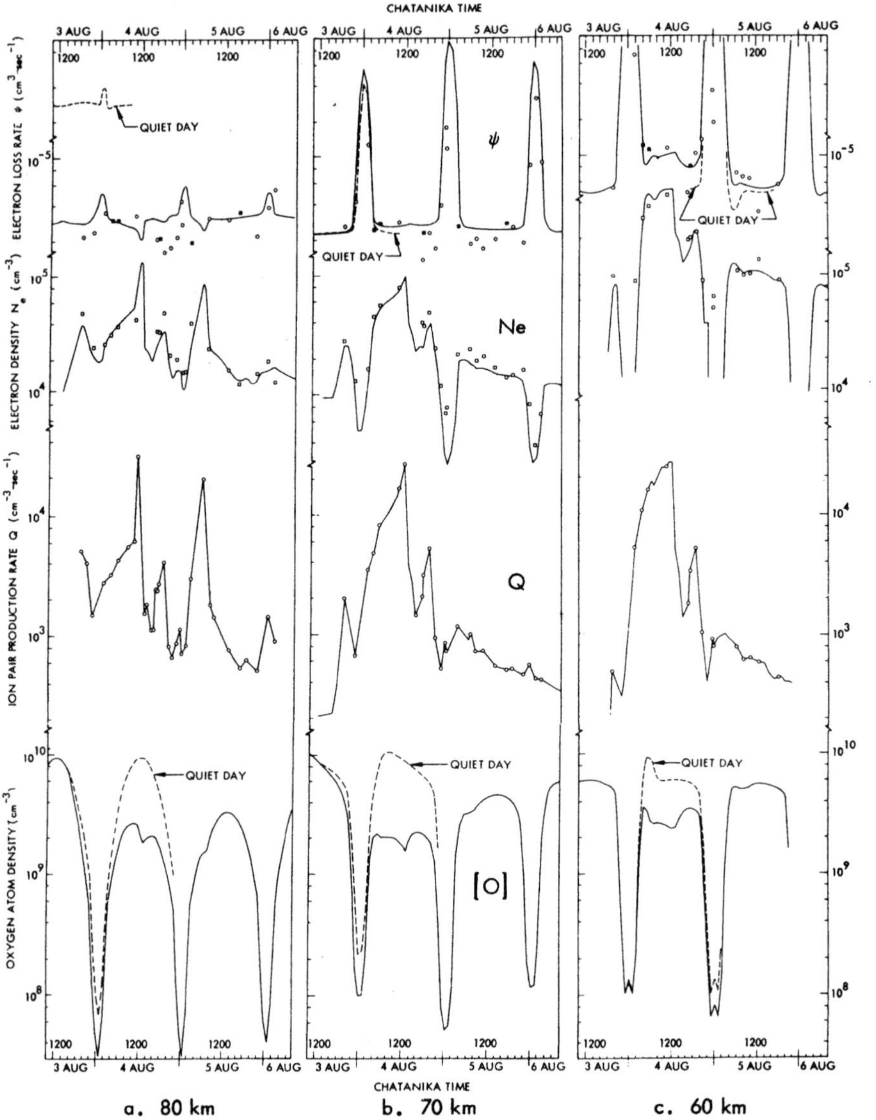

FIG. 10. Some results of a neutral and ion chemistry analysis of the lower atmosphere at altitudes of 80, 70, and 60 km during the August 1972 solar particle event as taken from Gunton et al. (1977).

points obtained from the measured production rates and electron densities. Reasonable agreement is obtained throughout the diurnally varying conditions of the event. In the second panel from the top, the electron densities obtained from the analysis (solid lines) are found to be in good agreement with the data measured by the radar. The production rate profiles are shown in the third panel.

One of the more important results of the analysis is shown in the lower panel of Figure 10. The diurnally varying atomic oxygen concentration is shown for a quiet day (dashed line) and as calculated for the SPE. At the time of the peak ionization, the calculations show that the atomic oxygen concentration was reduced by approximately a factor of 5 at 70 and 80 km altitudes and over 2.5 times at 60 km altitude. Atomic oxygen, although a very minor species in the mesosphere and stratosphere, plays an important role in determining the ultimate electron densities. Thus, we see the important coupling of the neutral and ionized atmosphere and the effect that changes in the former, due to solar particle bombardment, have on the latter.

Several other minor neutral species were altered by the August 1972 SPE. The ozone concentrations in the upper stratosphere were substantially reduced (Crutzen et al. 1975; Reagan and Imhof 1976). The ozone reduction was principally due to the catalytic destruction by the odd hydrogen and odd nitrogen molecules which were greatly increased by the proton bombardment. The odd hydrogen molecules produce short-term destructive effects, but the creation of odd nitrogen molecules, such as nitric oxide and nitrous oxide, produced long-term reductions. The worldwide climatic effects of stratospheric ozone reductions during solar particle events will be a subject of considerable interest in the future.

Figure 11, taken from Reagan and Imhof (1976), shows the correspondence between the latitudinal profile of the energetic protons that penetrated deep into the stratosphere on 4 August 1972 and the corresponding profile of the ozone concentrations measured in the northern polar regions. The ratio of the ozone concentrations measured by the Nimbus 4 satellite (Heath et al. 1976) two days after the peak of the event (day 219) to one day before the event (day 216) is shown for altitudes near 48 and 34 km. Substantial decreases in the ozone coinciding with the proton bombardment region are clearly evident.

4. CONCLUSIONS

Electron precipitation and the associated bremsstrahlung have been quantitatively shown to be important and often the dominant ionization sources in the middle and high latitude atmosphere

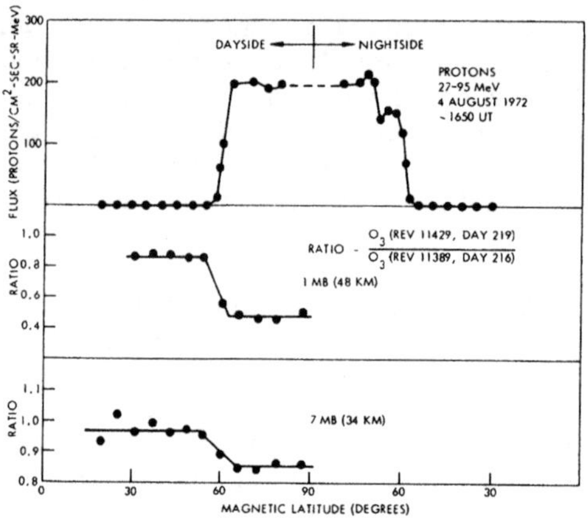

FIG. 11. Correspondence between the proton bombardment region and the reduction of stratospheric ozone as measured with the Nimbus 4 satellite during the August 1972 SPE.

below 100 km altitude. Electron precipitation may also be important in altering the neutral chemistry of the upper stratosphere. Solar particle events not only produce greatly enhanced ionization but alter the chemistry of key minor neutral species in the atmosphere such that long-term changes in the atmosphere/ionosphere persist after the events are over.

5. ACKNOWLEDGMENTS

The author wishes to acknowledge the important contributions of Drs. William L. Imhof and Trygve R. Larsen to this review. The support of the Lockheed Independent Research program in preparing this review is appreciated. A great deal of the results presented were obtained from satellite experiments sponsored by the Office of Naval Research. Special acknowledgments are made to Mr. R. Gracen Joiner of that office for his continued support and encouragement.

6. REFERENCES

Aiken, A. C., 1971, Review presented at COSPAR Symposium on D and E Region Ion Chemistry, Urbana, Illinois, 6-9 July 1971.

Bailey, D. K., R. R. Brown, and M. H. Rees, 1970, J. Geophys. Res., 75, 7161.

Banks, P. M., and G. Kockarts, 1973, Aeronomy, Academic Press, New York.

Bauer, S. J., 1973, Physics of Planetary Ionospheres, Springer-Verlag Publ. Co., New York.

Belrose, J. S., and L. Thomas, 1968, J. Atmos. Terr. Phys., 30, 1397.

Coffey, H., 1973, Collected Data Reports on August 1972 Solar-Terrestrial Event, Parts I, II, and III, U.S. Dept. of Commerce, Environmental Data Service, Ashville, N. Car.

Crutzen, P. J., I. S. A. Isaksen, and G. C. Reid, 1975, Science, 189, 457.

Gunton, R. C., R. E. Meyerott, and J. B. Reagan, 1977, Lockheed Report D556351, Ion and Neutral Chemistry of the D-Region during the Intense Solar Particle Event of August 1972.

Heath, D. F., A. J. Krueger, and P. J. Crutzen, 1976, NASA-GSFC Report X-912-76-172, Influence of a Solar Proton Event on Stratospheric Ozone.

Heaps, M. G., 1976, USA Ballistic Research Laboratories Report 2641, The Ion Pair Production Rate Due to Cosmic Rays.

Imhof, W. L., G. H. Nakano, E. E. Gaines, and J. B. Reagan, 1975, J. Geophys. Res., 80, 3622.

Imhof, W. L., T. R. Larsen, J. B. Reagan, and E. E. Gaines, 1976, Lockheed Report D502063, Analysis of Satellite Data on Precipitating Particles in Coordination with ELF Propagation Anomalies.

Imhof, W. L., J. B. Reagan, and E. E. Gaines, 1977, EOS, 58, 464.

Isaksen, I. S. A., 1974, ELF-VLF Radio Wave Propagation, ed. J. Holtet, D. Reidel Publ. Co., Dordrecht-Holland.

Johnson, R. G., and W. L. Imhof, 1973, NASA Report SP-366, Possible Relationships Between Solar Activity and Meteorological Phenomena, 89.

Larsen, T. R., J. B. Reagan, W. L. Imhof, L. E. Montbriand, and J. S. Belrose, 1976a, J. Geophys. Res., 81, 2200.

Larsen, T. R., W. L. Imhof, and J. B. Reagan, 1976b, J. Geophys. Res., 81, 3444.

Lauter, E. A., and R. Knuth, 1967, J. Atmos. Terr. Phys., 29, 411.

Lyons, L. R., and D. J. Williams, 1975, J. Geophys. Res., 80, 3985.

Maehlum, B., 1967, J. Geophys. Res., 72, 2287.

Manson, A. H., and M. J. W. Merry, 1971, J. Atmos. Terr. Phys., 33, 413.

Manson, A. H., 1972, Radio Science, 7, 57.

Newkirk, L. L., W. E. Francis, and M. Walt, 1974, Lockheed Report D407007, BREMGAT, A Code for the Generation and Transport of Bremsstrahlung through the Atmosphere.

Paulikas, G. A., 1975, Reviews of Geophysics and Space Physics, 13, 709.

Potemra, T. A., and A. J. Zmuda, 1970, J. Geophys. Res., 75, 7161.

Potemra, T. A., 1973, Physics and Chemistry of the Upper Atmospheres, ed. B. M. McCormac, D. Reidel Publ. Co., Dordrecht-Holland.

Potemra, T. A., and T. J. Rosenberg, 1973, J. Geophys. Res., 78, 1572.

Potemra, T. A., 1974, ELF-VLF Radio Wave Propagation, ed. J. Holtet, D.Reidel Publ. Co., Dordrecht-Holland.

Reagan, J. B., and T. M. Watt, 1976, J. Geophys. Res., 81, 4579.

Reagan, J. B., and W. L. Imhof, 1976, EOS, 57, 972.

Rees, M. H., 1969, Space Science Reviews, 10, 413.

Spjeldvik, W. N., and R. M. Thorne, 1975, J. Atmos. Terr. Phys., 37, 777.

Swider, W., and T. J. Keneshea, 1973, Planet. Space Sci., 21, 1969-1973.

Thomas, L., 1971, J. Atmos. Terr. Phys., 33, 157.

Thorne, R. M., E. J. Smith, R. K. Burton, and R. E. Holzer, 1973, J. Geophys. Res., 78, 1581.

Thorne, R. M., 1977, Science, 195, 287-289.

Ulwick, J., 1972, ed., AFCRL Report 72-0474, Proceedings of COSPAR Symposium on Solar Particle Event of November 1969.

Walt, M., W. M. McDonald, and W. E. Francis, 1968, Physics of the Magnetosphere, ed. R. Carovillano and J. F. McClay, Reinhold Publ. Co., New York, New York.

Weeks, L. H., R. S. Cuikay and J. R. Corbin, 1972, J. Atmos. Sci., 29, 1138.

Zmuda, A. J., 1966, J. Geophys. Res., 71, 1911.

DISCUSSION

Witt: What is the reason and consequence of the pronounced change in the shape of the energy spectrum at 100 keV during storm ?

Reagan: The shape of the precipitated spectrum is probably determined by the interaction of the electron spectrum near the edge of the loss cone in the equatorial regions and the existing wave spectrum, but since we have never yet measured both parameters simultaneously we cannot be certain.

Thorne: The characteristic break in the shape of the spectrum of trapped electrons near 100 keV can qualitatively be understood in terms of a different source mechanisms. Below 100 keV adiabatic compression of solar wind injected particles can explain the observed fluxes. Above 100 keV we need an additional acceleration process.

Witt: What method is used to calculate the energy deposition ?

Reagan: The electron deposition in the atmosphere was calculated using the computer code AURORA developed by Walt at Lockheed (Walt et al. 1968, in Physics of the Magnetosphere, ed. by R.L.Carovillano, J.F.McClay and H.R.Radoski, D. Reidel, Dordrecht, Netherlands).

Piggott: Electron densities shown are consistent with ionospheric evidence over North America. Over Europe these are much smaller by factors of order 5. Are these data from the US sector ? Europeans should note the large longitude change in all these phenomena, eg Es at 20 MHz over North America corresponds to Es at 6 mHz over Europe.

Reagan: All of the data shown were for the North America sector. The presence of the South Atlantic anomaly at longitudes between North America and Europe can have profound effects on the particle precipitation present over ionospheric stations on the two continents at the same magnetic latitude.

Rawer: How many orders of magnitude is the increase of electron production during strong storm conditions as compared with quiet conditions ?

Reagan: At night time the measured ion-electron production leads to electron densities that are approximately 100 to 1000 times higher than the quiet time densities at altitudes between 70-100 km.

INFLUENCE OF RELATIVISTIC ELECTRON PRECIPITATION ON THE LOWER IONOSPHERE AND STRATOSPHERE

Richard Mansergh Thorne

Department of Atmospheric Sciences, University of California, Los Angeles, California 90024 U.S.A.

INTRODUCTION

Under geomagnetically quiet conditions there is a pronounced deficiency in energy deposition capable of ionizing the lower mesosphere and upper stratosphere. The precipitation of energetic outer radiation zone electrons can therefore provide the dominant ionization source at such altitudes, particularly during geomagnetically disturbed periods. The general importance of this energy input has already been stressed in the review talk by J. B. Reagen (1977). At relativistic energies ($\gtrsim$ 500 kev) the precipitating electrons deposit their energy near the base of the mesosphere; and the concomitant Bremsstrahlung X-rays provide additional energy deposition deep into the stratosphere.

This paper describes a preliminary attempt to assess the relative importance of relativistic electron precipitation on the ionization state and long term Nitric Oxide production in the lower mesosphere and upper stratosphere. Comparisons will be made with previously identified sources due to bombardment of the atmosphere by galactic cosmic rays (GCR) and energetic solar protons (during SPE). Potential mechanisms leading to the rapid precipitational loss of trapped relativistic electrons will also be discussed.

IONIZATION DURING RELATIVISTIC ELECTRON PRECIPITATION

Long distance radio communication is known to be severely disturbed when intense fluxes of electrons penetrate to D-region altitudes. Indirect information on the electron precipitation can therefore be inferred by monitoring such signals. Figure 1

B. Grandal and J. A. Holtet (eds.), Dynamical and Chemical Coupling, 161-168.

shows an example of two techniques which provide a particularly sensitive measure of relativistic electron bombardment. The top panel shows the temporal change in the absorption of 23.2 Mhz signals transmitted on the Alaskan forward scatter link from Barrow to Anchorage. Under daytime conditions the reflection height of such waves is generally near 70 km. The abrupt increase in absorption results from the precipitation of energetic particles leading to enhanced ionization below this altitude. The central panel shows the phase of VLF signals transmitted between Aldra and Tromsø during the same time interval. The normal reference quiet day diurnal variation is indicated by dashed lines. During this interval the nocturnal phase showed an abrupt phase advance

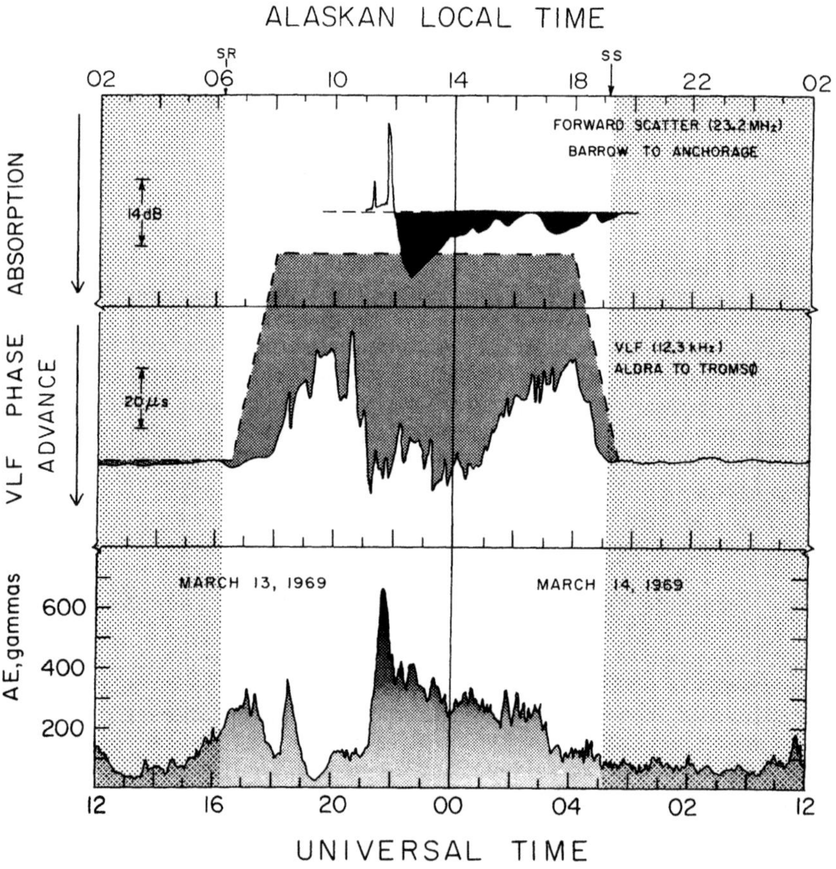

Fig. 1. Simultaneous observation of REP events during daytime over Alaska and at night over Norway (Thorne and Larsen 1976)

at 2100 UT, which extended past the normal daytime level. This may be interpreted as a lowering in the VLF wave reflection altitude due to intense ionization of the lower mesosphere. The inferred onset of nocturnal energetic electron precipitation occurred in approximate coincidence with the onset of intense substorm activity here identified through the AE index. The onset of dayside precipitation over Alaska was delayed by approximately one hour. An extended study of such correlated precipitation events (Thorne and Larsen, 1976) has established this as a characteristic response to intense substorm activity. Nightside precipitation generally begins with the substorm onset whereas dayside events are less frequent, of shorter duration and generally delayed by up to a few hours. Earlier studies (e.g. Bailey and Pomerantz, 1967; Bailey, 1968; Rosenberg et al., 1972; Larsen and Thomas, 1974) have already stressed the correlation with substorms and established that REP events are usually restricted to sub-auroral latitudes, $4 \lesssim L \lesssim 8$.

Unfortunately there is little published data on the energy

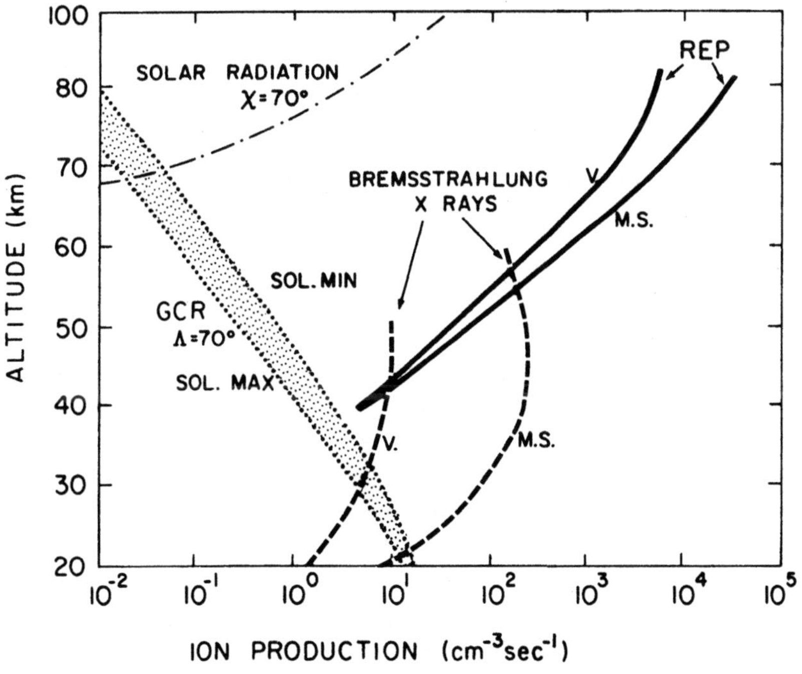

Fig. 2. Ionization rates during two REP events compared to that by quiet-time solar radiation and galactic cosmic rays.

spectrum of precipitating electron measured on satellite during REP events. Two specific examples (Vampola, 1971; Mathews and Simons, 1973) have been used in Figure 2 to compute altitude profiles of ion pair production in the lower atmosphere. The direct electron energy deposition was computed using a modified version of the program described by Walt et al. (1968) and the Bremsstrahlung X-ray ionization was computed using the results of Berger et al. (1974). When compared with the quiet-time sources of ionization due to solar radiation and galactic cosmic rays it is clear that REP events should locally control the ionization in the upper stratosphere and throughout the mesosphere.

Recent studies have also established that between one to two molecules of NO are produced for each generated ion pair (e.g. Delgarno, 1967; Crutzen et al., 1972). Nitric oxide has a relatively long lifetime in the lower atmosphere and since it is now thought to be an import, if not dominant, catalytic agent in the removal of stratospheric ozone it becomes imperative to establish whether the NO produced during relativistic electron precipitation events is competitive with other natural sources. In order to assess the long term influence of REP events one must have information on the statistical occurrence of the precipitation and its global coverage. Such information has yet to be deduced from satellite measurements. The ground based radio probing techniques however, indicate that dayside REP events occur as much as 10% of the time and the nightside events may be even more frequent. The precipitation of course is limited to sub-auroral latitudes which cover only 7% of the globe. Using this as a guide Figure 3 shows an estimate of the annual REP production of ion pairs (or NO) assuming either a 1% or 10% occurrence. This has been compared with both the annual production from GCR and the integrated effect during the August 1972 SPE, which was the most intense solar proton event ever recorded by satellite instruments.

It is clear that even with a 1% occurrence rate, REP events should dominate over GCR in the upper stratosphere and throughout the mesosphere. It is also clear that the extreme August 1972 SPE dominated over both annual sources except in the upper mesosphere where electron precipitation should remain predominant. One should however recall that the August 1972 event represents an extreme SPE, more than an order of magnitude larger than any other recorded by satellite data.

The majority of atmospheric ozone of course occurs at altitudes below the cross over between REP and GCR domination as indicated in Figure 3. Nevertheless, below 60 km the height integrated NO production from a 1% REP occurrence is comparable to that from GCR's and this suggests that one must therefore resort to models which include vertical transport effects in order to realistically assess the relative importance of the various

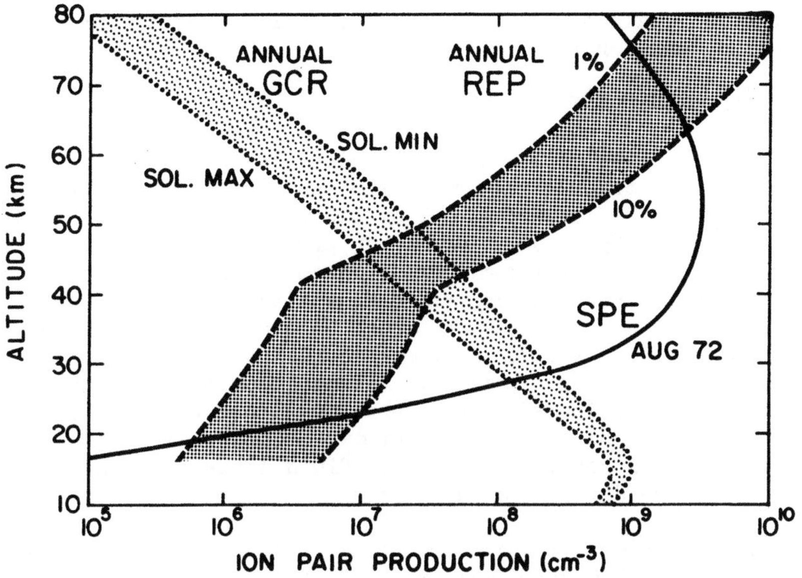

Fig. 3. Height profile of annual production of ion pairs due to 1 or 10% REP occurrence is compared to annual GCR and total production during the intense SPE of August 1972.

sources of NO on establishing the natural level of stratospheric ozone. Crudely one should expect REP events (or SPE) to mainly influence the upper stratosphere whereas GCR should predominate in the lower stratosphere where vertical transport is greatly inhibited. In the mesosphere, ozone chemistry is further complicated by the inclusion of water compounds (particularly OH) which becomes the dominant O_3 destruction element at higher altitudes (Crutzen, 1971). The production of hydroxyl radicals however will accompany ionization so that enhanced mesospheric O_3 loss should still result from relativistic electron precipitation.

In view of the well established correlation between magnetospheric substorms and solar activity (or solar wind sector structure) it is tempting to speculate on the role that substorm related relativistic electron precipitation might play in the realm of Sun-Weather Relationships. The following scenario might provide one channel for establishing a physically acceptable causal link: "A modulation of stratospheric ozone will cause changes in both the thermal structure and radiative damping properties of the middle atmosphere. This in turn should influence both the tropospheric energy budget and the reflection characteristics of atmospheric waves which are involved in the development of tropospheric weather systems." But whether such a speculative chain of reasoning can withstand quantitative scrutiny remains a task for the future.

POTENTIAL MECHANISMS FOR REP

The intensity of the precipitation flux during REP events indicates that trapped electrons are scattered from the radiation belts at the limit imposed by strong pitch-angle diffusion (e.g. Kennel, 1969). The scattering process must involve a resonant interaction with plasma waves which are naturally generated in the magnetosphere during substorm activity. During the resonant interaction the electrons experience a small change in direction relative to the ambient magnetic field. Under strong diffusion the scattering becomes so intense that the electrons are forced to be distributed isotropically. Those with a velocity vector close to the magnetic field direction, namely within the atmospheric loss cone, stream along field lines into the atmosphere and there deposit their energy.

The condition for first order cyclotron resonant scattering is simply that electrons Doppler shift the fluctuating wave field to a frequency equal to their rate of gyration about the field

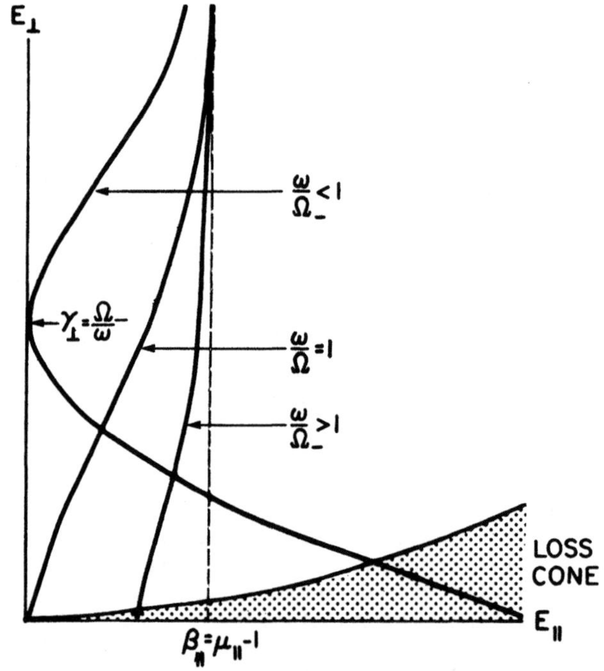

Fig. 4. Schematic plot of first order cyclotron resonance surfaces for relativistic electrons.

lines; namely

$$1 - \mu_{\shortparallel}\beta_{\shortparallel} = \frac{\Omega_-}{\omega\ \gamma}$$

where $\mu_{\shortparallel} = k_{\shortparallel}c/\omega$ is the wave refractive index and $\beta_{\shortparallel} = v_{\shortparallel}/c$ the normalized electron velocity along the ambient field direction, ω is the wave frequency, Ω_-/γ is the relativistic electron gyrofrequency with $\gamma = (1-\beta^2)^{-\frac{1}{2}}$. The general morphology of the surfaces in velocity space for resonance with a wave of prescribed frequency and refractive index is shown in Figure 4. For precipitation of relativistic electrons the resonance surfaces must intersect the loss cone at relativistic energies. Since $\mu_{\shortparallel}$ is generally much larger than unity, it is clear that this can only occur for waves with frequency below the electron gyrofrequency. In addition the waves must be sufficiently intense to cause scattering near the strong diffusion limit.

Four distinct classes of magnetospheric plasma waves have now been identified in the outer radiation zone. Their characteristics and the conditions needed to cause the observed strong diffusion loss of relativistic electrons are listed in Table 1. One can of course rule out the socalled electron electrostatic $(n + \frac{1}{2})\Omega_-$ waves since they can only resonate with low energy electrons. Our current knowledge of the ambient magnetospheric properties of the other three wave modes is not yet sufficiently advanced to distinguish between the competing processes. Further satellite measurement and indepth analysis will be required to answer this tantalizing question.

Table 1. Potential Mechanisms for REP

Wave Mode	Typical Intensity	Comments on Resonance and Strong Diffusion Requirements
Whistler mode chorus $1/10 < \frac{\omega}{\Omega_-} < 1$	10m γ	Resonance in tenuous plasma trough. Need B' ~ 100 to 10^3 m γ
E. M. Ion Cyclotron $\frac{\omega}{\Omega_+} < 1$	few γ	Strong diffusion loss in regions of high plasma density
Electrostatic Cyclotron Waves $(n + \frac{1}{2})\Omega_-$	10-100 mv/m	Resonance only with low energy precipitating electrons
Broad band electrostatic turbulence $\Omega_+ < \omega < \omega_{LHR}$	≈10 mv/m (but not well established)	Effective resonance in tenuous plasma trough. Need E' ~ 100 mv/m

ACKNOWLEDGEMENTS

Numerical computation of the ionization during REP events was performed by L. Andreoli. The research was supported in part by ONR Contract N 00014-76-C-0307 and NSF DES-13792.

REFERENCES

Bailey, D.K. 1968, Rev. Geophys., 6, 289.
Bailey, D.K., and Pomerantz, M.A. 1965, J. Geophys. Res., 70, 5823.
Berger, M.J., Seltzer, S.M., and Maeda, K. 1974, J. Atmos. Terr. Phys., 36, 591.
Crutzen, P.J. 1971, J. Geophys. Res., 76, 7311.
Crutzen, P.J., Isaksen, I.S.A., Reid, G.C. 1975, Science, 189, 457.
Dalgarno, A. 1967, Space Res., 7, 849.
Kennel, C.F. 1969, Rev. Geophys., 7, 379.
Larsen, T.R., and Thomas, G.R. 1974, J. Atmos. Terr. Phys., 36 1613.
Mathews, D.L., and Simons, D.J. 1973, J. Geophys. Res., 78, 7539.
Reagan, J.B. 1977, This volume.
Rosenberg, T.J., Lanzerotti, L.J., Bailey, D.K., and Pierson, J.D. 1972, J. Atmos. Terr. Phys., 34, 1977.
Thorne, R.M., and Larsen, T.R. 1967, J. Geophys. Res., 81, 5501.
Vampola, A.L. 1971, J. Geophys. Res., 76, 4685.
Walt, M., MacDonald, W.M., and Francis, W.E. 1968, p. 534 in Physics of the Magnetosphere, (Ed. B.M. McCormac) Reidel Dordrecht, Holland.

DISCUSSION

Hargreaves: Auroral radio-absorption events, observed with riometers, show a movement from night to day, giving a time delay of 30 mins to 1 hour between midnight and noon. No doubt some of these events contain a relativistic component. The REP events identified by forward scatter or VLF are a daytime phenomena. Is there evidence for a night-day movement in REP events, and, if so, what are the implications for the precipitation mechanism ?

Thorne: The simultaneous dayside forward scatter and nightside VLF radio wave observation do indeed indicate a delay on the order of one hour between the nightside and dayside relativistic electron precipitation. This delay is too long to be simply caused by gradient drifting of relativistic electrons injected into the nightside outer radiation zone during substorms. Rather the delay probably represents a movement of the zone of strong diffusion scattering. It is important to establish how this zone moves in order to distinguish the various potential mechanisms.

COHERENT WAVE INDUCED PARTICLE PRECIPITATION INTO THE UPPER ATMOSPHERE*

U. S. Inan, T. F. Bell and R. A. Helliwell

Radioscience Laboratory, Stanford University,
Stanford, Calif., 94305, USA

ABSTRACT. Particle precipitation induced by coherent VLF waves in the magnetosphere is simulated with a computer. It is found that energy fluxes of the order of 0.5 ergs/cm^2-sec are precipitated by 10 mγ waves. It is also shown that significant density enhancements will be produced in the D and E regions by these incoming fluxes.

Energetic electrons in the earth's radiation belts are precipitated into the ionosphere through their interaction with coherent VLF waves in the magnetosphere. Examples of such coherent waves are natural whistlers (Helliwell 1965), triggered emissions (Stiles and Helliwell 1975), signals that are injected by VLF ground transmitters (Helliwell and Katsufrakis 1974) and large scale power grids (Helliwell et al 1975; Park 1976). Future experiments may involve transmissions of VLF signals from laboratories carried on the Space Shuttle.

The particles interact with the waves in the cyclotron resonance mode, resulting in amplification of the waves (up to 30 db) and pitch angle scattering of the particles. As a result energetic electrons enter the loss cone and precipitate into the lower ionosphere. There they produce secondary electrons creating impulsive enhancement of ionization throughout the volume of the

*This work was supported by the National Aeronautics and Space Administration. The computer simulations were made on the CDC7600 Computer of the National Center for Atmospheric Research (NCAR). The use of this facility was made possible by a computer time grant from NCAR.

B. Grandal and J. A. Holtet (eds.), Dynamical and Chemical Coupling, 169-174.

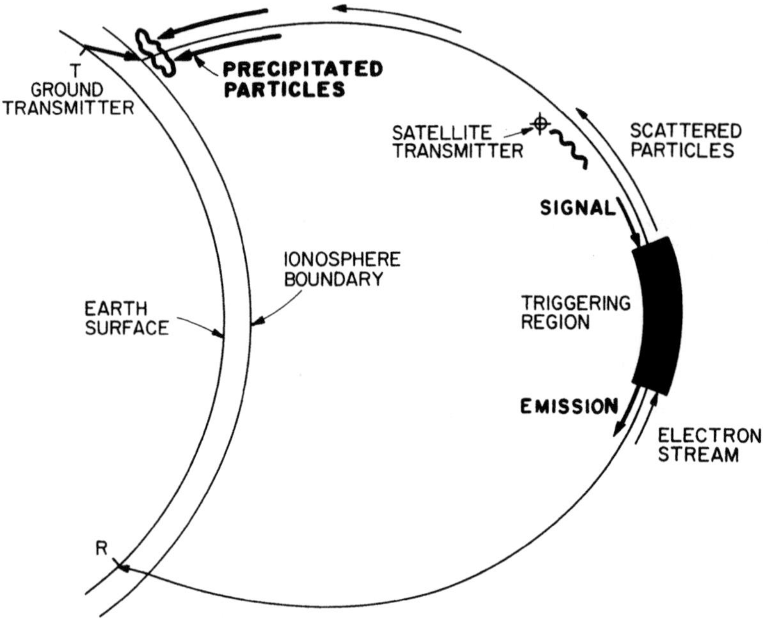

FIG. 1. Schematic representation of the wave particle interaction.

precipitation region. Precipitated energy deposition rates approaching that of a moderate aurora have been measured (Rosenberg et al 1971). The interaction and the resulting precipitation are schematically illustrated in Fig. 1.

Recently, the cyclotron resonance interaction was simulated on a computer (Inan et al 1977) primarily to obtain theoretical estimates for fluxes of electrons precipitated by VLF signals from ground transmitters such as that at Siple Station, Antarctica. Complete equations of motion are employed in the simulation and extensive study of scattering of single particles, sheets of particles and a full distribution of particles has clarified many features of the interaction (Inan et al 1977).

The computer simulation approach has, for the first time, determined the limits of validity of the linear theory under given conditions. Previous studies of pitch angle scattering by coherent waves (Ashour-Abdalla 1972) have used the linear theory approximation without establishing its limits. Figure 2 compares the root mean square scattering obtained from linear theory and the full nonlinear analysis. It can be seen that the linear theory is inaccurate for $B_w > 3$ mγ. Inan et al (1977) present a quantitative criterion for determining the validity of linear theory under a wide range of conditions.

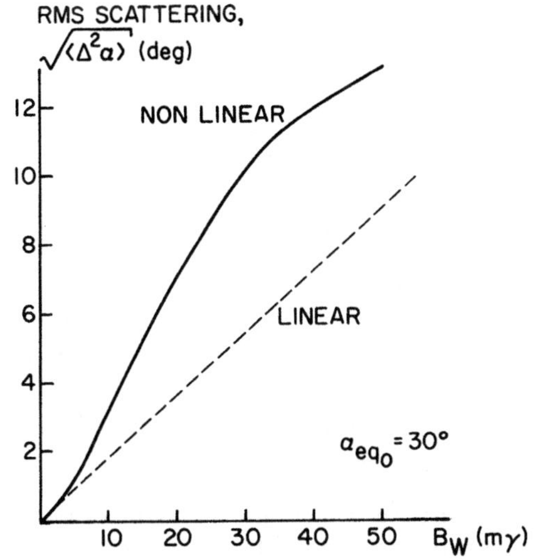

FIG. 2. RMS scattering predicted by linear and nonlinear analyses vs. wave intensity. This result is for equatorial scattering at L = 4.

The full distribution of particles is simulated by test electrons appropriately distributed in energy and pitch angle. The scattering of each test particle is computed by integrating the equations of motion over the course of time that the particle is in resonance with the wave. After the nonlinear gyroresonance interaction has taken place the perturbed particles are placed into new locations in the distribution in accordance with their total scattering in pitch angle and energy. To accurately simulate a full distribution, typically 40,000-50,000 test electrons are used for each computation.

Figure 3 shows the result of a computation for a 10 mγ (1 mγ = 10^{-12}wb/m^2) wave amplitude. The dashed line represents the unperturbed distribution which in this case is assumed to be uniform with a sharp cutoff at the loss cone. The solid curve represents the perturbed distribution after one pass of the wave. The loss cone half-angle at L ~ 4 is ~ 5.5 wide. More than 20,000 test particles with energies ranging from 1 to 10 keV were used for computing the new distribution. We see that the loss cone is partly filled with particles scattered from higher pitch angles.

The normalized precipitated flux can be computed by integrating the distribution over the loss cone. We have used Shield and Frank (1970) measurements of low energy electron fluxes in converting from normalized flux to energy deposition rates. Table 1 gives the precipitated flux for three different energies. The

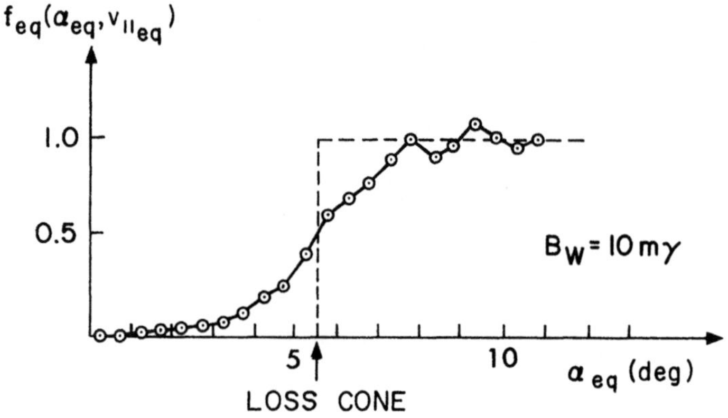

FIG. 3. Normalized equatorial distribution (integrated over energy) vs. pitch angle.

upper values are obtained assuming an isotropic pitch angle distribution of the trapped particles whereas the lower values are obtained assuming a highly anisotropic pitch angle distribution. Hence the values given in Table 1 could be viewed as upper and lower bounds to the flux. Note that for 1.5 keV the energy deposition rate is about 0.8 erg/cm^2-sec, almost as intense as a moderate aurora. We conclude that significant energy is deposited by a wave of 10 mγ intensity. Such a wave amplitude is representative of highly coherent VLF wave types that are found in the magnetosphere.

Energy (keV)	1.5 ± 0.5	7 ± 0.5	30 ± 0.5
Flux(ergs/cm^2-sec)	0.73	0.13	0.02
	0.10	0.02	0.004

Table I. Incoming fluxes precipitated by a CW signal of 10 m intensity, propagating along the L = 4 field line.

The ionospheric density enhancements produced by the incoming fluxes given in Table 1 are estimated using the results given by Banks et al (1974) and Bailey (1968). These authors give the ion pair production rate as a function of altitude for precipitated fluxes with an exponential form with different energies. The precipitation induced density enhancements for the fluxes of Table 1 are shown in Fig. 4 for a precipitation pulse of 1 second duration. The average nighttime electron density profile (Hanson 1961) is also shown for comparison. We see from the figure that a 1 second precipitation pulse produces significant perturbation in the nighttime electron density over the 60-250 km altitude range. Enhanced photo emission due to the precipitated flux should be easily measured by photometers. The low

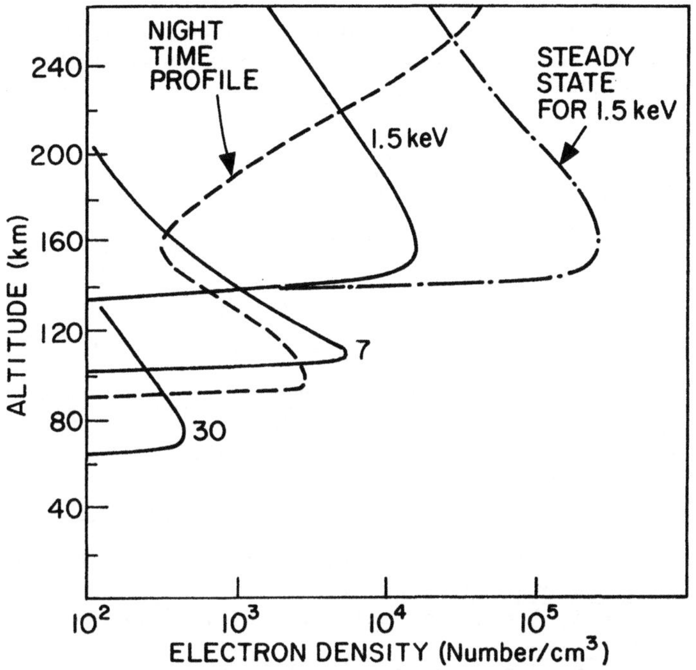

FIG. 4. Precipitation induced density enhancements produced by the incoming fluxes of Table 1. The solid lines represent the enhancements due to a precipitation pulse of 1 sec duration. The steady state due to the 1.5 keV flux is also shown. (The solid curves in the above should be scaled down by a factor of 8 in electron density.)

energy perturbations could also be measured by ionosondes. The enhancements caused by the 30 keV flux occur at low altitudes and could therefore cause detectable modifications in the subionospheric VLF propagation (Helliwell et al 1973). These density enhancements also result in significant enhancements in the ionospheric conductivity and could conceivably give rise to micropulsation activity (Bell 1976).

Past attempts to measure the energetic particle flux precipitated by the Siple transmitter using photometers and high and low altitude satellites have been frustrated by excessive cloud cover, insufficient angular resolution of satellite particle detectors, and lack of knowledge of the location of the precipitation region. However precipitation effects induced by natural whistlers are reported (Helliwell et al 1973; Rosenberg et al 1971). In one case (Jan 2 1971 event reported by Rosenberg et al 1971), natural whistlers were seen to cause precipitation of particles in the 30-70 keV range with an energy flux of approximately 2 ergs/cm^2-sec.

In the light of these results it appears that controlled VLF wave injection in the magnetosphere could be an important and useful tool to study the coupling processes between the atmosphere and magnetosphere.

REFERENCES

Ashour-Abdalla, M., 1972, Planet. Space Sci., 20, 639.
Bailey, D.K., 1968, Rev. of Geophys., 6, 289.
Banks, P.M., Chappell, C.R and Nagy, A.F., 1974, J. Geophys. Res., 79, 1459.
Bell, T.F., 1976, J. Geophys. Res., 81, 3316.
Hanson, W.B., 1961, Satellite Environment Handbook, Stanford University Press, Stanford, Ca.
Helliwell, R.A., 1965, Whistlers and Related Ionospheric Phenomena, Stanford University Press, Stanford, Ca.
Helliwell, R.A., Katsufrakis, J.P. and Trimpi, M.L., 1973, J. Geophys. Res., 78, 4679.
Helliwell, R.A. and Katsufrakis, J.P., 1974, J. Geophys. Res., 79, 2577.
Helliwell, R.A., Katsufrakis, J.P., Bell, T.F.,and Raghuram, R., 1975, J. Geophys. Res., 80, 4249.
Inan, U.S., Bell, T.F., and Helliwell, R.A., 1977, submitted to J. Geophys. Res.
Park, C.G., 1976, Proc. International Symp. on Solar-Terres. Phys., Boulder, Colo., D. Williams (ed.).
Rosenberg, T.J., Helliwell, R.A. and Katsufrakis, J.P., 1971, J. Geophys. Res., 76, 8445.
Stiles, G.S. and Helliwell, R.A., 1975, J. Geophys. Res., 80, 608.

DISCUSSION

Thorne: I am somewhat surprised that 10 mγ waves can produce strong diffusion scattering since much larger (1000 mγ) signals are required for such scattering by incoherent waves.

Inan: For an incoherent signal a wave amplitude of 1000 mγ would be the integrated amplitude over the whole spectrum. The scattering induced by such a signal would cover a wide range of energies. The interaction is of course much stronger for the coherent signal and the scatterings encountered by the particle at a single encounter are at least an order of magnitude greater.

THERMAL ELECTRON COLLISIONS AND ENERGY TRANSFER IN ATMOSPHERIC GASES

M. Mentzoni

Department of Physics, University of Oslo, Norway

There has been a number of review articles dealing with the subject covered by the title of this paper, for example Banks (1966), Dalgarno (1969) and Takayanagi and Itikawa (1970). Electron collisions and energy transfer in our ionosphere has been of great interest for many years, particularly in view of the cross-modulation phenomenon used in ionospheric research as a diagnostic tool for probing electron density, collision frequency and the energy loss coefficient (G-factor). A summary of all the work that has been done in this field up to 1969 has been given by Fejer (1969). Whereas the earlier cross-modulation experiments were mainly aimed at perturbing the ionosphere by moderately increasing the electron temperature the interest in recent years has shifted towards making substantial changes of the ionosphere by exposing it to the radiation from high power ground based transmitters.

The various manifestations of such ionospheric modification are discussed by Meltz and Perkins (1974) who's paper is to be found in a special issue of Radio Science devoted entirely to the theme "Ionospheric Modification by High Power Transmitters", particularly dealing with the Plattville and Arecibo experiments. One should note that ionospheric heating experiments also are planned for the multi-national EISCAT-facility currently being completed in Tromsø, Norway. It is clear by now that the heating of electrons in the lower ionosphere by powerful radio waves is limited primarily by rotational excitation of N_2 in the D-region and that electron excitation of fine-structure transitions of atomic oxygen is important in the E-region (Holway and Meltz 1973). The role of the rotational excitation of O_2 in the con-

B. Grandal and J. A. Holtet (eds.), Dynamical and Chemical Coupling, 175-183.

text of electron heating (or cooling) in the D-region is presently still not quite clear, even though it has been pointed out earlier (Mentzoni 1968) that O_2 may play a significant role in the energy balance for thermal electrons. A further discussion on this point is given by Hagenbuch (1973).

Up to present there exists conflicting experimental results with regard to the electron collision frequency for momentum transfer and the energy transfer in O_2 at thermal energies. It will be the main purpose of this paper to attempt to assess the influence of molecular oxygen on the G-factor in air.

THEORETICAL BASES

The laboratory experiments designed to measure electron collision frequency and energy transfer directly are based on microwave diagnostic techniques and indirectly on measurements of transport coefficients in drift tube studies (Gilardini 1972). The G-factor enters into the time averaged energy balance equation as follows

$$\frac{dU}{dt} = \sigma_r E_o^2/2 - G\,\nu_{eff}\,(U - U_o) \tag{1}$$

where the electron energy $U = (3/2)k\,T_e$; the real part of the conductivity

$$\sigma_r = N_e\, e^2\, \nu_{eff}/m(\omega^2 + \nu_{eff}^2)$$

U_o = electron equilibium energy,
E_o = electric field peak amplitude,
k = Boltzmann constant,
e, m, and N_e = electron charge, mass and number density respectively,
ω = disturbing wave angular frequency.

The effective electron collision frequency, ν_{eff}, is defined as follows for a Maxwell velocity distribution

$$\nu_{eff} = \frac{8}{3\pi^{\frac{1}{2}}}\, N \left(\frac{m}{2k\,T_e}\right)^{5/2} \int_o^\infty v^5\, Q_m(v) \exp\left(-\frac{mv^2}{2k\,T_e}\right) dv \tag{2}$$

Here N is the neutral number density, v the electron velocity and Q_m (v) the collision cross-section for momentum transfer. The collision frequency $\nu(v) = N\,v\,Q_m\,(v)$.

If one can approximate $Q_m(v)$ in a polynomical formula

$$Q_m\,(v) = \sum_\alpha Q_\alpha\, v^\alpha$$

by using a least-square method one obtains

$$\nu_{eff} = \frac{4}{3\pi^{\frac{1}{2}}} N \sum_{\alpha} Q_{\alpha} \left(\frac{2k\,T_e}{m}\right)^{(\alpha+1)/2} \Gamma(3 + \frac{\alpha}{2}) \quad (3)$$

If we can put $\nu_{eff} = \beta\, U^{\gamma}$ one obtains by insertion into (1) and assuming G energy independent

$$\frac{d\nu_{eff}}{dt} = a\, \nu_{eff}^{\frac{2\gamma-1}{\gamma}} + b\, \nu_{eff}^{2} \quad (4)$$

with

$$a = \gamma\, \beta^{\frac{1}{\gamma}} \left((e^2\, E_o^2/2m\omega^2) + G\, U_o\right)$$

and $b = -\,G\,\gamma$

For $\gamma = 1$, which is an important case since it applies for both N_2 and O_2, the solution of (4) reads (dropping the subscript)

$$\nu(t) = \nu_o \frac{\nu_1}{\Delta\nu} \exp\,(G\nu_o t)/((\frac{\nu_1}{\Delta\nu}) \exp\,(G\nu_o t) - 1) \quad (5)$$

, Here ν_1 is the steady state value during the heating pulse and ν_o the anbient value. Eq. (5) is for the cooling phase of the crossmodulation, i.e. t = 0 of the cessation of a rectangular heating pulse. For $\Delta\nu/\nu_o = (\nu_1 - \nu_o)/\nu_o << 1$ eq. (5) reads

$$\nu(t) = \nu_o + \Delta\nu \exp\,(-\,G\nu_o t) \quad (6)$$

which is an important result used in laboratory cross modulation experiments.

N_2

In N_2 the collision cross-section found by Engelhardt et al. (1964) seems to be quite well established. By using the procedure outlined above Itikawa (1971, 1973) has computed ν_{eff} from eq. (3). The electron energy transfer due to rotational excitation is given by (Mentzoni and Row, 1963)

$$\frac{dU}{dt} = 3.79 \times 10^{-8}\, N(N_2)\, k(T_e - T_o)/(kT_e)^{\frac{1}{2}},\ \text{erg sec}^{-1} \quad (7)$$

Combining eq. (7) with eq. (1), the latter without the source term on the right side, one obtains

$$G = \frac{2.15 \times 10^{-10}}{\nu_{eff}\, T_e^{\frac{1}{2}}} + \frac{2m}{M} \quad (8)$$

where M = mass of N_2-molecule.

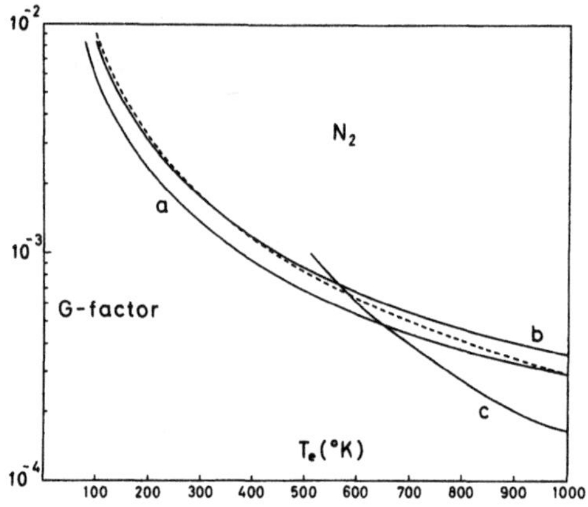

Fig. 1. G-factor in N_2.

This relation is plotted in fig. 1 as curve b, curve a being based upon the results published by Dalgarno and Moffett (1962). For both curves the recently published values of ν_{eff} have been used (Itikawa 1973). The dotted line is computed on the basis of a $T_e^{-3/2}$-dependence of G normalized to the value of 300°K of curve b. It can be seen that the $T_e^{-3/2}$-dependence is well obeyed except at the highest temperatures where the 2m/M-correction becomes increasingly important. It must be emphasized that the $T_e^{-3/2}$ temperature dependence has been experimentally verified in the laboratory over the temperature range 300°K - 735°K (Mentzoni and Row 1963) thus justifying the extrapolation towards lower temperatures. Curve c is taken from the recently published results of Dean and Smith (1975).

O_2

The effective collision frequency in O_2 has been computed from eq. (3) by Itikawa (1973) using the results of Mentzoni (1965) and Veatch et al. (1966) at low electron energies ($\leq$ 0.3 eV) and Hake and Phelps (1967) for higher energies. The results are shown in fig. 2 as curve b. It should be mentioned that laboratory results using microwave diagnostics in general yield a lower value of ν_{eff} than do those derived from transport coefficients (Gilardini 1972; Massey 1969). Based upon the earlier calculations of Itikawa (1971) using the cross-sections of Hake and Phelps (1967), the curve a of fig. 2 has been plotted. The dotted line follows a linear approximation which

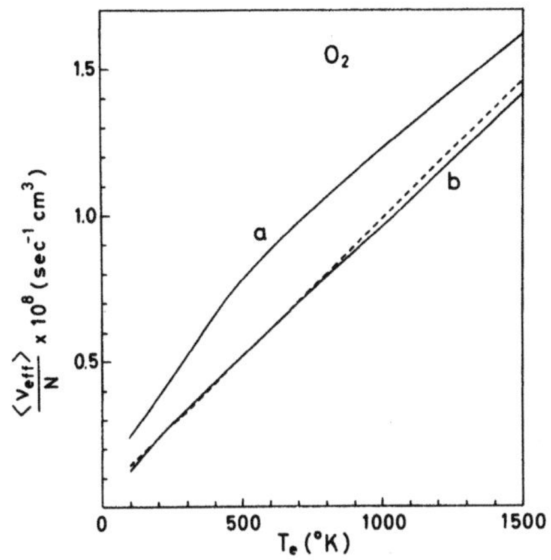

Fig. 2. Effective collision frequency in O_2

reads

$$\nu_{eff} = 9.3 \times 10^{-12} N(O_2) T_e \quad (9)$$

which is seen to give a good fit to curve b, particularly for $T_e \leq 800^oK$.

The electron energy transfer in molecular oxygen has been investigated in the laboratory using cross-modulation technique (Mentzoni & Rao 1965), these results are supported by earlier preliminary microwave data (Rao and Goldstein 1963; Formato and Gilardini 1962). Hake and Phelps (1967) analysed data pretaining to drift velocities and characteristic energies measured by a number og investigators. These data correspond to an electron energy transfer rate smaller than the microwave results by a factor of 2.3. On theoretical grounds rotational excitation in O_2 may yield a large cross-section because of the presence of short range interaction (Geltman and Takayanagi 1965). On the basis of the Born approximation cross-section for rotational excitation (Gerjuoy & Stein 1955) the energy loss rate should have been much smaller than what has been found experimentally from microwave and drift tube measurements.

The following relations apply

$$\frac{dU}{dt} = 9.17 \times 10^{-18} k(T_e - T_o)/(kT_e)^{\frac{1}{2}} \text{ (Hake and Phelps 1967)} \quad (10)$$

and $$G = \frac{5.2 \times 10^{-10}\, N(O_2)}{\nu_{eff}\, T_e^{\frac{1}{2}}} + \frac{2m}{M}$$

and $$\frac{dU}{dt} = 2.14 \times 10^{-17}\, k(T_e - T_o)/kT_e)^{\frac{1}{2}}$$

(Mentzoni and Rao 1965) (11)

$$G = \frac{1.21 \times 10^{-9}\, N(O_2)}{\nu_{eff}\, T_e^{\frac{1}{2}}} + \frac{2m}{M}$$

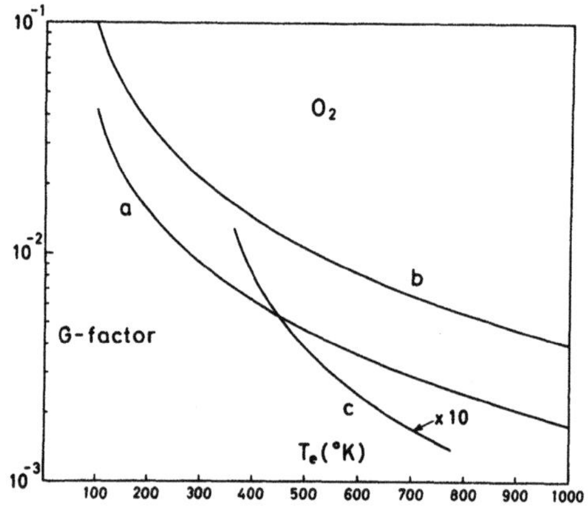

Fig. 3. G-factor in O_2

The G-factor based upon formulas are shown in fig. 3, curve a and b, respectively, ν_{eff} is taken from fig. 2 curve b. Curve c in fig. 2 is obtained by Dean and Smith (1975) with the gas being at 300°K. It should be noted that curve c is multiplied by a factor of ten. The extrapolation of curves a and b towards lower temperatures is based on the $T_e^{-3/2}$-dependence which was found to be obeyed resonably well between 300°K and 1000°K (Mentzoni and Rao 1965).

AIR

In order to compute the G-factor in air from the values obtained for N_2 and O_2 the following formula has been used (Mentzoni 1968)

$$G = \frac{[G\, \nu_{eff}]_{N_2} \times 0.78 + [G\, \nu_{eff}]_{O_2} \times 0.21}{\nu_{eff}(N_2) \times 0.78 + \nu_{eff}(O_2) \times 0.21} \qquad (12)$$

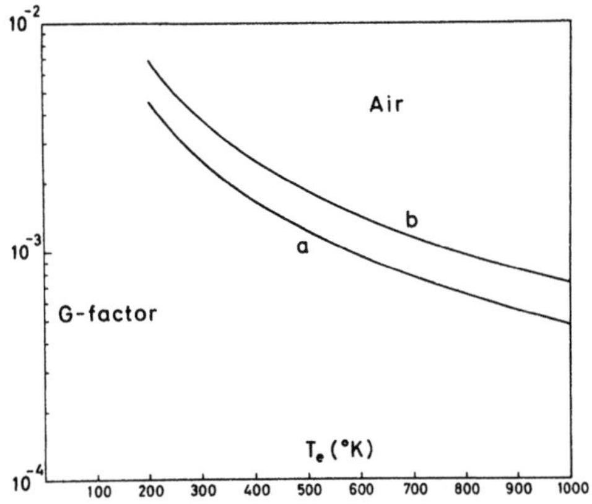

Fig. 4. G-factor in air

Complete mixing between N_2 and O_2 is assumed which is a reasonable assumption in the D-region. The results of this computation is shown in fig. 4 where curve a is based upon curve a of fig. 3 whereas curve b of fig. 3 yields curve b of fig. 4. The results of fig. 4 can be expressed with an accuracy of better than 5% by the following relations up to 500°K.

$$G = 2.52 \times 10^{-3} \, (300/T_e)^{3/2} \quad \text{, curve } \underline{a} \tag{13}$$

$$G = 3.85 \times 10^{-3} \, (300/T_e)^{3/2} \quad \text{, curve } \underline{b} \tag{14}$$

With the very low G-factor for O_2 reported by Dean and Smith their G-factors for air would essentially be that of N_2.

At this point it would be natural to compare the laboratory results presented here with those obtained by ionospheric cross-modulation measurements in the D-region. This is done in Table I.

Table I G-factor in air

Ionospheric measurements	$2.5 \times 10^{-3} - 12.5 \times 10^{-3}$	Thrane (1966)
	$(6.8 \pm 0.3) \times 10^{-3}$	Smith et al. (1967)
	$5 \times 10^{-3} - 8 \times 10^{-3}$	Showen (1972)

Laboratory measurements ($T_e \approx T_{gas} = 200^oK$)	4.5×10^{-3}	transport coefficients
	6.8×10^{-3}	microwave

It can be seen from the table that the ionospheric G-factor measurements are in good agreement with the laboratory data as presented here.

DISCUSSION

Figures 1 and 3 show that the results recently obtained by Dean and Smith (1975) differs markedly with the earlier data. This is particularly true for oxygen where the discrepancy is very large. For O_2 the G-factors measured by Mentzoni and Rao, Hake and Phelps, and Dean and Smith are in the ratios 30:13:1. In other words whereas microwave and drift tube data differ by a factor of 2.3, the flowing afterglow studies of Dean and Smith yield a G-factor more than one order of magnitude smaller than the earlier results. These authors claim that the microwave results may be in error because the molecules may be internally excited by the discharge pulse contrary to the situation in their experiment.

It is very difficult to understand how this could cause such a large discrepancy. Furthermore, an explanation would also have to be found for the drift tube results used by Hake and Phelps, who obtain a G-factor 13 times larger than Dean and Smith. Unfortunately it seems therefore that no satisfactory explanation exists at the present time for explaining the G-factor discrepancy in O_2. It should be mentioned that the ionospheric G-factor calculated by Hagenbuch (1973) using the same model as used here is from 50% to 100% higher than the microwave and drifttube results presented in fig. 4. He also finds that by using the results for O_2 calculated by Geltman & Takayanagi (1966) the G-factor will increase significantly and vary as $T_e^{-3/2} + C\,T_e^{\frac{1}{2}}$, with C being a constant. Clearly, more work is needed, both theoretical and experimental, with regard to electron energy transfer in O_2 at thermal energies.

REFERENCES

Banks, P. 1966, Planet. Space Sci., 14, 1085.

Dalgarno, A. 1969, Can. J. Chem., 47, 1723.

Dalgarno, A., and Moffett, R.J. 1962, Planet. Space Sci., 9, 439.

Dean, A.G., and Smith, D. 1975, Atmos. Terr. Phys., 37, 1419.

Engelhardt, A.G., Phelps, A.V., and Risk, C.G. 1964, Phys. Rev., 135, A 1566.

Fejer, J.A. 1970, J. Atmos. Terr. Phys., 32, 597.

Formato, D., and Gilardini, A. 1962, Proc. Vth Interntl. Conf. on Ionization Phenomena in Gases, North-Holland Publishing Co., Amsterdam.
Geltman, S. and Takayanagi, K. 1966, Phys. Rev., 143, 25.
Gilardini, A. 1972, p. 89 and 397 in Low Energy Electron Collisions in Gases. John Wiley & Sons, New York.
Hagenbuch, K.M. 1973, J. Geophys. Res., 78, 8345.
Hake, R.D. Jr., and Phelps, A.V. 1967, Phys. Rev., 158, 70.
Holway, L.H. Jr., and Meltz, G. 1973, J. Geophys. Res., 78, 8402.
Itikawa, Y. 1971, Planet. Space Sci., 19, 993.
Itikawa, Y. 1973, Phys. Fluids, 16, 831.
Massey, H.S.W. 1969, p. 787 in Electron Collisions with Molecules-Photoionization. Oxford at the Clarendon Press.
Meltz, G., and Perkins, F.W. 1974, Radio Science, 9, 885.
Mentzoni, M.H. 1965, Radio Science, 690, 213.
Mentzoni, M.H.,1968, J.Geophys. Res., 73, 4444.
Mentzoni, M.H., and Row, R.V. 1963, Phys. Rev., 130, 2312.
Mentzoni, M.H., and Narasinga Rao, K.V. 1965, Phys. Rev. Letters, 14, 779.
Rao, K.V., and Goldstein, L. 1963, p. 63 in Proc. VIth Interntl. Conf. on Ionization Phenomena in Gases, (Ed. P. Hubert) S.E.R.M.A., Paris, 1964, Vol. 3.
Showen, R.L. 1972, J. Geophys., 77, 1923.
Smith, R.A., Loch, R.G., Setty, G.S.G.K., Coyne, T.N.R., Barratt, P.H., and Orasad, B.S.N. 1967, In Proc. DR TE Conf. on Ground-Based Radio Wave Propagation Studies of the Lower Ionosphere.
Takayanagi, K., and Itikawa, Y. 1970, Space Sci. Rev., 11, 380.
Thrane, E.V. 1966, Experimental studies of the structure of the ionospheric D-region, NDRE Rep. 54, Kjeller, Norway.
Veatch, G.E., Verdeyen, J.T., and Cahn, J.H. 1966, Bull. Am. Phys. Soc., Ser. II, 11, 496.

AN ATTEMPT TO INTERPRETE COLLISION FREQUENCY MEASUREMENTS IN THE D-REGION

K. Torkar[1] and M. Friedrich[2]

[1]Space Rec. Inst. of the Austrian Academy of Sciences, c/o
[2]Dept. of Communications and Wave Propagation, Techn. Univ. Graz, Graz, Austria

1. INTRODUCTION

Effective electron collision frequencies, ν, in the ionosphere's D-region are usually derived from some kind of radio wave experiment. Hence, ν is not measured, but rather deduced from the effect it has upon radio waves. This interaction between collision frequency and wave propagation is described by a magneto-ionic theory and the complex refractive indices expressed by it. The most general formulation commonly used today is the one derived by Sen and Wyller (1960) which is based on laboratory measurements by Phelps and Pack (1959). They found a linear relationship between electron energy and collision frequency for nitrogen, the most abundant gas in the D-region. This linearity which was later also found for O_2, the second most abundant constituent, implies as a further consequence a proportionality of the pressure, p, and the collision frequency, ν.

2. MEASUREMENTS

Thrane and Piggott (1966) demonstrated the typical seasonal variation of the collision frequency (summer/winter, winter/spring etc.) and showed the relatively good agreement with the pressure from a model atmosphere. In an attempt to derive a more general seasonal behaviour, a number of collision frequency profiles were analyzed. All values were obtained by the same programmes (Torkar and Friedrich 1976; Torkar *et al.* 1976 *et loc. cit.*) from Faraday rotation and differential absorption data of rockets flown from high latitudes.

B. Grandal and J. A. Holtet (eds.), Dynamical and Chemical Coupling, 185-188.

Nine of them were launched from the Andøya Rocket Range, Norway, and three from Esrange, Sweden, both ranges being very near to 70°N latitude. The weighted collision frequency values were fitted by a least-mean-square method at any altitude to a sinusoidal seasonal variation and at any day of the year to a profile allowing a quadratic dependence on height, h, on a logarithmic scale of the following form:

$$\log\nu = A+Bh+Ch^2+(E+Fh+Gh^2)\cos\{\frac{360}{365}(t-\tau)\} \quad (1)$$

A to G constant
t day of the year
τ day of the maximum at one height

The result in Figure 1 shows the altitude variations of various constant collision frequencies with the time of the year and the height regions covered by the different rocket launches.

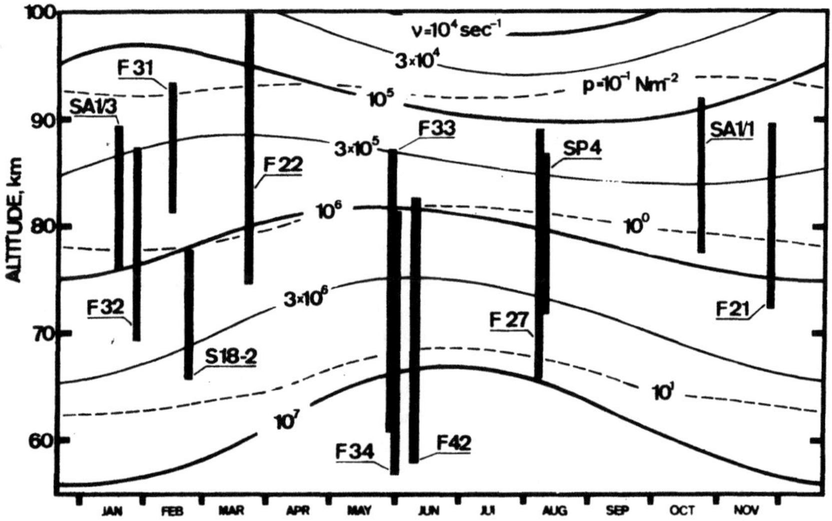

Fig. 1 Seasonal variation of constant collision frequencies at high latitudes derived from twelve rocket flights

These curves display a seasonal behaviour similar to that of neutral pressure taken from CIRA (1972) at 70°N (dotted lines), as already pointed out by Thrane and Piggott (1966). The next tempting step is to divide the two quantities and thus derive the proportionality factor

$$K = \frac{\nu}{p} \quad (2)$$

in order to remove seasonal and latitudinal variations.

According to the laboratory measurements this factor should be about $6 \cdot 3 \times 10^5$ m^2/N sec for the gas mixing ratio in the D-region. In Figure 2 the variation of this factor K with height which was derived from all rocket flights (plus one from mid-latitudes) is depicted. Individual K-values were obtained by dividing ν by the CIRA (1972) model. The profile in the figure is again a least-mean-square fit on the logarithmic scale

$$\log K = H + Jh \tag{3}$$

H, J constants

Also shown is the best fit by Mechtly (1974), however, based on the CIRA (1965) reference atmosphere. Two extreme values found in literature ($6 \cdot 4 \times 10^5$ from Benson 1964 and $9 \cdot 9 \times 10^5$ m^2/N sec from Bennet *et al.* 1972) are also indicated.

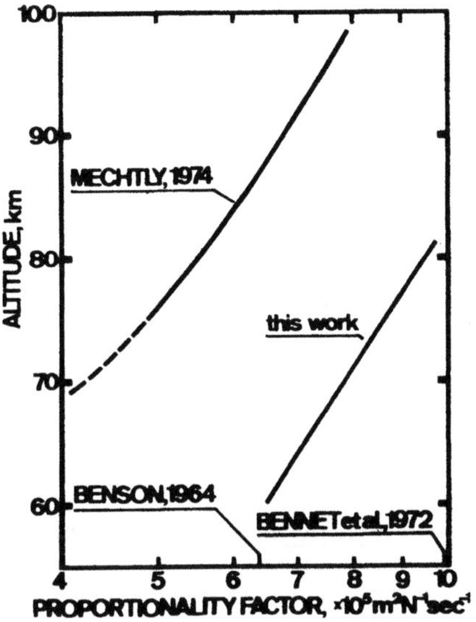

Fig. 2 Altitude dependence of the proportionality factor K between ν and p

3. INTERPRETATION

Several authors have reported this discrepancy with the laboratory measurements, some have simply explained it as inaccuracies (Hall 1970). If one, however, trusts the measurements and accepts the Sen and Wyller equations - which all of the above authors have done - the consequence is that the pressure at the times of the different rocket launches was higher than what one would expect from the reference atmosphere. The other feature is the height dependence of K, which means larger pressure scale heights and hence higher temperatures. The deviation, ΔT, from the reference can be derived by:

$$\Delta T = - \frac{H_O}{H_O + H_K} \cdot T_O \tag{4}$$

H_O reference atmosphere pressure scale height
T_O reference atmosphere temperature
H_K scale height of K (here ca. -50 km)

The temperatures thus derived are on average between 20 and 30° above CIRA (1972). The agreement with temperatures derived by measuring the extinction of the solar Lyman-α line is fairly good for all cases where the two experiments were on the same payload (Thrane and Johannessen 1975; Thrane *et al.* 1977). Generally speaking, temperature enhancements appear to be associated with ionospheric disturbances (riometer absorption, winter anomaly etc.).

REFERENCES

Bennett, F. D. G., Hall, J. E., and Dickinson, P. H. G. 1972, J. Atmosph. Terr. Phys., 34, 1321

Benson, R. F. 1964, J. Res. Nat. Bur. Stand., Sect. D., 68D, 1123

CIRA 1972, Cospar International Reference Atmosphere, Akademie Verlag Berlin

Hall, J. E. 1970, J. Atmosph. Terr. Phys., 32, 1849

Mechtly, E. A. 1974, Radio Sci., 9, 349

Phelps, A. V., and Pack, J. L. 1959, Phys. Rev. Lett., 3, 340

Thrane, E. V., and Piggott, W. R. 1966, J. Atmosph. Terr. Phys., 28, 721

Thrane, E. V., and Johannessen, A. 1975, J. Atmosph. Terr. Phys., 37, 655

Thrane, E. V., Grandal, B., Hagen, O., and Ugletveit, F. 1977, Z. f. Geophys., 34, in press

Torkar, K., and Friedrich, M. 1976, Techn. Univ. Graz, Internal Report INW 7602

Torkar, K., Jacobsen, T. A., Bauer, A., and Friedrich, M. 1976, Techn. Univ. Graz, Internal Report INW 7603

ACKNOWLEDGEMENT

Messrs. L. Björn and B. Holback from the Uppsala Ionospheric Observatory were most generous to let us also include the data from their rocket flights. The participation in the other flights was made possible by several grants of the Austrian Council for Scientific Research. The discussions with Dr. E. V. Thrane were both stimulating and encouraging. The numerical calculations were carried out at the Graz Computer Centre (RZG).

PART III: COUPLING BETWEEN NEUTRAL AND IONIZED ATMOSPHERE

INFLUENCE OF IONIZATION ON THE NEUTRAL ATMOSPHERE

George C. Reid

Aeronomy Laboratory, NOAA, Boulder, Colorado 80302

INTRODUCTION

Ionizing radiation can affect both the dynamics and the chemical composition of the neutral atmosphere through macroscopic heating effects and through microscopic chemical reactions. The dynamical effects, which take place mainly in the thermosphere, will be described in other papers in this collection, and this paper will concentrate on the chemical aspects. Specifically, we will be concerned with the stratosphere, where residence times of certain species tend to be very long, and where effects can take place that are potentially important to our surface environment. The chief objective will be to explore the consequences of very intense ionizing events, more intense than any that have occurred within the brief period of man's interest in such effects, but well within the range of intensities that may have occurred in the distant past, or that might occur in the future. The latter part of the paper will be concerned with a tentative study of the potential effects of such events on the global climate, which is a topic of major current interest.

PRODUCTION OF NO_x BY IONIZATION EVENTS

Recently Crutzen et al. (1975) showed that solar-proton events could produce large quantities of nitric oxide (NO) in the stratosphere, and that during periods of high solar activity this source could become comparable in intensity with the principal known source of stratospheric NO, i.e., the oxidation of nitrous oxide (N_2O) generated by denitrifying bacteria at the surface. A brief review of the mechanism that creates this NO is worthwhile.

B. Grandal and J. A. Holtet (eds.), Dynamical and Chemical Coupling, 191-202. *All Rights Reserved.*

A solar proton requires a minimum energy of about 30 Mev in order to reach the top of the stratosphere, at about 50 km altitude. Due to the concentration of its energy loss toward the end of its track, coupled with the exponentially increasing density of the atmosphere with depth, virtually all of the ionization produced by such protons will appear within a few kilometers of their stopping point. Thus all protons with incident energies greater than about 30 Mev will cause only stratospheric ionization. The bulk of the ionization is actually created by secondary electrons, rather than by the protons themselves, and these secondary electrons can also dissociate molecular nitrogen quite efficiently.

Figure 1 shows cross sections for production of N_2^+ by direct ionization of N_2, of N^+ and N by dissociative ionization, and of N + N by dissociation. The curves cover the range of energies to be expected in the case of primary energetic protons, and are taken from the work of Tate and Smith (1932) (total ionization), Rapp et al. (1965) (dissociative ionization), and Winters (1966) (dissociation). Since each dissociation event produces two N atoms, and since the N^+ ion produced in a dissociative ionization event will rapidly charge transfer with O_2 to produce another neutral N atom, it is obvious from the data that the total number of N atoms produced in the atmosphere must be comparable to the total ion-pair production rate. In fact, a recent detailed study of the problem of energy transfer by protons carried out by Porter et al. (1976) has shown that about 1.3 N atoms are produced for each ion pair. Since a proton loses about 36 ev per ion pair created, we can thus estimate roughly that 1 erg of incident ionizing radiation results in the formation of about 2×10^{10} N atoms in the atmosphere. This is a useful number to bear in mind, and is large enough that a substantial fraction of the energy lost by the protons ultimately goes into dissociating N_2 molecules, i.e., into the fixation of atmospheric nitrogen.

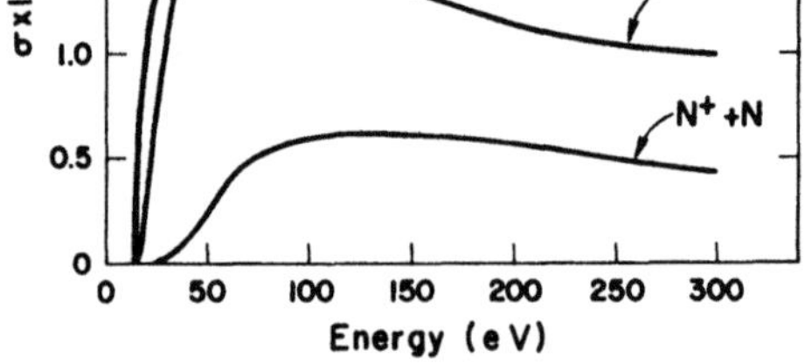

Figure 1. Electron impact cross sections in nitrogen.

The principal chemical reactions undergone by N atoms in the stratosphere are

$$N^* + O_2 \rightarrow NO + O \qquad (k_1)$$

$$N + O_2 \rightarrow NO + O \qquad (k_2)$$

$$N + NO \rightarrow N_2 + O \qquad (k_3)$$

where N^* is used to denote one of the metastable (^2D or ^2P) states of the N atom. The N atoms formed in association with ionizing events may be in one of these excited states or in the ground (^4S) states, and laboratory evidence on the branching ratio is not yet clear. This branching ratio, however, is of crucial importance to the question of compositional changes during very intense ionizing events. We define it here as $B = q(N^*)/[q(N^*) + q(N)]$, where the q's are production rates in $cm^{-3}\,sec^{-1}$.

The N atoms can be assumed to be in photochemical equilibrium at all times, and if we also assume that the duration of an ionizing event is much less than the lifetime of NO in the stratosphere, an analytical expression can be derived for the concentration of NO in terms of Q, the time-integrated ionization rate. This expression is not particularly simple, but the important point is that there is always a finite amount of NO that can be produced in the atmosphere by this mechanism. One cannot increase the total ionization indefinitely and expect the NO concentration to build up indefinitely. The reason is clear from the reactions as written above: as long as there is comparatively little NO present, the N atoms will react with O_2 to produce NO; when the NO content becomes large, however, the N atoms destroy NO instead. This dual role of N atoms has become well known in recent years in connection with studies of NO in the thermosphere, but the interest here is to determine how large the limiting NO concentration can become in the stratosphere, since we are going to be concerned with very intense ionization events.

The limiting ratio of NO and O_2 concentrations can be written as

$$\frac{[NO]}{[O_2]}_{lim} = \frac{[(2B-1)^2 k_1^{\,2} + 4k_1k_2]^{1/2} + (2B-1)k_1}{2k_3} \qquad (1)$$

$k_1 = 6 \times 10^{-12}\,cm^3\,sec^{-1}$ (Black et al. 1969)

$k_2 = 1.5 \times 10^{-11} \exp(-3580/T)\,cm^3\,sec^{-1}$ (Clark and Wayne 1970)

$k_3 = 2.2 \times 10^{-11}\ cm^3\ sec^{-1}$ (Phillips and Schiff 1962)

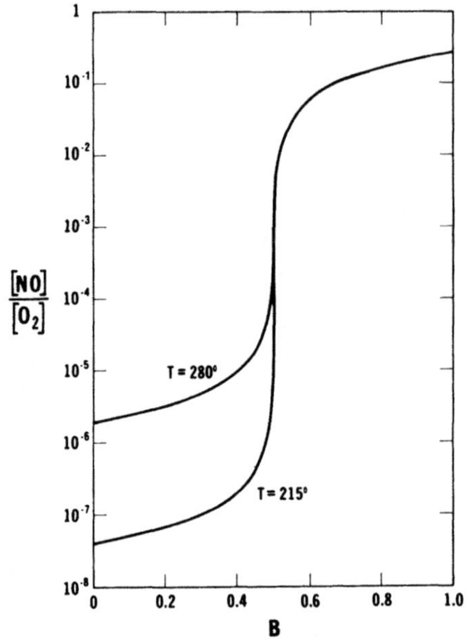

Figure 2. Limiting NO mixing ratios

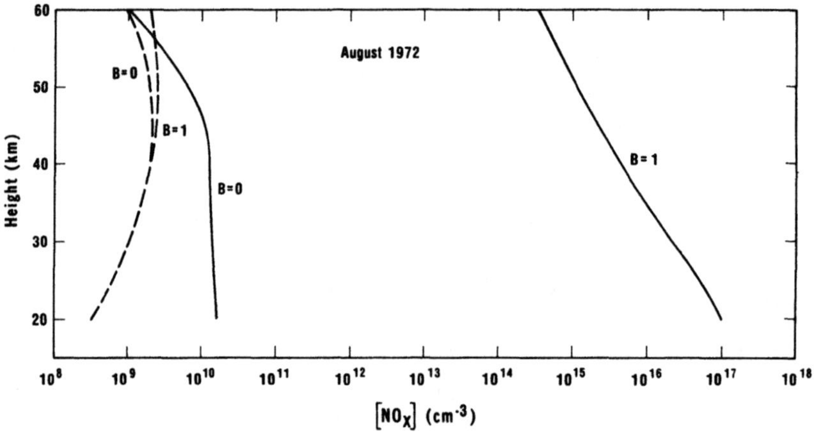

Figure 3. Maximum possible NO production (solid lines) and amount produced by August 1972 events (broken lines)

Since k_1 is always very much larger than k_2, the limiting value is very sensitive to whether B is larger or smaller than 0.5. This is illustrated in Figure 2, which shows the limiting mixing ratio as a function of B for two temperatures: 215°K, corresponding to 20 or 70 km, and 280°K, corresponding to the stratopause, near 50 km. The limiting ratio changes by six orders of magnitude according to the value chosen for B, and the obvious random selection of B = 0.5 would be a very poor one indeed.

Figure 3 shows the maximum amount of NO that could be created by a very intense ionizing event in an actual model atmosphere for the two extreme cases B = 0 and B = 1. It appears that if B = 0 the August 1972 events created nearly the maximum possible amount of NO_x in the lower mesosphere, and an event only ten times more intense would effectively saturate the stratosphere with NO_x. If B = 1, however, we are still a very long way from producing the maximum attainable amount of NO_x.

All available evidence suggests that B is considerably closer to 1 than to 0, and may be in the neighborhood of 0.8. Thus we can imagine that intense ionizing events are capable of producing much more NO_x in the atmosphere than anything we have yet knowingly experienced. The remainder of the paper will consist of an attempt to estimate some of the consequences of such very large amounts of NO_x, particularly as they might affect the global climate.

The NO_x will exist in a variety of forms, such as NO, NO_2, NO_3, N_2O_5, and will also produce such species as HNO_2 and HNO_3 by reaction with odd-hydrogen species. I shall concentrate here on NO_2, since this is the species with the highest potential for altering surface climate.

NO_2 is produced by reaction with ozone:

$$NO + O_3 \rightarrow NO_2 + O_2 \qquad (k_4)$$

and is destroyed principally by reaction with atomic oxygen:

$$NO_2 + O \rightarrow NO + O_2 \qquad (k_5)$$

and by photodissociation;

$$NO_2 + h\nu \rightarrow NO + O \qquad (J_1,\ \lambda < 398\ \text{nm})$$

(Again, these are only the principal reactions, which are used here to illustrate the mechanism).

These reactions take place on a time-scale short compared to the lifetime of NO_x in the stratosphere, and we can assume $d[NO_x]/dt = 0$, leading to

$$\frac{[NO_2]}{[NO_x]} = \frac{k_4[O_3]}{k_4[O_3] + k_5[O] + J_1} \tag{2}$$

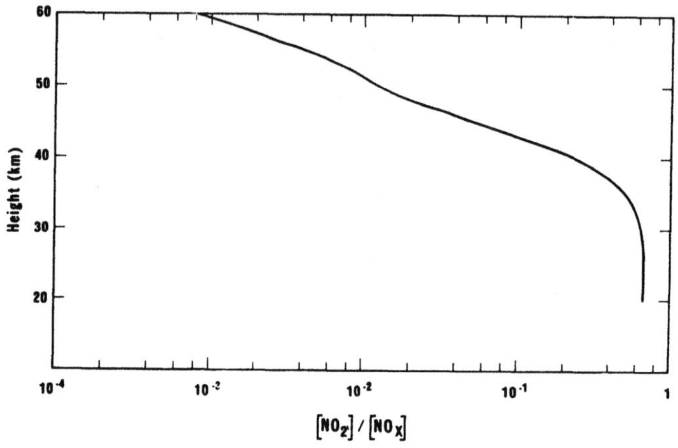

Figure 4. Fraction of NO_x in the form of NO_2

A problem arises in attempting to calculate this ratio without a full photochemical model, since the ozone content will be severely reduced below its normal value. However, for the sake of rough estimates, Figure 4 shows the ratio for the normal situation, in which we have used currently accepted values for the concentrations of O and O_3 and the following reaction rates:

$k_4 = 1.7 \times 10^{-12} \exp(-1310/T)\ cm^3\ sec^{-1}$ (Schofield 1967)

$k_5 = 9.1 \times 10^{-12}\ cm^3\ sec^{-1}$ (Davis et al. 1973)

$J_1 = 7 \times 10^{-3}\ sec^{-1}$ (Crutzen 1971)

In the upper stratosphere, most of the NO_x remains in the form of NO, but below about 40 km there is substantial conversion to NO_2, even in daytime conditions. Allowing for the effects of ozone depletion mentioned above, it seems reasonable that about 1/4 of the total NO_x content might be expected to be in the form of NO_2.

CLIMATIC EFFECTS OF NO_2

The unique and important feature of NO_2 is its strong absorption spectrum in the visible and near ultraviolet regions of the

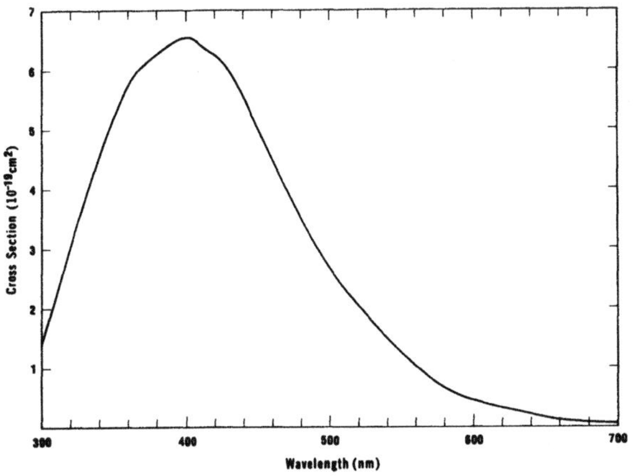

Figure 5. Absorption cross section of NO_2

solar spectrum. Figure 5 shows the results of laboratory measurements of this spectrum (Dixon 1940; Hall and Blacet 1952), indicating a peak cross section of about 6.5×10^{-19} cm^2 near 400 nm wave length. The absorption spectrum extends with appreciable intensities through the blue and green portions of the visible spectrum and well into the near ultraviolet. At wavelengths below 398 nm the NO_2 molecule dissociates, and the absorbed energy is converted directly into heating. There is a considerable amount of uncertainty regarding what happens at longer wavelengths, but laboratory evidence suggests that the NO_2 molecule is raised into one or more excited states with relatively long lifetimes that would be collisionally deactivated in the atmosphere, again leading directly to heating.

The net result of the injection of large quantities (of order 10^{-17} cm^{-2} column) of NO_2 into the atmosphere would be to reduce the amount of solar radiation reaching the surface of the earth, i.e., to reduce the effective solar constant. The orange and red parts of the spectrum would be little affected, and the sky would appear reddish in color, identical in fact to the color seen in the photochemical smog layers often present over large cities, where NO_2 is produced as a by-product of car exhausts.

The temperature of the surface of the earth is determined by a delicate balance between three factors: the incoming solar radiation (reduced by the fraction scattered back into space - the effective albedo), the infrared radiation escaping to space, and the quantity of heat exported and imported by the general circulation of the oceans and atmosphere. The infrared radiation that does not escape is radiated back to the ground, causing the well-known greenhouse effect, and if the NO_2 were to contribute appreciably to this effect it would help cancel out its tendency to reduce surface temperatures. The quantities of NO_2 we are concerned with, however, are so small that they would have little impact by comparison with water vapor and CO_2. The large increase in stratospheric temperature, however, would affect the radiation characteristics of the stratosphere, and we have not taken this into account in what follows.

Most discussions of variations in the solar input to the earth have centered on increases of CO_2 content, leading to increases in temperature caused by the greenhouse effect, and increases in dust due to anthropogenic and volcanic activity, leading to decreases in temperature through the increased albedo. To this I would like to add a third ingredient, i.e., the production of NO_2 by intense ionizing events.

As anyone acquainted with the field of climate modelling knows, it is an extremely difficult task to try to estimate the effects of changing environmental parameters on climate, since what we call climate is the averaged result of a complex network of interacting effects. As a first step, however, it seems reasonable to follow the lead of Budyko (1969) and of Sellers (1969), who independently constructed fairly simple one-dimensional radiative-balance climate models. In Budyko's model, the starting point is the relation

$$Q(1-\alpha) - I = A \tag{3}$$

where Q is the solar radiation incident at the top of the atmosphere during the course of a year at any given latitude, α is the average albedo for that latitude, I is the infrared radiation escaping to space, and A is the heat lost through circulation of the atmosphere and oceans. Budyko adopted a simple linear relationship between I and the surface temperature, derived empirically on the basis of a large number of observations. This can be written in the form

$$I = C_1 + C_2 T \tag{4}$$

where T is the surface temperature and C_1 and C_2 are coefficients that depend weakly on average cloudiness at the latitude under consideration.

For the earth as a whole, the quantity A must be zero, which allows a 'planetary mean' temperature, T_p, to be calculated. From observations of T as a function of latitude, Budyko calculated A as a function of latitude, and showed that it could be written as a simple linear function of $T - T_p$, i.e., the departure of the local temperature from the planetary mean.

Gathering all this together, it is possible to calculate T as a function of latitude for varying solar constants, albedos, or both. Both Budyko (1969) and Sellers (1969), who developed a rather more sophisticated model, reached the conclusion that the earth's climate is very sensitive to small reductions in the effective solar constant. In fact, a reduction of 1.5 to 2% led to a completely glaciated earth, a situation from which there would be little hope of recovery. The reason for this sensitivity is the strong positive feedback between temperature and albedo, since an ice-covered surface sends much more solar radiation back into space than an ice-free surface.

Both of these models, however, predict only equilibrium temperatures, and the time needed to reach equilibrium following a change in external conditions is likely to be at least some tens of years, due to the large quantities of heat stored in the oceans. Since the lifetime of NO_x in the stratosphere is likely to be only a few years, the surface temperatures calculated represent only limiting values toward which the actual temperature will tend. The situation is further complicated, however, by the fact that stratospheric residence times are determined by the thermal structure of the stratosphere, and this will be greatly modified by large amounts of NO_2.

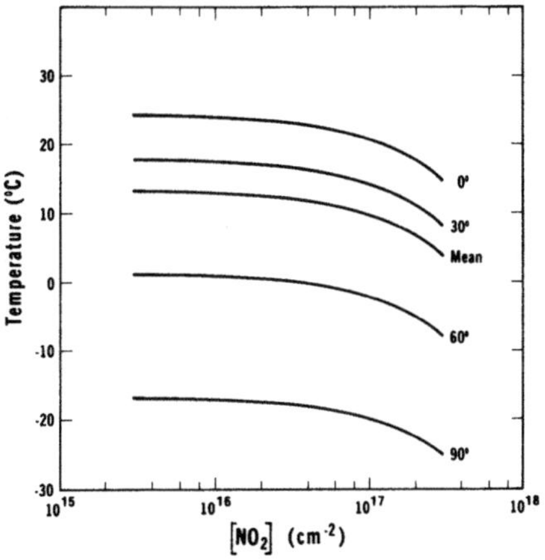

Figure 6. Equilibrium mean temperatures as a function of NO_2 vertical column content

Bearing in mind these reservations, it is still instructive to calculate the changes in surface equilibrium temperature brought about by large NO_2 concentrations. Figure 6 shows the results predicted by a model similar to that of Budyko for the planetary mean temperature and the temperature at latitudes 0°, 30°, 60° and 90°

as a function of vertical NO_2 column content. Since an advance of the polar ice cover would presumably require times longer than the residence time of NO_2, these calculations have not included the coupled ice-albedo effect described above, and the ice cover has been assumed to remain at its present limits, i.e., at latitudes above 72°.

Using these results, together with the results discussed earlier, we can construct Table 1, which relates decreases in global mean temperature to energy input in the form of ionizing radiation that penetrates the stratopause.

T(°C)	$[NO_2]$ (cm^{-2})	$[NO_x]$ (cm^{-2})	Energy flux (erg cm^{-2})
1	3×10^{16}	1.2×10^{17}	6×10^{6}
5	1.5×10^{17}	6×10^{17}	3×10^{7}
9	3×10^{17}	1.2×10^{18}	6×10^{7}

For comparison with these energy fluxes, Bostrom et al. (1972) estimated that the flux of protons with energies above 30 Mev during the August 1972 events was about 8×10^{9} cm^{-2}. Allowing an average energy for these protons of about 40 Mev, the corresponding energy flux is about 6×10^{5} erg cm^{-2}, which is only one order of magnitude less than the flux needed to produce an equilibrium global temperature decrease of 1°C. Also for comparison, recent measurements by Noxon (1975) have shown that the normal ambient column content of NO_2 in the atmosphere in an unpolluted situation is about 3×10^{15} cm^{-2}, which is again only one order of magnitude less than that needed to cause a 1°C temperature drop.

One apparent result of these calculations is the suggestion that very intense solar particle events could have a very significant effect on global climate.

Another important fact not discussed so far is the confinement of the direct effects to the polar regions by the geomagnetic field. This would initially have the effect of reducing the temperature at high latitudes, while leaving lower latitudes unaffected, resulting in a general increase in the intensity of the atmospheric circulation in response to the increased temperature gradient, and a general increase in storminess. As the NO_x gradually spread to lower latitudes, temperature gradients would again be reduced, but the global mean temperature would be lowered.

DISCUSSION

We have said little about the possible sources of very intense ionizing events, other than a brief mention of solar flares. Two other sources deserve mention, i.e., glactic cosmic rays and supernovae. Galactic cosmic rays should produce about 3×10^7 NO_x molecules cm^{-2} sec^{-1} at high magnetic latitudes. If we take the stratospheric residence time as 3 years, we find that the quantity produced within the residence time is about 3×10^{15} cm^{-2}, about 1×10^{15} cm^{-2} of which would be in the form of NO_2. Comparing Table 1, we see that at present-day levels, galactic cosmic rays fall short of causing interesting climatic effects by quite a large factor, and it does not seem likely that variations in cosmic-ray flux could be the cause of the various suggested relationships between climate and the sunspot cycle, at least through the NO_2 mechanism. We have to be careful, however, about extending this conclusion to the distant past.

Turning to a more exotic possibility, it is worth considering the effects of supernovae, which have been suggested as the cause of some of the dramatic extinction events that have punctuated the history of life on earth. Estimates of the energy released by a 'typical' supernova in the form of energetic ionizing radiation are not easy to obtain with any degree of confidence. A total energy release of 10^{48} to 10^{50} erg is often assumed, however, and perhaps 10^{47} erg might be in the form of gamma rays and hard x rays that would ionize the stratosphere (Ruderman 1974). At a distance of 30 light-years (1 ly = 10^{18} cm), the energy flux at the earth would be about 10^7 erg cm^{-2}, and at 10 ly about 10^8 erg cm^{-2}. Comparing with Table 1, we see that events of this magnitude could certainly produce severe climatic effects, with important consequences for the biosphere.

In conclusion, it appears that ionizing radiation can produce important climatic effects on earth through the peculiar properties of the NO_2 molecule. The possibility that this mechanism is responsible for the sun-weather relationships that have been widely discussed in recent years does not look too hopeful, since the variation in ionizing radiation is not large enough. We do not yet fully understand the various chemical and dynamical influences at work in the stratosphere, however, and we should keep an open mind to some extent. It is worth noting that the apparent association between the 'little ice age' and the Maunder minimum in solar activity (Eddy 1976) is in the correct sense to be explained by this mechanism. When solar activity is very low, cosmic-ray intensities are high, leading to an increased production of NO_2 in the stratosphere, and a global cooling. I should also emphasize that the climate analysis presented here is very crude, and there may well be more subtle influences at work.

REFERENCES

Black, G., Slanger, T.G., St.John, G.A., and Young, R.A. 1969, J. Chem. Phys., 51, 116.

Bostrom, C.O., Kohl, J.W., McEntire, R.W., and Williams, D.J. 1972, The Solar Proton Flux - August 2-12, 1972, Applied Physics Laboratory, Johns Hopkins University, preprint.

Budyko, M.I. 1969, Tellus, 21, 611.

Clark, I.D., and Wayne, R.P. 1970, Proc. Roy. Soc. London, A316, 539.

Crutzen, P.J. 1971, J. Geophys. Res., 76, 7311.

Crutzen, P.J., Isaksen, I.S.A., and Reid, G.C. 1975 Science, 189, 457.

Davis, D.D., Herron, J.T., and Huie, R.E. 1973, J. Chem. Phys., 58, 530.

Dixon, J.K. 1940, J. Chem. Phys., 8, 157.

Eddy, J.A. 1976, Science, 192, 1189.

Hall, T.C., Jr., and Blacet, F.E. 1952, J. Chem. Phys. , 20, 1745.

Noxon, J.F. 1975, Science, 189, 547.

Phillips, L.F., and Schiff, H.I. 1962, J. Chem. Phys., 36, 1509.

Porter, H.S., Jackman, C.H., and Green, A.E.S. 1976, J. Chem. Phys., 65, 154.

Rapp, D., Englander-Golden, P., andBriglia, D.D. 1965, J. Chem. Phys., 42, 4081.

Ruderman, M.A. 1974, Science, 184, 1079.

Schofield, K 1967, Planetary Space Sci., 15, 643.

Sellers, W.D. 1969, J. Appl. Meteor., 8, 392.

Tate, J.T., and Smith, P.T. 1932, Phys. Rev., 39, 270.

Winters, H.F. 1966, J. Chem. Phys., 44, 1472.

DISCUSSION

Rusch: What is the feedback effect when the NO_2 increases and depletes the visible radiation in the following respect ? The O_3 will decrease and allow at least a partially compensating effect by increasing the UV radiation which reaches the earth.

Reid: The enhanced heating due to the ozone depletion has not been considered in these preliminary calculations. The effect near the surface will be negligible and will be confined to the lower stratosphere and upper troposphere. There will also be a small compensating effect on visible radiation through the decrease in ozone absorption in the Chappuis bands.

ENERGY BALANCE OF THE ATMOSPHERE UNDER THE INFLUENCE OF A DISTURBED SUN AND MAGNETOSPHERIC PROCESSES

K.D. Cole

Division of Theoretical and Space Physics, La Trobe University, Bundoora, Victoria, Australia. 3083

This paper reviews physical processes which may control the energy density of the atmosphere particularly between the altitudes of 30 and 160 km. Changes of composition and atmospheric parameters which follow from magnetospheric processes and concommitant processes on the sun are emphasized to present a framework for discussing sun-weather relationships. An outline of a theory of sun-weather relationships is presented, and a related quasi-bienniel forcing function identified.

The energy of a packet of air depends on its position in the atmosphere and its composition. The sources at its disposal include electromagnetic radiation of the sun, terrestrial radiation, radiation from other parts of the atmosphere, rotation of the earth, gravitational and thermal tides, winds, heat conduction, viscous dissipation, internal atmospheric waves, electromagnetic radiation from the universe at large (especially x-rays, cosmic rays), energy inputs to the atmosphere from solar wind interaction with the geomagnetic field, plasma processes within the geomagnetic field and energetic particles from the sun. The paper will discuss the various energy sources in order to present a framework for discussion of the physics of sun-weather relationships - a subject reactivated recently by scientists who have been concerned principally with disturbances on the sun, the modulation of the interplanetary medium, plasma processes within the geomagnetic field and the apparent relationships of these things with the modulation of the weather, climate and various parameters of the atmosphere (see King 1975, for a review of these relationships). Our understanding of the lower and middle atmosphere has been based principally on the concept of a steady quiet sun. Advances in knowledge, in the last

B. Grandal and J. A. Holtet (eds.), Dynamical and Chemical Coupling, 203-216.

thirty or so years, of processes associated with solar activity and affecting the upper and middle atmosphere (particularly composition and winds in polar regions) now point strongly to the need for research into the ways in which solar activity may affect the dynamics of the atmosphere. The amount of energy deposited in the earth's atmosphere as the result of solar activity is much smaller than that associated with the solar constant, i.e., 1-100 ergs cm^{-2} sec^{-1}, averaged over the globe, compared to perhaps 2×10^5 ergs cm^{-2} sec^{-1}. It follows that any physical mechanism for a solar activity effect on the weather is likely to be caused by a solar activity effect on some important factor such as albedo, or variation in minor constituent concentrations able to cause significant variations in the amount of absorption of the solar radiation or the planetary reradiation. We must therefore examine all the factors which contribute to the energy balance of the atmosphere at all levels.

THE ATMOSPHERE UNDER A QUIET SUN AND IN THE ABSENCE OF SOLAR AND GEOMAGNETIC PLASMA

The quiet sun radiates approximately as a black body at a temperature of about 5750°K according to the Stephan-Boltzmann law or about 6100°K according to Wien's law (e.g. Hess 1959). "The irradiation at mean solar distance - the solar constant - is approximately 2.0 Langleys (cal cm^{-2} min^{-1}) giving a flux of solar energy perpendicular to the earth's surface of 0.5 L" (Goody 1964). The principal known variation of solar irradiation (± 3.5%) is due to variation of the sun-earth distance which also causes a north-south hemisphere asymmetry in the irradiation of the earth - the southern hemisphere receives more radiation in its summer than does the northern hemisphere. Radiations at $\lambda \leq 3000$Å are absorbed above 50 km altitude. The so-called accessible radiation at $\lambda \gtrsim 3000$Å penetrates to the earth's surface virtually unattenuated in a dust-free atmosphere. Here it is reradiated to space according to the Planck distribution appropriate to the surface temperature.

The albedo varies on account of (i) the distribution of land and sea masses; (ii) the reflection quality of their surfaces which changes with geological movements, interaction with the atmosphere (e.g. by production of snow, ice and perturbation of the sea surface), and man's interferences, e.g. by agricultural, industrial and architectural practices; and (iii) variations in cloud cover, scattering by atmospheric molecules. Gross hemispherical asymmetry of the albedo may result from these causes and combinations of them.

Though the predominant molecules of the atmospheres (N_2, O_2) cause scattering of sunlight, they contribute directly little to the absorption of either solar or terrestrial radiation except for the Schumann-Runge and Hergberg bands which cause dissociation of O_2 in the upper atmosphere.

Table 1 lists constituents and the radiations they absorb or emit.

Table 1 (based on CIAP 1975)

Altitude (km)	Constituent	Radiation
30-75	O_3 (absorption)	2500-3000Å (Hartley)
	(major)	3200-3600Å (Huggins)
		4000-7500Å (Chappuis)
75-90	O_3 (absorption)	"
75-90	O_2 (dissociation)	1750-2000Å (Schumann-Runge
> 90	O_2 (dissociation)	1100-1750Å (Schumann-Runge)
< 60	O_2 (dissociation)	2000-2424Å (Herzberg)
ALL	Dust	ALL
60-100	CO_2	2.7μ, 4.3μ
Cooling	CO_2 (major)	15μ
30-110	O_3	9.6μ
ALL (mostly troposphere)	H_2O	80μ

The heating of the atmosphere from 20 to 90 km is dominated by absorption by O_3 and the production of O_3 is controlled by the photodissociation of O_2 by the Chapman sequence (see Johnston, 1975)

$$O_2 + h\nu \rightarrow O + O$$

$$O + O_2 + M \rightarrow O_3 + M$$

CO_2 is responsible for the major cooling of this region of the atmosphere. It is clear that the densities of minor constituents play a fundamental role in the energy balance of the middle atmosphere (CIAP 1975) and factors that control them are therefore of basic importance.

THE THERMOSPHERE (< 160 KM)

Sources of energy for the thermosphere are: (1) electromagnetic radiation from the sun principally in the wavelength range $\lambda \lesssim 3000$Å (Chandra and Sinha 1975); (2) joule dissipation of ionospheric electric currents (Cole 1962, 1971); (3) corpuscular bombardment of the atmosphere (Rees 1975); (4) viscous dissipation of atmospheric motion (Cole 1963; Wu and Cole 1974); (5) heat conduction from the magnetosphere (Cole 1965); (6) electromagnetic radiation from auroras (e.g., Swider 1974); and (7) acoustic radiation from auroras (Wilson 1976). The electromagnetic heating of the "quiet" thermosphere at low and mid-latitudes occurs principally through photo-dissociation of O_2 by the Schumann-Runge continuum (1300-1750Å) and by photoelectrons and this has been reported by Chandra and Sinha (1975) (see their Fig. 2). They estimate a heating efficiency of between 0.5 and 0.3. Stolarski et al. (1975) have examined the heating due to photoelectrons produced by photons of $\lambda \lesssim 1025$Å and find an efficiency of ~ 30-35% compared to 50% found by Chandra and Sinha (1973), and they attribute the difference to allowance for photoelectron transport which they made. As Chandra and Sinha point out the thermal energy available from the S-R continuum is proportional to the O_2 density. <u>It is suggested here that variations of $[O_2]$ caused by auroral processes and disturbance wind systems will couple with solar S-R to produce magnetic storm variations of the thermo-sphere in additional to those caused directly by the auroral processes</u>. Total heating rates as a function of altitude calculated by Chandra and Sinha (1973) range from 10^{-7} ergs cm^{-3} sec^{-1} at 200 km to 10^{-6} ergs cm^{-3} sec^{-1} at 120 km.

Joule Heating: Joule dissipation of which dissipation in terms of "ion-drag" is one special formulation can be represented by

$$Q_J = \sigma_1 (\underline{E}_s + \underline{V} \times \underline{B})^2 \qquad (1)$$

where σ_1 is the Pedersen conductivity (Cole 1962, 1971), $\underline{E}_s$ is the electrostatic field and $\underline{V}$ is the centre of mass velocity of the medium. At all altitudes above 100 km, in the absence of anomalous resistivity, $\underline{E}_s = -\underline{V}_e \times \underline{B}$ where $\underline{V}_e$ is the drift velocity of electrons so that equation (1) can be represented as

$$Q_J = \sigma_1 [(\underline{V} - V_e) \times \underline{B}]^2 \tag{2}$$

This formula may be useful for experiments which measure $\underline{V}_e$ and $\underline{V}$. At altitudes above 170 km, $V_e \approx V_i$ (the drift velocity of the ions) so that at these altitudes an alternative to equation (2) becomes

$$Q_J = \sigma_1 [(\underline{V} - \underline{V}_i) \times \underline{B}]^2 \tag{3}$$

At these altitudes also $\sigma_1 \to n_i m_i \nu_i / B^2$ so that equation (3) is the conventional "ion-drag" dissipation rate.

$$Q_J = n_i m_i \nu_i \; (\underline{V} - \underline{V}_i)^2 \tag{4}$$

where n_i, m_i and ν_i are the number density, mass and collision frequency of ions. Here ν_i is the number of times per second that an ion loses all its momentum to the neutral gas. In fact, formula (4) is a good approximation above 90 km (below which dissipation of electron motion becomes important). In formula (2) the ion contribution to σ_1 dominates and approximately (2) is given by

$$Q_J = \frac{n_i}{B^2} \frac{m_i \nu_i}{(1 + \nu_i^2/\omega_i^2)} [(\underline{V} - \underline{V}_e) \times B]^2 \tag{5}$$

The difference between equation (4) and (5) is only apparent and comes from the different formulation of Q_J in terms of conductivity and electric field on the one hand, and differences of ion and neutral drift velocities on the other.

Joule heating persists while current flows. However, the collisions of ions with neutrals accelerates the atmosphere and in the absence of significant mechanical force the atmosphere would take up the motion presented by $\underline{E}_s + \underline{V} \times \underline{B} = 0$ in a characteristic time of $\rho/\sigma_1 B^2$ where ρ is the gas density (Cole 1962, 1971). Electric fields (electrostatic or dynamo) which fluctuate with periods less than the quantity are therefore effective heat sources. To be compatible with heating (Q_{UV}) of the atmosphere by solar UV and EUV we require that

$$V - V_e \gtrsim \sqrt{\frac{Q_{UV}}{\sigma_1 B^2}} \tag{6}$$

Assuming an electron density of 2×10^5 cm^{-3} in the E region we then generate values of $V - V_e$ as in Table 1.

The entries in this table are proportional to $(n_e)^{-1/2}$ so that, in the equatorial electrojet and the auroral electrojet, this heating can be geophysically significant. Heating in model electrojets has been worked out and in some models the joule heating in the dark spaces between auroras is much greater than within the aurora itself (Cole 1971; Evans et al. 1976). Anomalies of the geomagnetic field cause corresponding asymmetries in the joule heating on account of the involvement of B in Q_J (Cole 1976a).

Table 2

Altitude (km)	150	130	120	110
Q_{UV} ergs cm^{-3} sec^{-1}	3×10^{-7}	5×10^{-7}	10^{-6}	2×10^{-6}
Mid-lat. $V - V_e$ m sec^{-1}	200	258	500	632
Auroral Zone $V - V_e$	165	215	420	527
Equator $V - V_e$	100	129	250	316

The Lorentz force $\underline{j} \times \underline{B}$ establishes motion of the thermosphere (Cole 1962). This is the same effect as the establishment of motion by "ion drag" or convection (Fedder and Banks 1972). The motion so established is eventually dissipated as heat by either joule dissipation or viscosity (e.g., Wu and Cole 1976).

Viscous Dissipation per Unit Volume (Q_{VIS}) is described by $\mu V \nabla^2 V$ where μ is the coefficient of viscosity (~ 5×10^{-4} poise for air) and is illustrated in Table 3 for velocity fields of scale L and amplitude V.

Table 3 (Q_{VIS} ergs cm^{-3} sec^{-1})

L(m) \ V(m sec^{-1})	10	100	1000	5000
100	5×10^{-6}			
1000	5×10^{-8}	5×10^{-6}		
10000	5×10^{-10}	5×10^{-8}	5×10^{-6}	1.25×10^{-4}
100000		5×10^{-10}	5×10^{-8}	1.25×10^{-6}

Viscous dissipation can become important in auroral atmospheric motions caused by Lorentz force (or "ion drag").

Corpuscular Bombardment produces heating of the atmosphere (Rees 1975) with an efficiency of 30 to 50%. The altitude (peak and range) of deposition of energy for electrons of varying energies is shown in Table 3, based on (Berger et al. 1970; Potemra and Rosenberg 1973).

Table 4

Energy (kev)	2	5	10	20	30	50	100
Altitude (km)	136 ± 20	115 ± 10	105 ± 5	97 ± 5	90 ± 10	85 ± 15	80 ± 20

It is to be noted that the altitude of deposition of energy by charged particles is variable on account of the variability of their energy spectrums and that, for the dominant energy range 5-40 kev, it is 115 to 97 km. On the other hand joule dissipation per ion-electron pair is a maximum at the altitude at which $\nu_i = \omega_i$ which may be between 125 and 135 km depending on the model atmosphere. Of course the actual joule heating is proportional to the electron density and may be expected to be at a different altitude. As illustrated in the calculation by Wu and Cole (1975) viscous dissipation may be as important as joule dissipation.

In terms of effects on the thermosphere (from 100-200 km) joule-cum-viscous dissipation appears to be greater than heating by corpuscular bombardment by electrons, in auroras (Boyd 1975), characteristically 4-12 kev, whose heating peaks at lower altitudes than the former heating. So that the heating per gram of air by joule-cum-viscous heating is much greater. At altitudes above 200 km soft electrons (energy < 1 kev) would appear to dominate in heating the auroral thermosphere.

HEAT CONDUCTION TO THE LOWER ATMOSPHERE

The motion produced in the thermosphere by electric fields is eventually dissipated as heat which is mostly conducted to the lower atmosphere. Joule heating alone of up to 100 ergs cm^{-2} sec^{-1} has been estimated from some measurements (Brekke 1976). Also within auroras energy deposition by electrons of up to 400 ergs cm^{-2} sec^{-1} has been reported (Frank and Akerson 1976). Evans et al. (1976) showed in one example that the energy flux into the atmosphere by auroral bombardment was similar to that in joule dissipation in the dark for space between auroras. The altitude down to which such additional heat from the thermosphere may perturb the lower atmosphere can be gauged by the downward heat fluxes as a function of altitude in a quiet sunlit atmosphere (Johnson and Gottleib 1970). Of course in a winter, sunless, atmosphere the perturbation depth of disturbance may be greater. This is illustrated in Table 5 for auroral latitudes.

Table 5

Altitude (km)	100	90	80	70	60	50
Heat Flux (ergs cm^{-2} sec^{-1})	3	7	11	70	200	800

It is apparent that magnetic disturbance of 400 ergs cm^{-2} sec^{-1} over a broad area could cause a thermal perturbation down to altitudes of about 55 km. This would be about the lower altitude limit for conduction heating from the disturbed thermosphere. We must invoke other processes to account for disturbances, of solar origin or due to processes in the magnetosphere, in the lower atmosphere. However, it is suggested here that this heating of the upper mesosphere may be expected to produce upwelling of the air providing support for aerosols (Reid 1975) such as exist in the summer polar cloud (Donahue et al. 1972). This could modulate albedo and contribute to variation of the lower atmosphere.

Radiation in the 1.27μ band from O_2 appears, at times, to be a significant energy sink for the lower thermosphere (see Swider 1974, and references therein). The energy flux in this radiation has on some occasions been comparable to the energy in the flux of energetic electrons in auroras. Observations of infrasonic radiation from those auroras (Wilson 1976) which move supersonically suggest that the flux of this radiation may at times be comparable to that dissipated in joule heating. The possible disturbance effect on the lower atmosphere of the 1.27μ electromagnetic radiation and the infrasound is not known.

Dust appears to contribute only in a small way to the heat balance of the middle atmosphere (Fiocco et al. 1975) except possibly in noctilucent clouds.

AN OUTLINE OF A THEORY OF SUN-WEATHER RELATIONSHIPS

Since the heat budget of the atmosphere depends critically on the concentrations of minor constituents and albedo, the likely origin of sun-weather relationships is in the variability of these quantities caused either by processes on the sun or magnetospheric processes.

Heath (1973) reports that the sun's emission of electromagnetic radiation in the wavelength range $\lambda < 3000\text{Å}$ is variable. The variability associated with the sun's rotation increases from about 1% at 3000Å to about 10% at 1750Å to about 80% at 1000Å. Other variations of UV and far UV irradiance of the sun have been found to be associated with (1) solar flares, (2) the solar cycle, and (3) a possible change of the sun over two years. However, I illustrate my discussion with the variation associated with solar rotation. This variability must cause a corresponding change in the rate of production of atomic oxygen in the atmosphere through dissociation of molecular oxygen. The altitude ranges in which this occurs, the responsible radiations, and their likely variabilities are shown in Table 6.

Table 6

λ (Å) Bands	Altitude Range (km)	Variability (%)
1100-1750 (Schumann-Runge)	> 100	70 - 7
1750-2000 (Schumann-Runge)	60 - 100	7 - 5
2000-2424 (Herzberg)	< 60	5 - 2

Changes in the production rate of O will change both the production rate and final height distribution of O_3 which is formed by the Chapman reaction

$$O + O_2 + M \rightarrow O_3 + M$$

At altitudes < 100 km this should cause variability in the local $[O_3]$ by amounts in the ranges shown in Table 6 together with changes in the distribution with altitude of $[O_3]$.

The heating rate of the mesosphere and stratosphere is dominated by absorption by O_3 of the radiation in the Hartley bands (2500-3000Å) in which there is a variability of about 1-2%, the Huggins bands (3000-3600Å) and the Chappuis band (4000-7500Å) in which there is no known variability. The heating rate of O_3 per atom of air is therefore expected to vary by about the amounts shown in Table 6. Now the nett heating rate of the mesosphere and stratosphere is determined principally by the difference of two large numbers, viz., the heating rate by O_3 (absorption) and the cooling rate by CO_2 (emission), the difference being about 1-10% of either (Newell and Gray 1972; Parks and London 1974). This is of the same order, in absolute terms, as the expected variability in heating rate due to O_3 absorption. It follows that the nett heating rate in these regions may vary by amounts in the range 10-100%, the particular amount being different at different altitudes, and determinable only by extremely complex (and model dependent) calculations. Without such calculations it seems possible to infer the large multiplicative effect on the nett heating rate due to variability of UV and far UV. This variability must in turn cause changes in the temperature profile of the stratosphere and mesosphere which is capable (Dickinson 1975) of modulating the temperature in the troposphere and the reflection coefficient of planetary waves from the troposphere causing geographical phase shifts of these waves. The phase shift of planetary scale waves in the atmosphere has been suggested by King (1975) to account for some sun-weather relationships, without specifying the mechanism by which this is achieved. Modulation of the temperature profiles in the manner suggested may also alter significantly the pattern of vertical winds and hence the concentration of nucleating particles for cloud formation, causing the modulation of albedo, an important factor for energetics of the lower atmosphere.

The solar variability reported by Heath (1973) is associated also with sector boundary crossings of the interplanetary magnetic field (Heath and Wilcox 1975). It has been noted earlier (Cole 1976) that the apparent relationship between the occurrence of sector boundary crossings and changes in the earth's atmosphere (Wilcox et al. 1974) may be traceable, in origin back to structures and their emission characteristics, on the sun. This theory gives such a physical mechanism to relate variability in the UV and far UV emissions of the sun with variations in the earth's atmosphere. In this global process of modulation of the atmosphere the interplanetary magnetic sector structure does not play a causitive role. The variability of the UV and far UV emission of the sun should produce, in the atmosphere, variability with the period of rotation of the sun and longer term variations such as solar cycle ones if these are present in the UV and far UV. The bombardment of the atmosphere in high magnetic latitudes can cause depletions of $[O_3]$ in the mesosphere (Reid et al. 1976). It is also likely that the intense production (Rusch and Barth, 1975) of NO in the auroral thermosphere will cause destruction of mesospheric O_3, following diffusion of NO downwards. Variations of $[O_3]$ in the mesosphere in the auroral zone of order 10% and more during severe disturbance can be expected to cause a multiplicative change of order 100% in the nett heating rate of a sunlit atmosphere. Since magnetic and auroral activity are related to the position of the earth in the sector structure of the interplanetary magnetic field (Wilcox and Ness 1965) there may be qualitative and quantitative differences in the response of the atmosphere related to the sector structure. This is in contrast to the non-auroral and the global response discussed above.

Somerville et al. (1976) have reported calculations of the response of an atmospheric general circulation model to drastic changes of the solar constant and of ozone amounts in the atmosphere. They conclude that "any causal relationships between solar variability and terrestrial weather, if they are to be of practical value for predictions on a time scale of 2 weeks or less will have to rely on changes in parameters other than the solar constant or the ozone content or on mechanisms not yet incorporated in our model". The top of their model corresponds to approximately 20-30 km altitude and they did report significant variations in the atmospheric temperature there of about +5°C, 10-12 days following a doubling of O_3 content. The authors showed that the globally integrated four basic forms of atmospheric energy are not significantly altered by ozone changes. However they do not comment on what appears to be significant changes in phase of the contours of the 500 mbar geopotential heights which are obscurred by orographic effects. The effect appears as a general straightening of the mid-latitude contours in their figure 1(d) as compared to 1(b). For example, at 160°W longitude the contours change from showing a latitude minimum (Fig. 1(b)) to showing a latitude maximum (Fig. 1(d)); also the maximum latitude shift shown at 60°W longitude in Fig. 1(b) is flattened out in Fig. 1(d). This effect may show up better on calculations for the southern hemisphere where orographic effects may be less.

Some correlations of disturbance in the atmosphere particularly in winter may be caused by propagation of planetary waves upwards to the polar ionosphere where electric fields generated may modulate the capture of solar wind plasma by the geomagnetic field (Cole 1976). This possibility is supported by the fact that on the middle atmosphere in winter westerly winds exist (CIAP 1975) and this is conducive to the transmission of planetary waves to the thermosphere (Dickinson 1968).

In future studies of apparent sun-weather relationships we must attempt to separate those that are due to energy sources in the lower atmosphere from those due to solar variability or magnetospheric processes.

A QUASI-BIENNIAL FORCING FUNCTION FOR THE ATMOSPHERE

It has been shown in the foregoing how the variability of the emission of the sun in the UV and far UV may produce a significant variation in the earth's atmosphere. The variability of interest here is that associated with the rotation of the sun and relatively fixed emission structures on it. Variations of order 100% in the heating rate in the mesosphere and stratosphere can be initiated by these UV variations. Their periodicity will be determined by some mean rotational speed of the inhomogeneous emission regions on the sun. Heath (1973) reports that from April 1969 to April 1972 two active UV regions on the sun about 180° apart persisted for 39 solar rotations. Their rotation period appeared to be about 27.4 days for both. The combination of a 27.4 day periodic heating of the mesosphere and stratosphere and the gravitational lunar tide which has a 29.56 day period should cause a beating of the two.

The moon returns to the same gravitational configuration with regard to the earth's atmosphere once every 29.56 earth rotations (the synodic period). It follows that a beating of the gravitational lunar tides and the UV heating of the atmosphere will take place with a period of 27.4 x 29.56 days or 26.5 months. The amplitude of the effect on atmospheric parameters is expected to be the product of the variation of the heating (order 10-100%) and the ratio of the gravitational lunar tide and the solar thermal tide which is about 10% (Chapman and Linzen 1970). A quasi-biennial variation of order up to 10% of many middle atmosphere parameters is expected to follow from this effect. This quasi-biennial oscillation would be dependent for its coherence and stability upon the stability of UV emission structures on the sun. An example of a study of a Q.B.O. is provided by Nastrom and Belmont (1975).

CONCLUSIONS

(1) The energy deposited as heat in the thermosphere during great magnetic storms may perturb the mesosphere thermally down to altitudes of 50 km. This may cause additional vertical winds in the polar upper mesosphere.

(2) Disturbance at lower altitudes depends upon alteration of concentrations of minor constituents either by solar variability in the UV and far UV or by auroral bombardment of the mesosphere by energetic protons or electrons. Changes in these concentrations can then change the absorption of solar radiation and hence the heating rate in the mesosphere and upper stratosphere.

(3) There is probably a class of weather systems which can launch planetary waves into the polar ionosphere capable of modulating the capture of solar wind plasma by the geomagnetic field and hence to modulate the production of geomagnetic disturbances.

REFERENCES

Berger, M.J., S.M. Seltzer and K. Maeda, J. Atmos. Terr. Phys. 32, 1015-1045, 1970.
Boyd, J.S., Rev. Geophys. & Space Phys. 13, 735-739, 1975.
Brekke, A., J. Atmos. Terr. Phys. 38, 887, 1976.
Chandra, S. and A.K. Sinha, Planet. Space Sci. 21, 593-604, 1973.
Chapman, S. and R. Linzen, "Atmospheric Tides", Gordon and Breach, Science Publishers, New York, 1970.
CIAP, The natural stratosphere of 1974. U.S. Dept. of Transport Publication No. DOT-TST-75-51, 1975.
Cole, K.D., Aust. J. Phys. 15, 223-235, 1962.
Cole, K.D., J. Geophys. Res. 68, 3231-3235, 1963.
Cole, K.D., J. Geophys. Res. 70, 1689-1706, 1965.
Cole, K.D., Planet. Space Sci. 19, 59-75, 1971.
Cole, K.D. in "Magnetospheric Particles and Fields", pp. 267-278, edited by B.M. McCormac, D. Riedel, Holland, 1976a.
Cole, K.D., Nature 260, 229-230, 1976.
Dickinson, R.E., J. Atmos. Sci. 25, 984-1002, 1968.
Dickinson, R.E., Can. J. Chem. 52, 1616-1624, 1975.
Donahue, T.M., B. Geunther and J.E. Belmont, J. Atmos. Sci. 29, 1205-1209, 1972.
Evans, D.S. et al., preprint 1976.
Fedder, J.A. and P.M. Banks, J. Geophys. Res. 77, 2328, 1972.
Fiocco, G., G. Grams and G. Visconti, J. Atmos. Terr. Phys. 37, 1327-1337, 1975.
Frank, L.A. and K.L. Akerson, J. Geophys. Res. 76, 3612-3643, 1971.
Goody, R.M., Atmospheric Radiation I, Oxford, Clarendon Press, 1964.
Heath, D.F., J. Geophys. Res. 78, 2779-2792, 1973.
Heath, D.F. and J.M. Wilcox, NASA Goddard Preprint No. X-901-74-156, pp. 116-120, 1974.
Hess, S.L., "Introduction to theoretical meteorology", Holt, Rurchart and Winston, New York, 1959.
Johnston, F.S. and B. Gottleib, Planet. Space Sci. 18, 1707-1718, 1970.
Johnston, H.S., Rev. Geophys. & Space Phys. 13, 637-649, 1975.
King, J.W., "Aeronautics and Astronautics" 13, 10-19, 1975.
Nastrom, G.C. and A.D. Belmont, J. Atmos. Sci. 32, 1715-1717, 1975.
Newell, R.E. and C.R. Gray - see CIAP, Chapter 1, pp. 22, 1975.
Nicolet, M., Rev. Geophys. and Space Phys. 13, 593-636, 1975.
Potemra, T.A. and T.J. Rosenberg, J. Geophys. Res. 1572-1580, 1973.
Rees, M.H., Planet. Space Sci. 23, 1589-1596, 1975.
Reid, G.C., I.S.A. Isakson, T.E. Holzer and P.J.Greutzen, Nature 259, 177-179, 1971.
Reid, G.C., J. Atmos. Sci. 32, 523-535, 1975.
Rusch, D.W. and C.A. Barth, J. Geophys. Res. 80, 3719-3721, 1975.

Parks, J.H. and J. London, J. Atmos. Sci. 31, 1898-1916, 1974.
Somerville, R.C.J., W.J. Quirk, J.E. Hansen, A.A. Lacis and P.H. Stone, J. Geophys. Res. 81, 1572-1576, 1976.
Stolarski, R.S., P.B. Hays and R.G. Roble, J. Geophys. Res. 80, 2266-2276, 1975.
Swider, W., J. Geophys. Res. 79, 3221-3224, 1974.
Wilcox, J.M. and N.F. Ness, J. Geophys. Res. 70, 5993, 1965.
Wilcox, J.M., P.H. Scherrer, L. Svalgaard, W.O. Roberts, R.H. Olson and R.L. Jenne, J. Atmos. Sci. 31, 581-588, 1974.
Wilson, C.R., R.D. Hunsucker and G.J. Romick, Planet. Space Sci. 24, 1155, 1976.
Wu, S.T. and K.D. Cole, Planet. Space Sci. 24, 727-730, 1976.

VARIATIONS OF THE MEAN MERIDIONAL CIRCULATION IN THE THERMOSPHERE

R. G. Roble

National Center for Atmospheric Research*
Boulder, Colorado 80307 USA

INTRODUCTION. Studies of the global circulation in the thermosphere have shown that the thermosphere is a dynamically active region with large variations about a basic state. This is due primarily to the variable nature of the main forcings for thermospheric dynamics--heating due to absorption of solar EUV and UV radiation, heating due to auroral particle precipitation, Joule heating due to the dissipation of ionospheric currents, momentum addition due to ionospheric drifts, and the effects of tides and other disturbances that propagate upward from the lower atmosphere and are dissipated in the thermosphere. Most of these energy and momentum sources have seasonal, solar cycle, and geomagnetic storm time variations that greatly influence the thermospheric structure and circulation. Several different theoretical models have been used to examine various aspects of the main global thermospheric variations. Some of these models are given by Geisler, 1967; Kohl and King, 1967; Dickinson et al. 1968; Dickinson, 1969; Mayr and Volland, 1972; 1973; Blum and Harris, 1975; Creekmore et al. 1975; Richmond and Matsushita, 1975; Hong and Lindzen, 1976; Forbes and Garrett, 1976.

In this paper the seasonal, solar cycle, and geomagnetic storm time variation of the mean meridional circulation and the latitudinal variation of temperature in the thermosphere are described. These variations have been determined by using a numerical model of the zonally symmetric thermosphere. The details of the model and a study of the mean meridional circulation and latitudinal variation of temperature in the thermosphere for equinox conditions during

*The National Center for Atmospheric Research is sponsored by the National Science Foundation

B. Grandal and J. A. Holtet (eds.), Dynamical and Chemical Coupling, 217-233.

solar cycle maximum have been given by Dickinson *et al*. 1975. A similar calculation for solstice conditions during solar cycle maximum has been described by Dickinson *et al*. 1977. Roble *et al*. 1977 examined the seasonal transition and the differences in the mean meridional circulation and latitudinal variation of temperature between solar cycle minimum and solar cycle maximum conditions. The heat and momentum sources that drive the mean meridional circulation in the model are solar EUV and UV heating, high latitude heating primarily due to auroral processes, and a momentum source that results from the correlation of diurnal variations of wind and ion drag. The highlights of these studies are reviewed and the results of some recent calculations describing the thermospheric response to geomagnetic substorms are presented in this paper.

NUMERICAL MODEL

The dynamic model that is used to calculate the zonal mean circulation and latitudinal variation of temperature in the thermosphere is a zonally averaged version of the pressure coordinate primitive equations of dynamic meteorology. The physical processes incorporated into the model are, however, those appropriate to thermospheric dynamics. A full description of the mathematical equations, reproduced in Table 1, is given by Dickinson *et al*. 1975. The model was primarily designed to examine the zonally symmetric motions above an altitude of about 150 km; however, it is extended down to about 80 km to minimize the influence of the lower boundary. The time-dependent thermodynamic equation, given in Table 1, includes the effects of vertical molecular heat conduction, radiative cooling, non linear meridional and vertical advection, adiabatic heating and cooling due to vertical motions, heating due to viscous dissipation, frictional heating due to ion drag, and a prescribed heat source. The time-dependent momentum equations include the inertial, viscous, coriolis, ion drag, pressure, and momentum forces in addition to transport of momentum by non linear advection. Continuity and hydrostatic equations and the equation of state complete the thermospheric dynamic system. The model calculates the perturbation temperature T' eastward velocity u, northward velocity v, vertical velocity w, and perturbation geopotential Φ'. The perturbations are departures from global mean reference values. The current thermospheric model does not calculate either the zonal mean distribution of composition or the global mean thermospheric temperature from global mean radiative heating. Rather, it prescribes basic state values determined from global means of the MSIS (Mass Spectrometer and Incoherent Scatter) empirical model of temperature and composition given by Hedin *et al*. 1977a, 1977b. The assumed boundary conditions are that vertical gradients of perturbation temperature and eastward, northward, and vertical velocities are zero at the upper boundary near 600 km and that the values of perturbation potential and

Table 1. Equations of Numerical Zonally Averaged Thermospheric Model

The thermodynamic equation

$$\frac{\partial T'}{\partial t}=\frac{ge^z}{p_0c_p}\frac{\partial}{\partial z}\left(\frac{K_T}{H}\frac{\partial T'}{\partial z}\right)-aT'-\frac{v}{r}\frac{\partial T'}{\partial \phi}-w\left(S+\frac{\partial T'}{\partial z}+\frac{RT'}{mc_p}\right)+Q'/c_p \quad (1)$$

The eastward momentum equation

$$\frac{\partial u}{\partial t}=\frac{ge^z}{p_0}\frac{\partial}{\partial z}\frac{\mu}{H}\frac{\partial u}{\partial z}+(f-\lambda_{xy})v-\lambda_{xx}u-\frac{v}{r\cos\phi}\frac{\partial}{\partial \phi}(u\cos\phi)-w\frac{\partial u}{\partial z}+M_x \quad (2)$$

The northward momentum equation

$$\frac{\partial v}{\partial t}=\frac{ge^z}{p_0}\frac{\partial}{\partial z}\frac{\mu}{H}\frac{\partial v}{\partial z}-(f-\lambda_{yx})u-\lambda_{yy}v-\frac{v}{r}\frac{\partial v}{\partial \phi}-w\frac{\partial v}{\partial z}-\frac{u^2}{r}\tan\phi-\frac{1}{r}\frac{\partial \Phi'}{\partial \phi}+M_y \quad (3)$$

The continuity equation

$$\frac{1}{r\cos\phi}\frac{\partial}{\partial \phi}(v\cos\phi)+e^z\frac{\partial}{\partial z}(e^{-z}w)=0 \quad (4)$$

The hydrostatic equation

$$\frac{\partial \Phi'}{\partial z}=\frac{RT'}{m}. \quad (5)$$

where the independent variables of the model are t = time, ϕ = latitude, z = $\log_e (p_o/p)$ where p = pressure and p_o = 5×10^{-4} mb. The dependent variables are u = eastward velocity, v = northward velocity, w = $\frac{dz}{dt}$ = "vertical motion," T' = perturbation temperature and ϕ' = perturbation geopotential. Other model parameters include g = gravity, c_p = specific heat at constant pressure, K_T = thermal conductivity coefficient, H = mean scale height, a = radiative damping coefficient, r = radius of earth, m = mean mass, Q' = perturbation heating, S = static stability = $\frac{\partial T_o}{\partial z} + \frac{RT_o}{mc_p}$, where $T_o(z)$ = global mean reference temperature, R = universal gas constant, μ = viscosity coefficient, f = coriolis parameter, (λ_{xx}, λ_{xy}, λ_{yx}, λ_{xy}) = ion drag tensor, M_x, M_y = momentum source. The coordinates x and y are positive eastward and northward respectively.

temperature and eastward and northward wind are zero at the bottom boundary near 80 km. The bottom boundary conditions are artificial and do not include the appropriate coupling with stratospheric and mesospheric dynamics. However, for zonally symmetric motions of a nearly inviscid, stratified compressible fluid, the influence of lower boundary conditions generally extends to at most a few scale heights above such a lower boundary. The numerical grid is in increments of 5° in latitude from the south to north pole, with five grid points per scale height in the vertical from about 80 to 600 km. Other details of the model and the numerical procedure for solving the equations are described in Dickinson et al. 1975.

PARAMETERIZATION OF ION DRAG

Above about 140 km, the ion gyrofrequency exceeds the ion - neutral collision frequency. Therefore, the ionization is constrained to move along geomagnetic field lines. A neutral wind flowing through the ions experiences a collisional drag that is proportional to the electron density and ion-neutral collision frequency and is dependent on the relative drift between the neutral and ionized constituents and the geomagnetic field line configuration. In the zonal mean dynamic model the geomagnetic pole is assumed to coincide with the geographic pole (Roble and Dickinson, 1974) and the geomagnetic field is represented by a centered dipole (Matsushita and Campbell, 1967). The components of the ion drag tensor that allow for the drift of ions due to neutral collisions are:

$$\lambda_{xx} = \sigma_p B^2 \rho_n^{-1}; \quad \lambda_{xy} = \sigma_H B^2 \rho_n^{-1} \sin I; \quad \lambda_{yy} = \sigma_p B^2 \rho_n \sin^2 I \quad (1)$$

where σ_p is Pedersen conductivity, σ_H is Hall conductivity, B is the magnitude of the geomagnetic field strength, ρ_n is neutral gas density, and I is the geomagnetic dip angle. The ion drag on the zonal wind, λ_{xx}, is proportional to the electron density values and has maximum values where the electron density is greatest. The ion drag on the meridional winds, λ_{yy}, is also proportional to the electron density but is weighted by $\sin^2$ I, where I is the geomagnetic dip angle and therefore goes to zero at the geomagnetic equator, where the geomagnetic field lines are horizontal. The Hall-type ion drag, λ_{xy}, is proportional to the Hall conductivity and sin I and therefore is asymmetrical about the equator because the dip angle is negative in the Southern Hemisphere. The Hall-type ion drag is relatively small at F-layer heights and also does not greatly influence the calculated motions in the thermosphere.

The zonal mean distribution of electron density is determined from the global empirical model of Ching and Chiu 1973. In addition, the global distribution of neutral atmospheric composition is determined from the MSIS empirical model of Hedin et al. 1977a;

1977b. Both the neutral composition and electron densities are used to calculate the Pedersen and Hall conductivities. Hence the ion drag components vary with the season, solar flux, geomagnetic activity, and other geophysical parameters inherent in the empirical models. Zonal mean quantities are considered diurnal averages within the model and interhemispheric differences are eliminated by averaging the properties of both hemispheres for the equivalent seasons.

SOLAR HEATING

Virtually all solar photons with wavelengths less than 1025 Å are absorbed in the thermosphere, leading to ionization of the absorbing species. Studies of this process by Stolarski et al. 1975 have shown that of the absorbed solar EUV energy within the thermosphere, approximately 30-40% of the energy appears as direct, local kinetic heating of the neutral gas, 40-50% of the energy is reradiated as airglow and lost from the region of direct excitation, and 20% appears as chemical energy of oxygen dissociation. The atomic oxygen is transported out of the region of direct production to the lower thermosphere, where it is regained as local heating in atomic oxygen recombination near 100 km.

The average energy of the solar flux in the Schumann-Runge continuum is about 7.5 eV (~ 1600 Å), and if it is assumed that only the excess energy over the O_2 dissociation energy (~ 5 eV) appears as neutral gas heating, the average neutral gas heating efficiency is about 33%. Thus, the solar EUV and UV heating in the thermosphere is determined by calculating the solar energy absorbed and multiplying this energy by a heating efficiency of about 33%. The zonal mean heating is calculated as a diurnal average at a given latitude. Heating due to the absorption of solar energy in the Schumann-Runge band system and the recombination of atomic oxygen in the lower thermosphere is also included in the calculation; and the details are described by Dickinson et al. 1977. Zonally averaged total heating rates and perturbation heating rates for various seasons are discussed in detail in Dickinson et al. 1975, 1977 and Roble et al. 1977a.

ZONAL MEAN MOMENTUM SOURCES

Thermospheric winds are smaller during daytime hours when the ion drag is large than at night when the ion drag is small. The zonally averaged momentum transfer between the neutral and ionized constituents of the earth's thermosphere exerts a "frictional" drag on the neutral winds that must be parameterized as a momentum source in the zonally symmetric model. By a correlation of the diurnal variation of ion drag and neutral wind the effective zonal mean zonal and meridional momentum sources are obtained:

$$M_x = -\overline{u'\lambda'_{xx}} \qquad M_y = -\overline{v'\lambda'_{yy}} \qquad (2)$$

The prime signs indicate departure from the zonally averaged values. Cross terms involving λ'_{xy} are negligibly small and are neglected. To calculate these momentum sources, a three-dimensional dynamic model of the neutral thermosphere (Roble and Dickinson, 1974) is used to calculate the global diurnal variation of thermospheric winds. Departures of these calculated winds from zonally averaged values are correlated with departures of ion drag from zonally averaged ion drag values determined from the empirical electron density model of Ching and Chiu 1973 and the MSIS model of Hedin et al. 1977a, 1977b. The zonally averaged momentum sources are not an important drive force for global thermospheric dynamics; however, they do alter the thermospheric structure slightly as described by Dickinson et al 1975; 1977.

HIGH LATITUDE HEAT SOURCE

It has been recognized for a long time that Joule heating due to ionospheric currents in the high latitude dynamo region is quite variable but could be an important part of the thermospheric energy budget (Cole, 1962). In addition, heating by direct particle precipitation in the auroral oval can also be an important source. Very little information on the global magnitude and distribution of this high latitude heat source is yet available. However, a recent study by Banks (1977) suggests that maximum total auroral heating occurs in the vicinity of the polar cusp and minimum occurs in the midnight sector, with the overall heating pattern having a horseshoe shape around the auroral oval. This heating appears to be equally divided between Joule heating of ionospheric currents and heating due to particle precipitation. The approach used in the studies of the mean meridional circulation is to multiply the estimated zonally averaged distribution of Joule heating due to time mean ionospheric currents (derived by Roble and Matsushita, 1975, and shown in Fig. 1) by an arbitrary factor until agreement between the calculated and observed circulations and latitudinal variation of temperature is obtained. In this manner, the magnitude of the total high latitude heat source due to auroral processes is estimated through studies of thermospheric dynamics.

SEASONAL VARIATION DURING SOLAR MAXIMUM CONDITIONS

The zonal mean circulation and the latitudinal variation of temperature in the thermosphere have been calculated by Dickinson et al. 1975 for equinox and by Dickinson et al. 1977 for solstice,

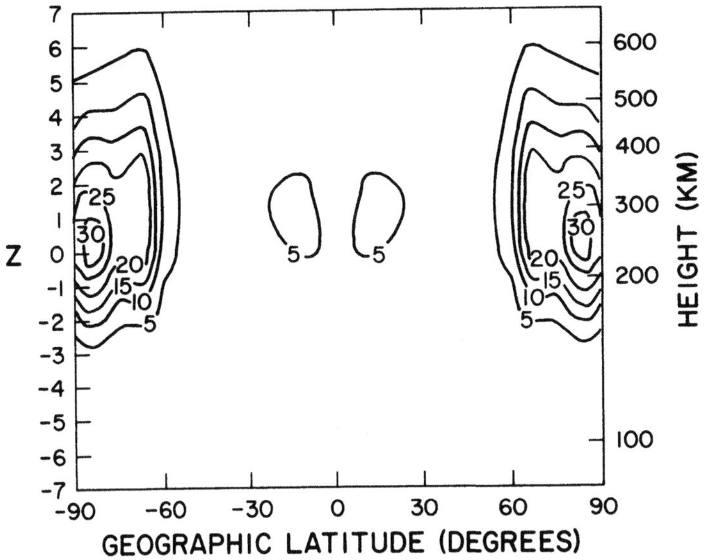

Fig. 1. Contours of the calculated zonal mean high latitude neutral gas heating used in model calculations. Units are ergs g^{-1} s^{-1} when divided by 10^{-4}. These values are derived from the study of Roble and Matsushita (1975); however, they are arbitrarily adjusted in each hemisphere by a constant factor to obtain agreement between the calculated and observed thermospheric temperature structure and circulation.

both for solar maximum conditions. These papers used the heat and momentum sources and parameterizations of ion drag and composition that were described in the previous section. The studies showed that the thermospheric circulation is driven primarily by heating due to the absorption of solar energy. The calculated mass flow stream functions for equinox and solstice conditions are shown in Figs. 2a and 2d, recpectively. Below about 150 km the circulation is from the summer pole to the winter pole during solstice and from the equator toward both poles during equinox. However, above about 150 km this circulation is modified by the high latitude heat source. During solstice the high latitude heat source reinforces the solar-driven summer-to-winter pole circulation in the summer hemisphere, but it reverses the circulation in the high latitude winter hemisphere at F-region heights, with transition from one cell to another occurring in the midlatitude winter hemisphere. During equinox the inferred circulation above 150 km is from both polar regions toward the equator, with the transition from one cell to another occurring near 20° latitude in both hemispheres for average geomagnetic quiet conditions.

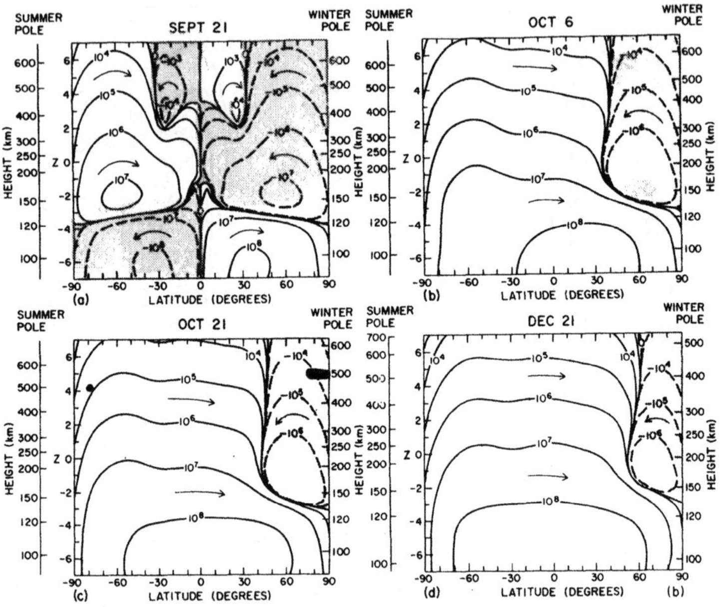

Fig. 2. Calculated mass flow stream function (g s^{-1}) for four days representing the seasonal variation during solar maximum, a) Equinox, 21 September, b) 6 October, c) 21 October and d) Solstice, 21 December.

The results of these studies suggest that the high latitude heat source is symmetrically distributed about the equator during equinox, with a total energy input of about 1.25 x 10^{18} ergs s^{-1} for average geomagnetic quiet conditions. During solstice the inferred high latitude heat source is about 75% greater--2 x 10^{18} ergs s^{-1}--but two and a half times more heating is required in the summer hemisphere than in the winter hemisphere to bring the calculated and observed thermospheric structure and circulation into agreement. The calculated distributions of perturbation temperature and meridional and zonal winds for equinox conditions are shown in Figs. 3a, 4a, and 5a, respectively, and for solstice conditions in Figs. 3d, 4d, and 5d, respectively.

The variation of the mean circulation and latitudinal variation of temperature between seasons were examined by Roble et al. 1977. They found a persistent solstice pattern of temperature and circulation that generally weakens in magnitude but persists as the season progresses to within a week or two of equinox. The circulation pattern then undergoes an abrupt breakdown with the establishment of the equinox circulation before the solstice

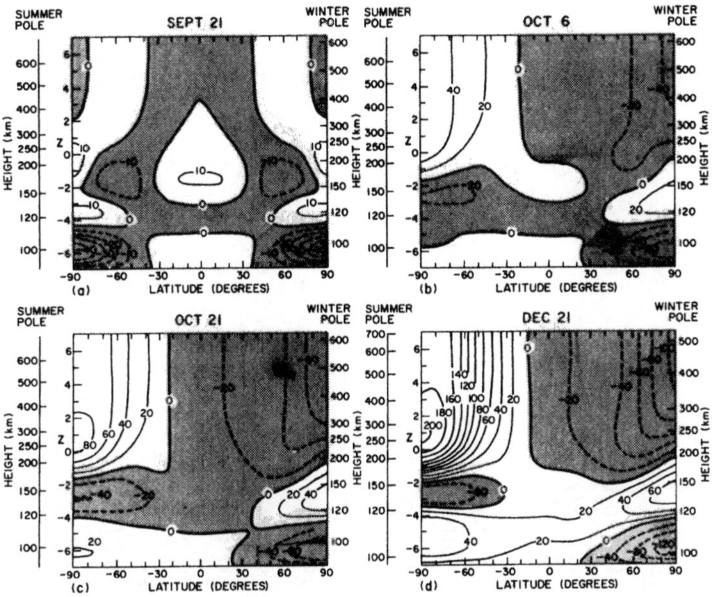

Fig. 3. The calculated zonal mean contours of perturbation temperature (K) for four days representing the seasonal variation during solar maximum, a) Equinox, 21 September, b) 6 October, c) 21 October and d) solstice, 21 December.

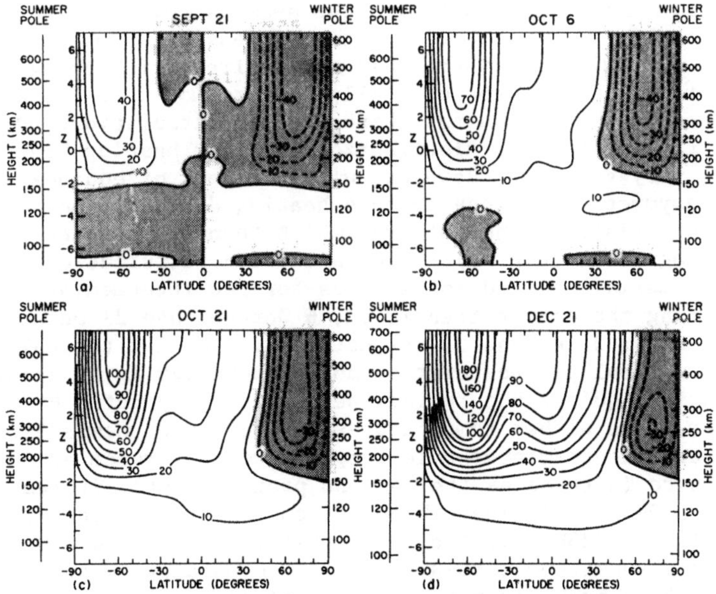

Fig. 4. The same as in Fig. 3 except the meridional wind component (ms^{-1}). Positive values are directed northward.

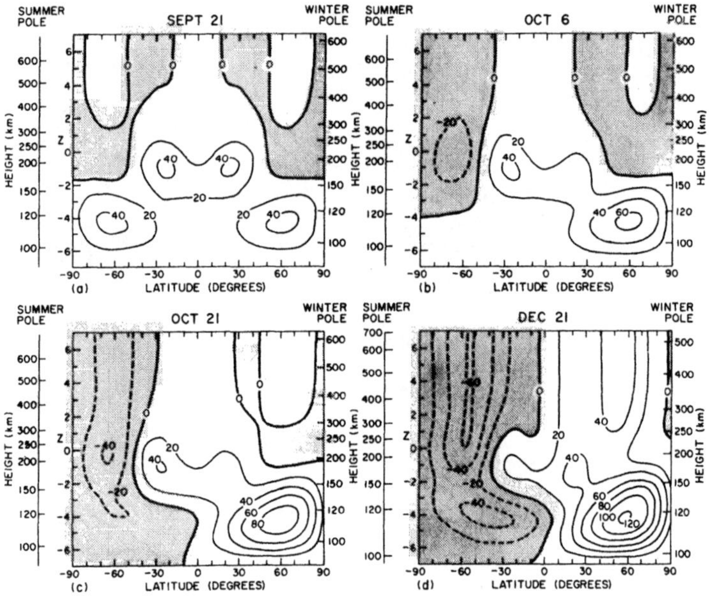

Fig. 5. The same as in Fig. 3 except the zonal wind component (ms^{-1}). Positive values are directed eastward.

pattern again develops, but with the seasons reversed in the hemispheres. Thereafter, the circulation intensifies towards solstice. Roble et al. 1977a attribute this rapid circulation change during equinox to heating in the optically thin region near 200 km, which is the main drive region for thermospheric circulation. During equinox the latitudinal distribution of solar heating hear 200 km does not vary greatly. However, within a week of equinox a pronounced asymmetry develops in this heating distribution that forces the solstice type pattern, which then intensifies as the season progresses (Fig. 2). The calculated perturbation temperature and meridional and zonal winds for two intermediate dates illustrating the abrupt transition, 6 October and 21 October, are given in Figs. 3, 4, and 5, respectively. The latitudinal variation of neutral gas temperature above about 120 km is relatively small during equinox but rapidly develops an asymmetry by 6 October, with higher temperatures in the summer hemisphere. The summer-pole-to-winter-pole temperature difference at 300 km continues to increase as the seasons progress until it reaches a value of 300K at solstice, which is consistent with the MSIS model value of 325K. In the lower thermosphere below 120 km there is a strong summer-pole-to-winter-pole temperature gradient; however, it is probably not realistic since coupling with mesospheric dynamics through the lower boundary is not included in the model calculations. Between 120-150 km the summer-pole-to-winter-pole temperature gradient reverses. The meridional winds are shown in Fig. 4. Below 150 km the winds are from the equator toward both poles during equinox. By 6 October the winds reverse in the summer hemisphere and a summer-to-winter hemisphere circulation develops. Above 150 km the effect of the high latitude heat source is to drive a circulation from both poles to the equator for equinox. However, as the season progresses, the summer hemisphere meridional winds penetrate into the winter hemisphere and force the transition between the two oppositely directed circulations to higher latitudes. This boundary is near 30° latitude on 6 October but moves to 50° latitude at solstice. The zonal winds, shown in Fig. 5, are eastward below 150 km during equinox, but as the seasons progress toward solstice they become westward in the summer hemisphere and remain eastward in the winter hemisphere. The winds below 150 km are in geostrophic balance. Above 150 km this pattern also develops, but it is primarily driven by the zonal momentum source. The magnitudes of these winds are consistent with the winds derived from midlatitude incoherent scatter radar observations by Roble _et al_. 1974; Amayenc 1974; Roble _et al_. 1977, and the 6300 Å optical measurements of Hernandez and Roble _et al_. 1977b. The seasonal transition shown in these figures is for average geomagnetic quiet conditions during solar maximum. The boundary between circulation cells is strongly dependent on the magnitude of the high latitude heat source and thus has considerable variability associated with geomagnetic activity. The schematic sequences shown in Figs. 6 and 7 illustrate this variability for equinox and solstice conditions. The top frames indicate the circulation for very quiet geomagnetic activity, with a total energy input of about 10^{17} ergs s^{-1}; the middle frames for average conditions of about 10^{18} ergs s^{-1}; and the bottom frames for geomagnetic storm conditions with an energy input of about 10^{19} ergs s^{-1}. Thus, the thermospheric circulation is strongly dependent on geomagnetic activity, with large variations about a basic state due to the intermittent nature of auroral activity.

SOLAR CYCLE VARIATIONS

Roble _et al_. 1977a calculated the zonal mean circulation and latitudinal variation of temperature for solar minimum conditions during solstice. The results of the calculation are shown in Fig. 8. To obtain agreement with observations, it was necessary to assume that the solar EUV radiation was about two times lower than at solar maximum and that the high latitude heat source was about four and a half times lower, 4.5×10^{17} ergs s^{-1}, than at solar maximum, with the bulk of the high latitude heat source in the summer hemisphere. The resulting latitudinal circulation is weaker, with a smaller summer-pole-to-winter-pole temperature dif-

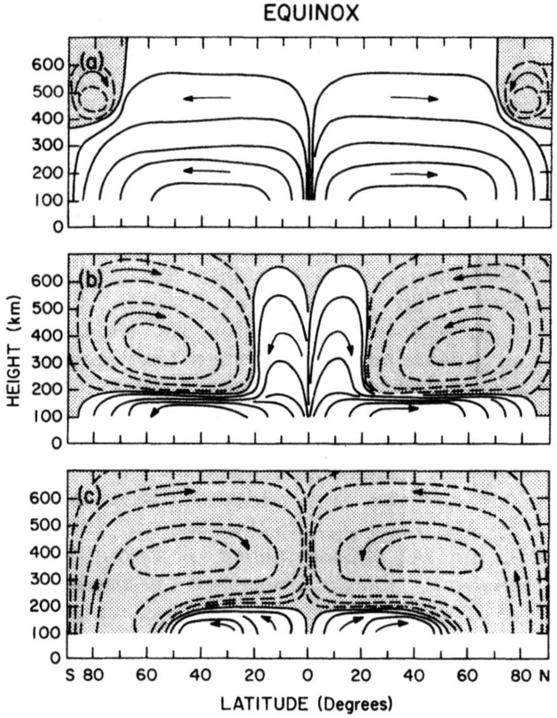

Fig. 6. Schematic diagram indicating the mass flow stream function variation with geomagnetic activity during equinox conditions during solar cycle maximum a) geomagnetic quiet conditions, b) average geomagnetic activity, and c) geomagnetic storms.

ference at F-region heights during solar minimum (200K) than at solar maximum (325K). These results are consistent with MSIS model predictions and the thermospheric temperature and wind measurements of Hernandez and Roble (1977). The seasonal variation during solar minimum conditions is similar to that previously described except that the magnitude of the high latitude heat source in the winter hemisphere is not sufficient to reverse the solar driven circulation (Fig. 9).

THERMOSPHERIC RESPONSE TO GEOMAGNETIC SUBSTORMS

The geomagnetic substorm has a characteristic time of one to two hours although the time interval, magnitude, and repetition rate are highly variable. To model the thermospheric response to a geomagnetic substorm, it is assumed that the high latitude heat source shown in Fig. 1 is superimposed on the basic heat sources derived for solstice conditions during solar minimum (Fig. 8).

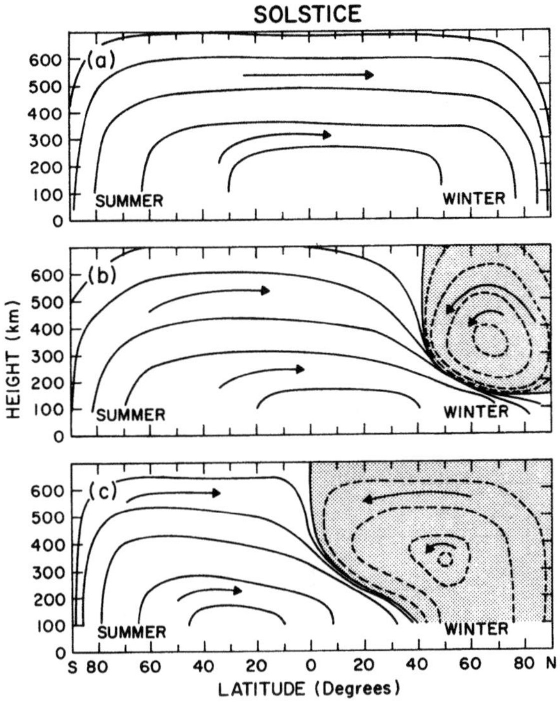

Fig. 7. Same as for Fig. 6 except for solstice conditions.

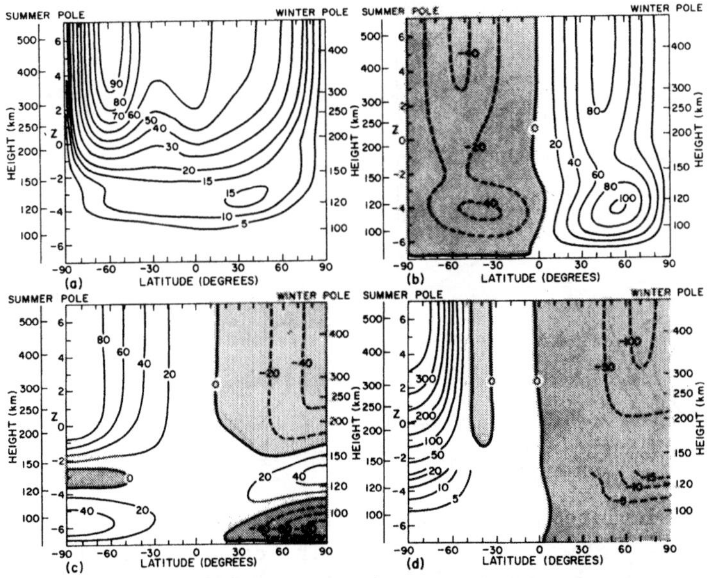

Fig. 8. Calculated zonal mean a) perturbation temperature (K), b) meridional wind (ms^{-1}), c) zonal wind (ms^{-1}), and d) vertical velocity (cm s^{-1}) for solstice conditions during solar cycle minimum.

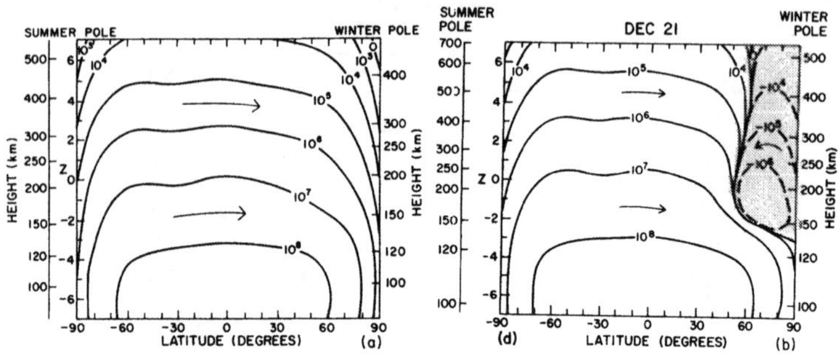

Fig. 9. Contours of mass flow stream function (g s^{-1}) for solstice conditions during geomagnetic quiet conditions during a) solar cycle minimum, and b) solar cycle maximum.

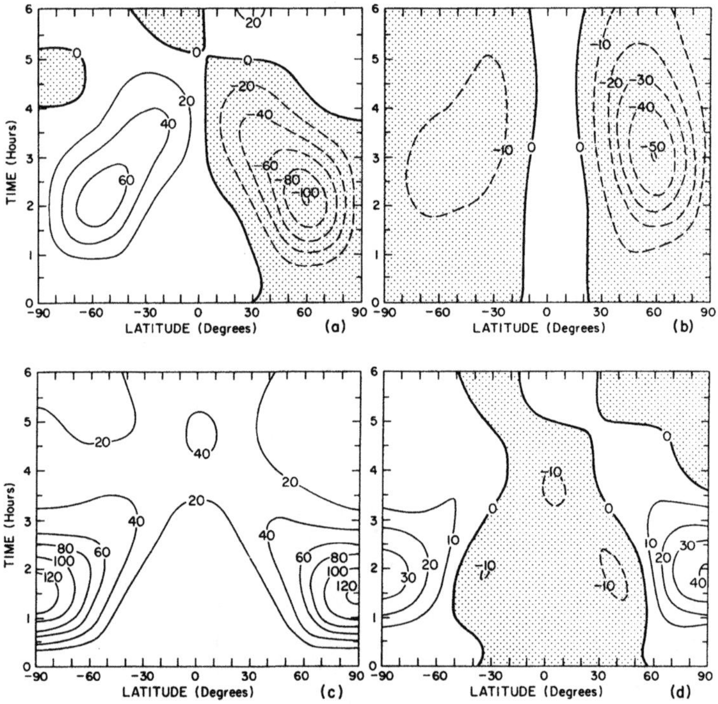

Fig. 10. Calculated mean thermospheric response to a geomagnetic substorm during solstice conditions. The contours are a) meridional wind (ms^{-1}), b) zonal wind (ms^{-1}), c) effective temperature difference (K), and d) vertical velocity (in cm s^{-1} when multiplied by 10) all at 250 km.

The substorm heat source is also assumed to be equally imposed on both the summer and winter hemispheres; the magnitude of the heat source is shown in Fig. 1. The time history of the substorm is arbitrary and for these calculations it is assumed to increase linearly with time from substorm onset, reach a maximum within one-half hour, and then decay linearly in time until it reaches zero two hours later. The total energy input during the substorm period is about 10^{22} ergs, and the substorm model is consistent with that described by Akasofu (1976). The calculated thermospheric response at about 250 km to the substorm is shown in Fig. 10. The temperature is not directly calculated since the model does not include coupling with composition. However, an effective temperature $\frac{T}{m} \times \bar{m}$ is shown, where T is the substorm perturbation temperature, m is the substorm perturbation mean mass, and $\bar{m}$ is the mean mass at a given latitude for the unperturbed solstice case. The effective thermospheric temperature at 250 km is seen to first increase at high latitudes, reaching maximum values one and a half hours after the initiation of the substorm. Maximum upward vertical velocities are also obtained on a similar time scale at high latitudes. The resulting pressure increase drives a thermospheric wave toward the equator in both hemispheres, with the differences in magnitude primarily due to variation of the mean background state for the solstice condition. Both the meridional and zonal wind components increase with time. The meridional wind is driven by the pressure force and the zonal wind develops through coriolis coupling with the meridional flow. The zonal wind is westward in both hemispheres. About four hours after the initiation of the substorm the downward motion in the equatorial region maximizes and the resulting temperature and pressure perturbations produce a pressure distribution that forces the wave to turn poleward at later times. The magnitude of the perturbation, however, is considerably reduced. The thermospheric circulation thus redistributes the energy of the geomagnetic substorm globally before it is conducted to the lower thermosphere, where in turn it is radiated to space by the optically active minor neutral constituents. Since the substorm is intermittent, thermospheric waves with various amplitudes and phasing can be launched intermittently and thus maintain the thermosphere in a constant state of agitation with these variations superimposed on the mean flow.

REFERENCES

Akasofu, S.-I. 1976, Space Sci. Rev. 19, 169.

Amayenc, P. 1974, Radio Sci. 9, 281.

Banks, P. M. 1977, J. Atmos. Terr. Phys., in press.

Blum, P. W. and Harris, I. 1975, Atmos. Terr. Phys. 37, 193.

Ching, B. K. and Chiu, Y. T. 1973, J. Atmos. Terr. Phys. 35, 1615.

Cole, K. D. 1962, Aust. J. Phys. 15, 223.

Creekmore, S. P., Straus, J. M., Harris, R. M., Ching, B. K., and Chiu, Y. T. 1975, J. Atmos. Terr. Phys. 37, 491.
Dickinson, R. E. 1969, J. Atmos. Sci. 26, 1191.
Dickinson, R. E., Lagos, C. P. and Newell, R. E. 1968, J. Geophys. Res. 73, 4299.
Dickinson, R. E., Ridley, E. C. and Roble, R. G. 1975, J. Atmos. Sci. 32, 1737.
Dickinson, R. E., Ridley, E. C. and Roble, R. G. 1977, J. Atmos. Sci. 34, 178.
Forbes, J. M. and Garrett, H. B. 1976, J. Atmos. Sci. 33, 2226.
Geisler, J. E. 1967, J. Atmos. Terr. Phys. 29, 1469.
Hedin, A. E., Reber, C. A., Newton, G. P., Spencer, N. W., Brinton, H. C. and Mayr, H. G. 1977b, J. Geophys. Res., in press.
Hedin, A. E., Salah, J. E., Evans, J. V., Reber, C. A., Newton, G. P., Spencer, N. W., Kayser, D. C., Alcaycle, D., Bauer, P., Cogger, L. and McClure, J. P. 1977a, J. Geophys. Res., in press.
Hernandez, G. and Roble, R. G. 1977, J. Geophys. Res., in press.
Hong, S. S. and Lindzen, R. S. 1976, J. Atmos. Sci. 33, 135.
Kohl, H. and King, J. W. 1967, J. Atmos. Terr. Phys. 29, 1045.
Matsushita, S. and Campbell, W. H. 1967, Physics of Geomagnetic Phenomena, Vol. II, Appendix 1, Academic Press, 1398 pp.
Mayr, H. G. and Volland, H. 1972, J. Geophys. Res. 77, 6774.
Mayr, H. G. and Volland, H. 1973, J. Atmos. Terr. Phys. 35, 669.
Richmond, A. D. and Matsushita, S., J. Geophys. Res. 80, 1975.
Roble, R. G., Emery, B. A., Salah, J. E. and Hays, P. B. 1974, J. Geophys. Res. 79, 2868.
Roble, R. G., Salah, J. E. and Emery, B. A. 1977b, J. Atmos. Terr. Phys., in press.
Roble, R. G., Dickinson, R. E. and Ridley, E. C. 1977a, J. Geophys. Res., in press.
Roble, R. G. and Dickinson, R. E. 1974, Planet. Space Sci. 22, 623.
Roble, R. G. and Matsushita, S. 1975, Radio Sci. 10, 389.
Stolarski, R. S., Hays, P. B. and Roble, R. G. 1975, J. Geophys. Res. 80, 2266.

DISCUSSION

Cole: Joule heating is given by $\sigma_p[(\underline{v}_e - \underline{v}) \times \underline{B}]^2$ where $\underline{v}_e$ is the electron velocity and $\underline{v}$ the centre of mass velocity of the whole gas. What have you assumed for v in your estimation of joule heating as a function of altitude ?

Roble: The joule heating was calculated by assuming v = 0. We found that our prescription of Joule heating from a direct calculation from time mean equivalent current systems inadequate. We, however, kept the spatial distribution defined by the equivalent current system and adjusted the overall magnitude of the high latitude heat source untill we could get agreement between the calculated and observed latitudinal variation of temperature and circulation. In this manner we have used neutral dynamics to define the overall magnitude of the solar wind energy entering the thermosphere through auroral processes.

Piggott: If a large part of the energy is associated with the midday cusp activity, the ionospheric evidence would suggest that this term should be very skew, mostly confined to the summer hemisphere. This is what is required to test the seasonal range.

While there is plenty of ionospheric evidence for minor phenomena which behave in the way suggested by the model, the large scale mid and low latitude quiet-storm behaviour is not consistent with it. Thus it looks as if a major term has been omitted. In general it is only storm centre phenomena with time constants of order 1 to 3 hours which behave as indicated. The big variations are usually restricted in longitude and can be delayed by several days relative to the substorm.

SOME RESULTS FROM THE EUROPEAN WINTER ANOMALY CAMPAIGN 1975/76

D. Offermann

University of Wuppertal, West Germany +)

Enhancements of D-region electron densities are known to be effected by solar or geomagnetic events as well as by aeronomic or meteorological events during winter. The basic goal of the western European winter anomaly campaign was the study of the atmosphere-ionosphere interaction, i.e. of the so-called meteorological type of winter anomaly. It therefore was intended to exclude effects due to solar flares, magnetic storms, and storm-after-effects. A number of possible explanations for this meteorological type of winter anomaly are presently discussed in the literature (Gregory 1965; Reid 1970; for a compilation of recent references see Sechrist 1975):

I. vertical and/or horizontal transport of atmospheric constituents with low ionization potential as NO

II. changes in water cluster ion chemistry as caused by transports of trace constituents (O, H_2O)

III. changes in neutral and ion chemistry by temperature variations.

To guarantee a broad coverage of atmospheric parameters, quite a number of atmospheric and ionospheric experiments for the upper mesosphere and lower thermosphere was integrated in the rocket campaign. These experiments were supplemented by measurements in the stratosphere and lower mesosphere by a number of meteorological rocket flights. The troposphere and the lower stratosphere were probed by balloon flights. The rocket and balloon launches took place at El Arenosillo, Spain, (37°6′N; 6°44′W) in the period December

---

+) formerly: Bonn University

B. Grandal and J. A. Holtet (eds.), Dynamical and Chemical Coupling, 235-252.

1975 to February 1976. The measurements were supported by an extended network of ground based stations in Spain, France, Italy, Great Britain, Austria and Germany. A total of 22 experimenters groups with 38 types of experiments participated in the campaign. The rocket experiments were allocated to 7 different types of payloads (called BI, BII,...,BVII), covering different altitude regimes. A total of 47 rockets and 23 balloons were successfully launched during the campaign. A detailed description of the campaign structure was given by Offermann (1977) (see also Dahl 1977). On two days of the campaign (January 4 and 21, 1976) there were large numbers of rocket launches, which were performed within a few hours. These so-called "salvoes" contained one experiment of all types available. The rest of the rockets (and balloons) was more or less evenly distributed over the period of January and begin of February 1976, mostly measuring the development of the wind and temperature field at medium and low altitudes (see Fig.1).

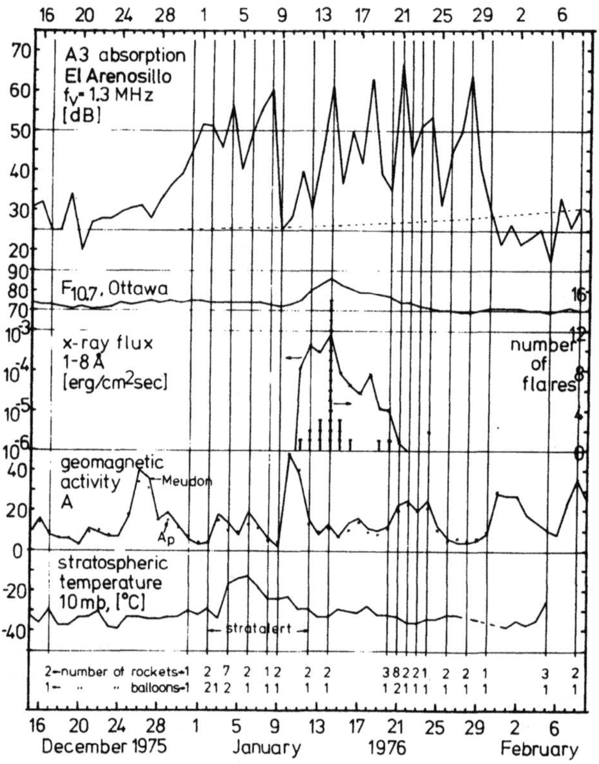

Fig. 1 : Geophysical data and rocket and balloon launches during the campaign

Some of the experiments participating in the campaign were used to forecast launch opportunities, to define launch criteria, and to check the verification of these criteria. A3 absorption measurements in Spain (supported by ionosonde measurements on the range) were used to define the launch days (absorption greater than 50 dB). Ursigrams and magnetic forecasts were obtained from the Observatoire de Paris, Meudon, to avoid magnetic storms and storm-after-effects. Days of solar x-ray activity and flares were avoided by use of information from the SMS satellite obtained through the Space Environment Center, NOAA, Boulder. Furthermore stratospheric weather reports and STRATALERT messages were obtained from the Institut für Meteorologie, Freie Universität, Berlin. Fig. 1 shows part of these data as they were received on the rocket range on a daily basis. (The stratospheric temperature given is the maximum temperature on the northern hemisphere of the respective day.) The vertical lines denote dates of rocket and balloon launches. The number of launches per day is given in the bottom of the picture. Fig. 1 shows that it was managed to launch the two rocket salvoes on days with strong winter anomaly activity, and nevertheless to avoid magnetic events, flares and periods of high x-ray emission. A number of particle and x-ray detectors flown as part of the two salvoes confirmed that there was only minor ion production due to particle and x-ray fluxes on these two days in the relevant altitude regime (Thrane et al. 1977).

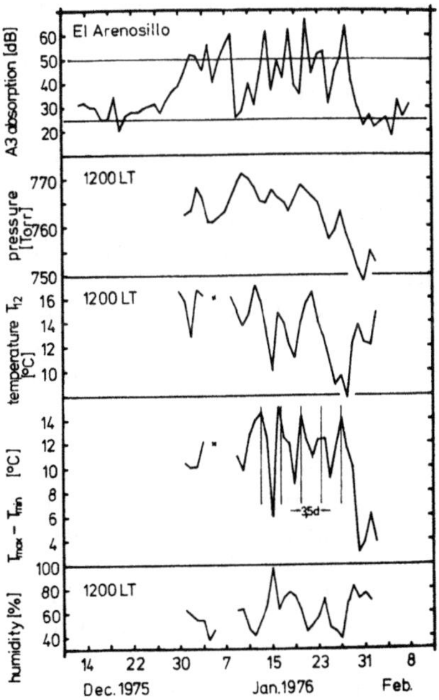

Fig. 2: Meteorological ground data

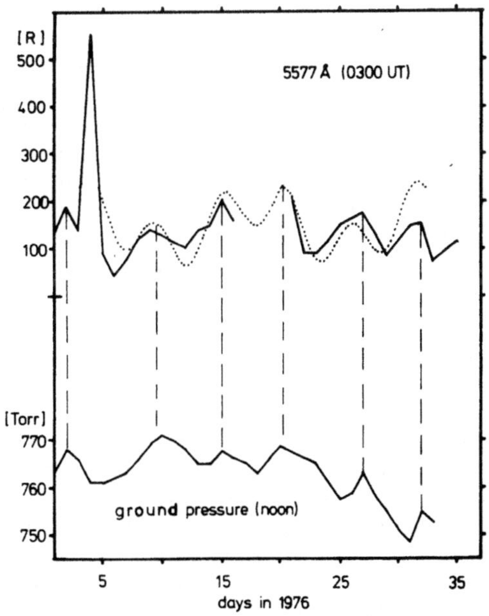

Fig. 3: Ground photometer measurements compared to ground pressure

In the following some experimental results obtained during the campaign are presented. It should be noted that most of these data are still preliminary. Fig. 2 shows some meteorological ground data as they were measured on the range (Offermann et al. 1977). A3 absorption is given for comparison. The central part of the campaign (January 1976) was characterized by high ground pressure. Superimposed on this high mean pressure there is a wavelike structure which is also found in a similar but time shifted fashion in the ground temperature. In an effort to analyze this wave structure, the daily maximum temperature difference (day/night) was determined. It shows a suspicious wavelike structure with a period of about 3.5 days (Fig.2). Longer periods may have been lost by the method of taking daily temperature differences. The ground pressure itself indicates a period somewhat smaller than 6 days. It should be mentioned that it also showed a diurnal and semidiurnal variation with peak to peak values up to 4 Torrs during most part of the campaign.

Fig. 3 shows ground based measurements of the 5577 Å emission from the night sky. (H.Lauche, private communication 1977). The ground pressure is also given for comparison. A periodic structure is seen in these night glow data, too. To demonstrate this, a superposition of two waves with 15 days and 5.5 days period and suitable amplitudes is given in Fig. 3 as a dotted line. The correlation of the maxima of the (interpolated) green line curve with the ground pressure maxima is very remarkable if one takes the very high emission on January 4, 1976, as a singular event. It indicates that during the campaign there were oscillations in the atmosphere with quite a spectrum of periods. At least part of these wavelike structures appears to have extended from the ground up to the atomic oxygen layer at around 95 km, where most of the 5577 Å line is emitted. Fig. 3 also shows that there was a remarkable difference in the night glow intensity on the two salvo days. In interpreting these data one has to take into account that the rocket launches took place half a day later than the photometer measurements. Nevertheless from this picture one clearly expects higher atomic oxygen densities in the 90 - 100 km regime on January 21, 1976.

Fig. 4 shows stratospheric temperatures at different altitude levels for the duration of the campaign (Offermann et al. 1977). The 50 mb, 30 mb, and 10 mb data (21 km, 24 km, and 31 km approx.) were taken from stratospheric weather maps of the Institut für Meteorologie of Freie Universität Berlin. In the upper part of Fig. 4 data of channel 2115 (40 - 45 km) of the PMR experiment on Nimbus 6 are given. They represent radiant fluxes in relative units. The temperatures at the lower altitudes are compared to atmospheric models (Cira 1972 and US Standard Atmosphere 1976). A common feature of all temperatures shown in Fig. 4 is that they vary strongly

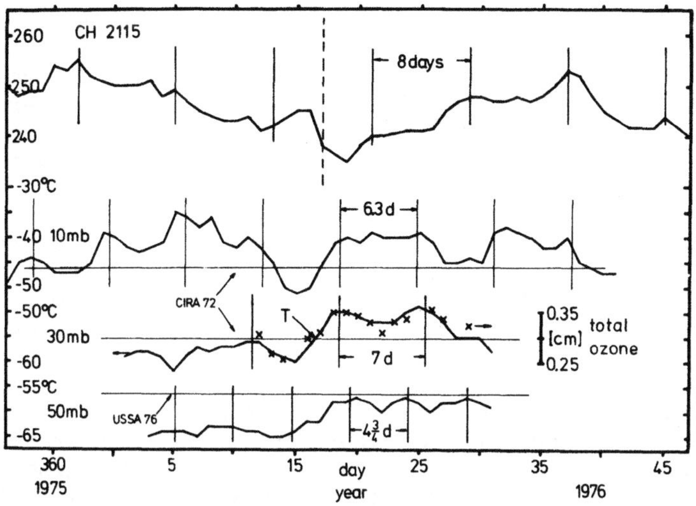

Fig. 4: Stratospheric temperatures and total ozone above the rocket range

during the campaign. At 50 mb they are generally too low as compared to the model, especially in the first part of January. At 30 mb the temperatures oscillate around the model value as a rough mean. At 10 mb they are generally too high as compared to the model.

In addition to the 30 mb temperatures total ozone data as measured by a Dobson spectrometer are also plotted in Fig. 4. (J. Cisneros and J. Satrustegui, private communication 1977.) The similarity of the variation of these ozone densities with those of the 30 mb temperatures is striking. This must be a purely dynamical effect since photochemistry would yield increase of ozone density with decreasing temperature.

The temperatures and the ozone densities shown in Fig. 4 again indicate wavelike structures. The periods of them are between 6 - 8 days (neglecting the 50 mb data). The periodic structure is least pronounced at the lower altitudes and best seen in the 10 mb data. These 10 mb data also indicate an oscillation with a periodic twice as long. Channel 2115 shows only a weak periodic structure, which may be due to the fact that this channel has rather poor altitude resolution and thus tends to smear out waves of short vertical wavelength. In part, however, this is a real phenomenon pertinent to this altitude regime during the campaign. This is because there is a node in the vertical wavelike structures around this altitude, as will be seen below.

The weakness of short term structures in the Channel 2115 data has the advantage, that other variations of smaller amplitude become visible. There is an obvious long term variation in these data with a time scale of the order of a month, yielding a temperature minimum around day 19 of 1976. The same feature with opposite phase appears to be present in the 10 mb temperatures, but it is masked by the strong short period variations at this altitude. The CIRA 1972 temperature model for 40°N shows an annual variation with a node around 60 - 65 km and extrema at 40 km and 85 - 90 km. The 40 km level shows a temperature minimum in winter, whereas the 80 - 95 km level reveals a temperature maximum at the same time. This opposite variation of stratospheric and mesospheric temperature is a well known phenomenon also for events of shorter duration, as for instance stratospheric warmings (K. Labitzke 1972). It may therefore be assumed that the temperature decrease during the campaign as shown by the Channel 2115 data corresponds to a temperature increase at the 85 - 90 km level, which is the altitude regime believed to be responsible for most part of the winter anomaly. (This temperature increase is actually there, as is indicated by high altitude data of the PMR experiments on Nimbus G; private communication by D. Rogers 1977). Fig. 5 shows the Channel 2115 data (40 - 45 km) together with the A3 absorption results. Indeed a correlation is found with long term periods of generally high absorption coinciding with low temperatures at the 40 - 45 km level, and vice versa. The correlation is still improved if one shifts the temperature by about 4 days, with respect to the A3 data. Then even detailed structures of similar form are found in both curves as is indicated in Fig. 5.

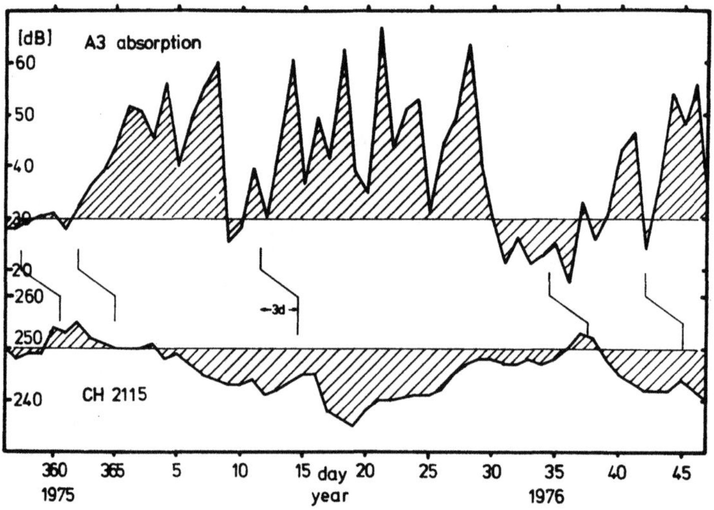

Fig. 5: A3 absorption and CH 2115 radiant fluxes (40 - 45 km)

Schwentek (1971) has proposed to consider the bulk of A3 absorption during winter as a normal behaviour, and take only the superimposed spikes as anomalous events. From Fig. 5 one could understand this "normal" winter anomaly as a consequence of a temperature decrease in the 40 - 45 km regime and a corresponding temperature increase in the 85 - 90 km regime with time scales of the order of a month.

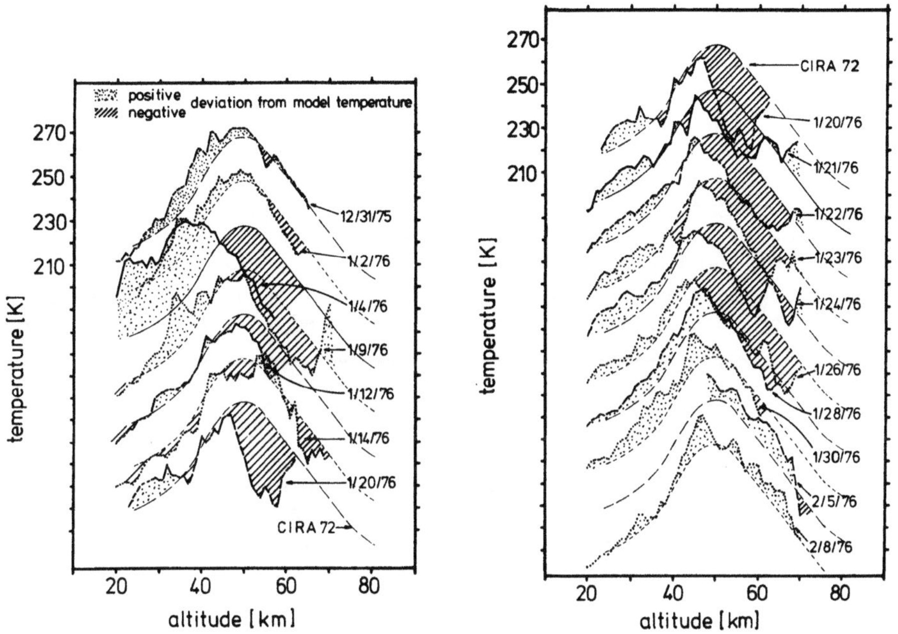

Fig. 6, 7: Rocket temperature measurements compared to CIRA 72 model profiles. Rocket flights are labelled by their dates

During the campaign 16 meteorological rockets were launched to measure temperatures and winds in the altitude regime 20 - 65 km. (The launches took place at nearly the same local time). (J.Cisneros and J. Satrustegui, private communication 1977). The temperature profiles obtained are shown in Fig. 6 and 7. They are shifted by 20 K with respect to each other. The respective model profiles (Cira 1972) are also given. The profiles exhibit strong wavelike structures both in time and in the vertical direction. Time variations are analyzed by making cuts in time at a given altitude. Vertical structures are analyzed by taking the difference of the measured profile and the Cira 72 curve (see below).

Fig. 8 shows the time variations of the temperature at 55 km altitude. The respective Cira 72 value is given for comparison. It shows that the temperature at this altitude is generally too low during the campaign. This is similar to the behaviour of the 40 - 45 km data, but not as indicative, because the 55 km level is not a node, and its behaviour might be masked by short period oscillations. It was therefore tried to fit the data by a superposition of two short period waves. The result for periods of 7 and 15 days, respectively, and equal wave amplitudes is given as a dotted line in Fig. 8. The fit is quite good and shows that if these short period oscillations were not present, the 55 km temperature would be low by up to 15 K. A further similarity to the data shown in Fig. 5 is the correlation with the A3 absorption data: at times when there is low A3 absorption the 55 km temperature is normal or higher than normal (after introduction of a suitable time shift of 1.5 days). Hence the temperatures at 40 - 45 km and 55 km show essentially the same gross features when compared to the A3 absorption.

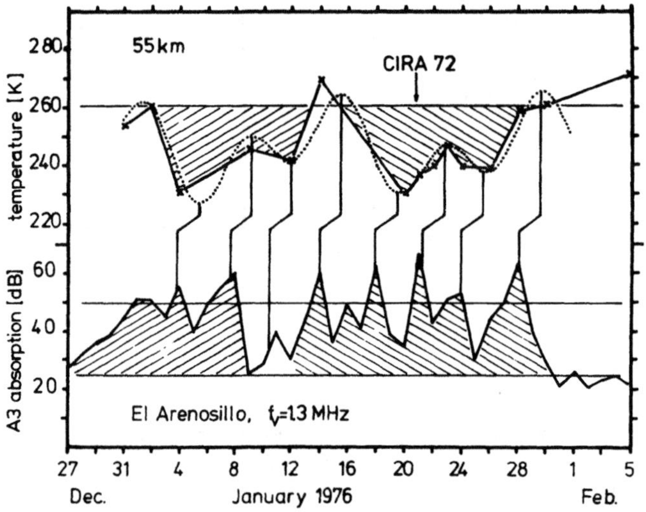

Fig. 8: Comparison of 55 km temperatures, measured by meteorological rockets, and A3 absorption.

Considering the fine structure of the temperature at 55 km, an intriguing correlation of the maxima and minima of the dotted curve and the A3 absorption data is found: each temperature extremum corresponds to an A3 absorption peak if one introduces the above mentioned time shift of 1.5 days. Hence not only the "normal" winter anomaly is correlated to the atmospheric temperature, but also the "anomalous" absorption peaks appear to be linked to wave-like structures in the temperature. Fig. 8 suggests that there might be strong temperature oscillation components at 80 - 90 km altitude with periods of the order of 7 and 3.5 days.

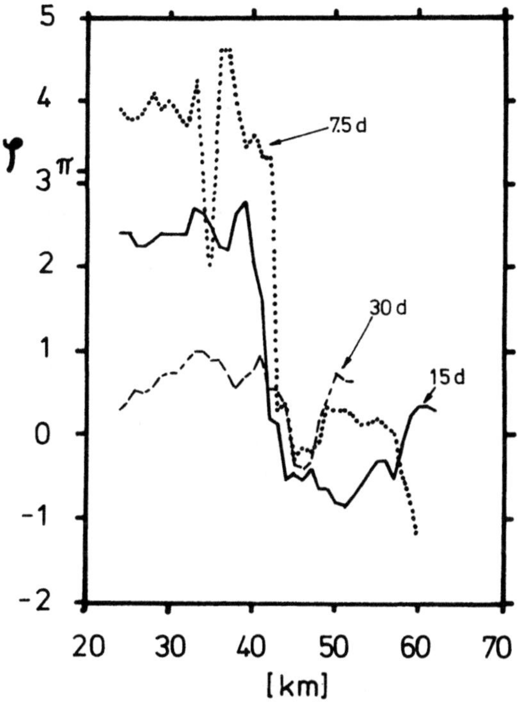

Fig. 9: Phases of the Fourier components with periods of 30, 15, and 7.5 days, respectively. Phases ϕ are given in radians.

The time variation of all of the temperature data given in Fig. 6 and 7 was analyzed similarly as in Fig. 8 at height increments of 1 km. Each set of data at a given altitude was Fourier-analyzed, starting from a basic period of 30 days, which was the main time interval covered by the measurements. Higher order periods up to the 5th harmonic (6 days) were taken into account. (Rose et al. 1977). The main result obtained from this analysis is that the dominating Fourier components are the 30 day and 15 day periods below 50 km. Between 50 and 60 km the 15 day and 7.5 day periods are most important. The regime between 40 - 45 km shows only small amplitudes and is therefore interpreted as a node in the vertical wavelike structure. The phases of the Fourier components with periods of 30 d, 15 d, and 7.5 d are given in Fig. 9. The 30 d phases are only shown up to 50 km, because at higher altitudes this component is of minor importance. It is interesting to see that the phases are about constant up to 40 - 45 km which is the location of the node. At this altitude the phases of the Fourier components dominant at higher altitudes change drastically by about 180° (Π in Fig. 9) and then again remain about constant. Thus the wavelike structures found in these temperature data resemble a standing wave. This may already be seen in the raw data shown in Fig. 6 and 7.

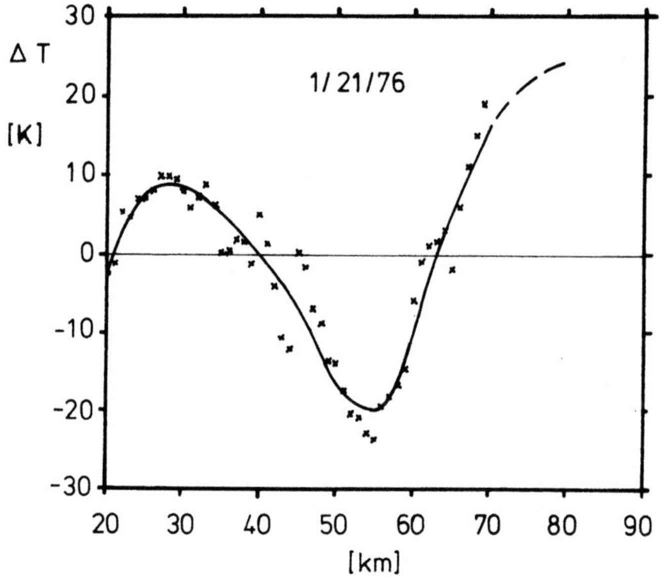

Fig. 10: Temperature deviations from the CIRA 1972 model profile on January 21, 1976.

To analyze the vertical wavelike structures in these temperatures the differences between the CIRA 1972 model temperatures and the measured data were calculated. As an example Fig. 10 shows these differences as obtained for one of the salvo days (Jan.21, 1976). The heavy curve is an approximate fit to the data points. The wavelike structure seen shows the above mentioned node at 40 - 45 km. It also shows a maximum at about 30 km and a minimum around

55 km, indicating a basic vertical wavelength of roughly 50 km. The 16 altitude profiles obtained during the campaign can be divided into two groups: one with strong wavelike structures and one with small variations (see also Fig. 6 and 7). All profiles with strong variations show the same features like Fig. 10 with similar wavelength and altitude of the node, though the magnitudes of the minima and the maxima vary. This again is what one would expect for a standing wave. The raw data in Fig. 6 and 7 show that the days with smaller variations occur in the begin, middle and end of January 1976. (In Feb. 1976 the wavelike activity appears to have disappeared.) This suggests that the vertical periodic structure with about 50 km wavelength might be correlated with the above found time period of about 15 days.

If one assumes that the wavelike structure shown in Fig. 10 extends to higher altitudes, one would predict enhanced temperatures in the 80 km altitude regime (dashed line in Fig. 10). This pertains also to the other days with strong oscillations. Hence such temperature increase would have a period of the order of 15 days. It would add to the above mentioned long term and possibly also short term temperature increases in the same altitude regime. It will be shown below that the assumed extrapolation is valid at least for the two salvo days (see Fig. 16).

Series measurements of selected upper atmosphere parameters were performed by a number of ground stations. Fig. 11 shows the results obtained for the eastward component of the prevailing wind in the altitude regime around 90 km as measured by the Garchy meteor wind radar (Bernard et al. 1977). There is considerable variation in these D-region winds during the central part of the campaign. The A3 absorption data measured at Balerma (Spain) were Fourier-analyzed. A synthesis of the first 12 Fourier components is given in Fig. 11 for comparison with the winds measured above France. A striking similarity in the time variation of both atmospheric parameters is found, if a time shift of about 2 - 3 days is introduced. Obviously there is a common reason (disturbance) in the atmosphere which becomes active first in Spain and later in France. It is difficult to say what the exact direction of movement of this disturbance is. If one assumes that it is horizontally sufficiently extended to cover Spain and France, and moves essentially eastwards, a velocity of 2 - 3 m/sec is obtained. Similar velocities are obtained from the analysis of measurements taken by various A3 stations in western Europe.

Meteor wind data measured at Sheffield at three time intervals during the campaign are similar to those obtained at Garchy (Bernard et al. 1977). At Sheffield, however, not only zonal but also meridional wind components could be measured. They were found to be generally small (5 - 15 m/sec) and mostly directed northward.

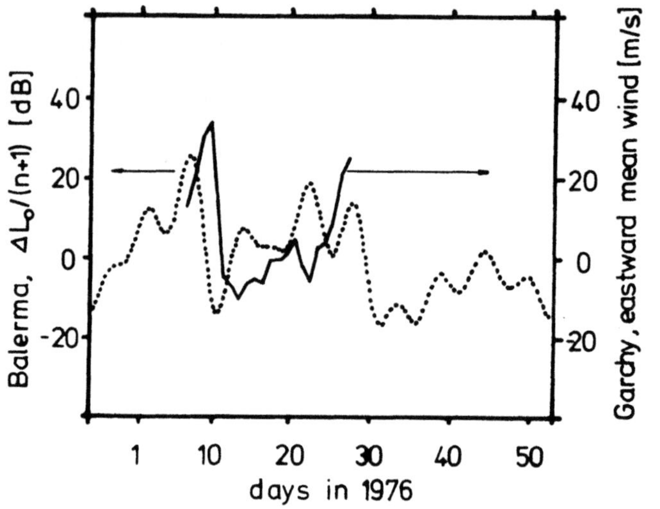

Fig. 11: Prevailing zonal winds above France (90 km) and A3 D-region absorption above Spain

Therefore horizontal transport of trace constituents like NO does not appear to be a primary reason for the winter anomaly activity during the campaign. The full wind analysis, however, is still under way.

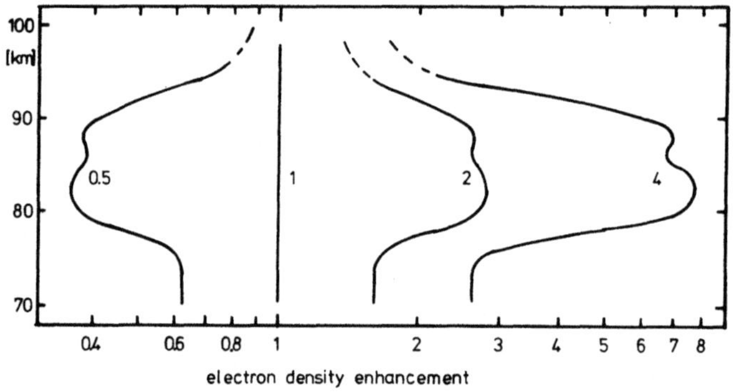

Fig. 12: Altitude distribution of relative electron density enhancement

Another series of measurements was taken by 15 small rockets, which determined - besides other parameters - electron densities in the altitude regime 70 - 95 km. From these profiles a "normal" electron density profile was determined (Friedrich et al. 1977). Days of high and low A3 absorption were then referred to this normal profile. Preliminary results for the relative electron density enhancements are shown in Fig. 12. The straight line labelled "1" represents the normal day. The labels of the other curves are a measure for the strength of A3 absorption. Low absorption days have labels smaller than 1, high absorption days labels larger than 1. The profile of the first salvo day (January 4, 1976) is near to the curve labelled "2", that of the second salvo (January 21, 1976) between the curves "2" and "4". This demonstrates that the two salvo days were not only characterized by high A3 absorption, but also by considerable electron density enhancement in the D-region. Fig. 12 furthermore shows that the electron density variations essentially occur in the altitude regime 75 - 95 km.

Rocket measurements other than of winds and electron densities were made at higher altitudes only on the two salvo days. On these days - besides many other experiments - mass spectrometers were flown which determined the composition of the neutral atmosphere. Fig. 13 shows densities of the major atmospheric constituents as they were obtained by a shockfreezing cryo-mass spectrometer (Bonn University, payload B II). As there are strong wavelike structures in the lower and middle atmosphere, it is not surprising to find similar variations in these mass spectrometer data, too. The deviations from the respective Cira 1972 curves are rather large in certain altitude regions. They are, however, supported by pitot tube measurements made on the same rockets (Becker et al. 1977).

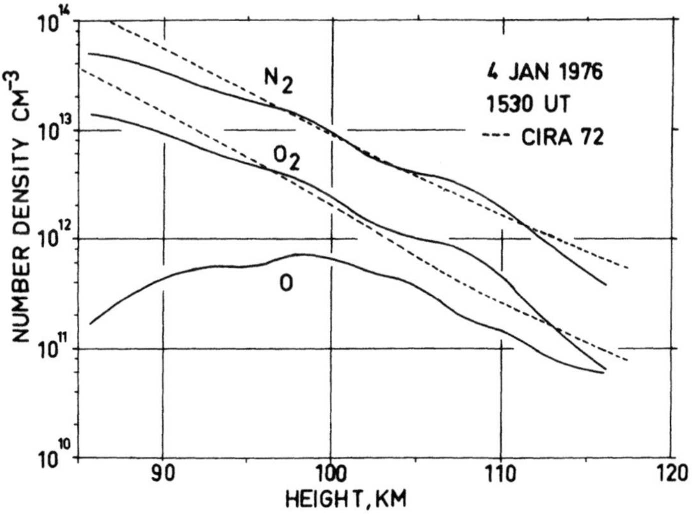

Fig. 13: Major atmospheric constituents as determined by mass spectrometer (Bonn University)

The results obtained on the second salvo day were similar. On this day, however, atomic oxygen was not measured, because it was below the detection limit of the spectrometer. The sensitivity of the peculiar cryo-mass spectrometer flown on that day was rather low. Nevertheless it would have measured atomic oxygen if the number densities had been larger than 1 - 2 x 10^{11} cm^{-3} around 100 km. Therefore it must be concluded that on January 21, 1976 the atomic oxygen density was lower by at least a factor of 5 than on January 4, 1976. A more specific result is obtained from two other mass spectrometers which were flown on different rockets (payloads B IV, cryopumped spectrometers of MPI Heidelberg, Krankowsky et al. 1977). They show (Fig. 14) that on January 21, 1976 the O density was lower by a factor of 5 - 10, depending on the altitude. The Bonn results are given for comparison in Fig. 14. The agreement of the data of the two different types of mass spectrometers is quite encouraging, though part of the data is still preliminary. There is an indication of this strong atomic oxygen variation already in the ground photometer measurements of the 5577 Å line (Fig. 3). Also more indirect hints are found, as a difference in suprathermal electron fluxes on the two days, which could be attributed to differences in atomic oxygen density (Spenner 1977).

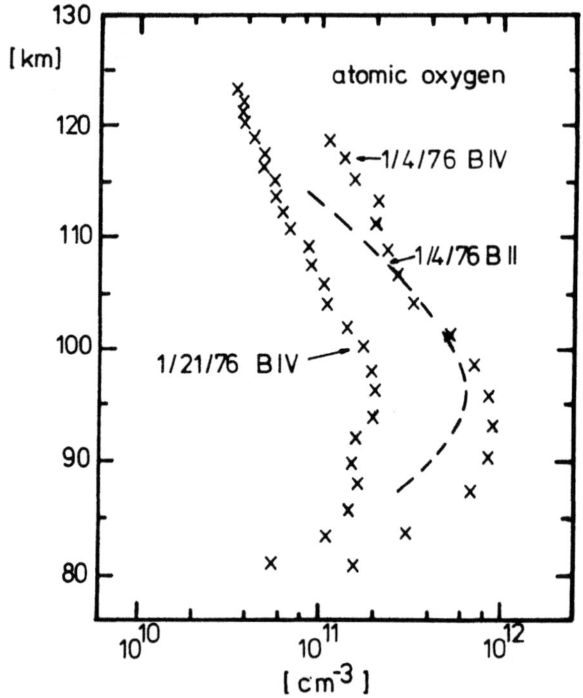

Fig. 14: Atomic oxygen densities measured by mass spectrometers (MPI Heidelberg). Dashed line is the O profile taken from Fig. 13 for comparison.

NO photometers were flown on the same rockets as the cryo-mass spectrometers (Beran et al. 1977). Preliminary results indicate that NO densities do not differ appreciably on the two salvo days, and that they are about a factor 2 - 3 higher than the profile of Meira (1971).

On the two salvo days there were temperature measurements in the whole altitude region 0 - 110 km. Fig. 15 shows the results obtained on Jan. 21, 1976. R denotes a radio balloon, B VII a meteorological rocket. B IV is a profile obtained from Ly-α measurement (see Becker et al. 1977). B II denotes a temperature profile obtained by integration of the total neutral density as measured by the Bonn cryo-mass spectrometer and the above mentioned pitot tubes (Becker et al. 1977). The model profile given is the respective Cira 1972 profile (40°N, January). Below 25 km the model profile was taken from the US Standard Atmosphere 1976. The agreement of the temperatures obtained by different methods and on different rockets is in general quite encouraging. There are, however, certain discrepancies. These may be due to experimental difficulties, but also to the

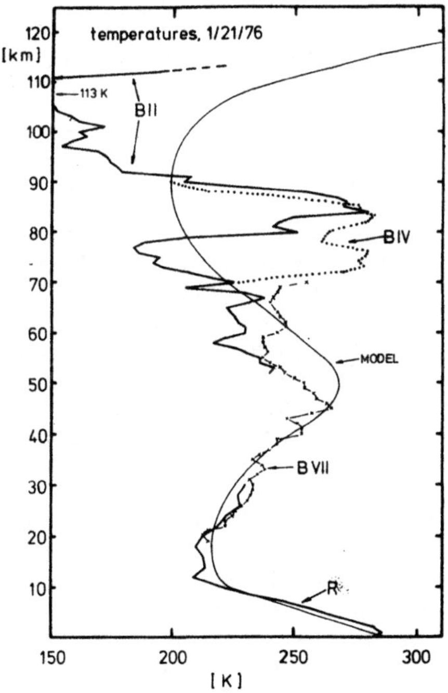

Fig. 15: Temperature profiles measured by different experiments on Jan. 21, 1976 (see text)

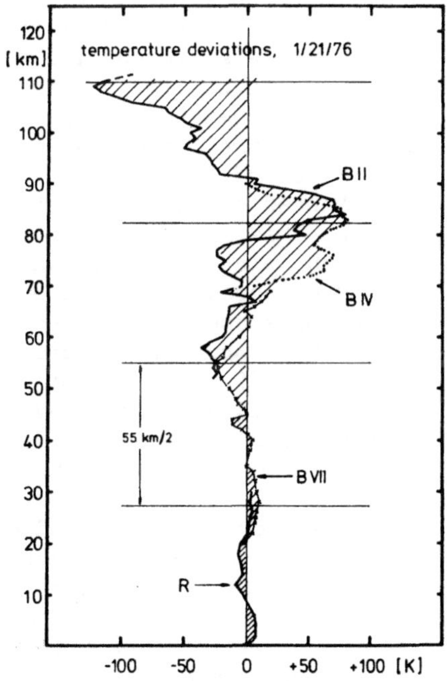

Fig. 16: Deviations of measured temperatures from model profile on Jan. 21, 1976

difference in horizontal location and in time of the measurements taken on different rockets (Offermann 1977). It is especially intriguing that the B II and B IV profiles show very similar structures, the details of which, however, do not fit together as long as one uses one common altitude and temperature scale.

Fig. 16 shows the deviations of the measured temperature profile from the model profile. It is clearly seen that a wavelike structure extends through the whole altitude regime. Above an altitude of 25 km it can be characterized by a wavelength of about 55 km. This is very close to what was deduced above from medium altitude temperature measurements. Fig. 16 shows also the expected node at 40 - 45 km. It shows furthermore a temperature maximum in the altitude regime around 80 km, as it was predicted from both the low and medium altitude temperatures. Very similar results are obtained for the other salvo day (Jan. 4, 1976).

The results of the temperature measurements at different altitudes may be summarized as follows: There were strong wavelike structures in the atmospheric temperature above Spain during the campaign, extending to above 100 km at least on two days. Medium altitude temperature measurements indicate, however, that at least part of this structure was a persistent feature during all of January, 1976, varying with a time period of the order of 15 days. Temperature oscillations were also found with shorter periods (6 - 8 days, 3.5 days) and much longer time scales (of the order of a month). Most of these variations seem to predict high temperatures in the 80 km region for those time periods when the winter anomaly activity was high. Comparison of Fig. 12 and Fig. 16 shows that the altitude regime of enhanced electron density coincides with that of increased temperature. This leads to the tentative suggestion that during the campaign the winter anomaly was caused (directly or indirectly) by temperature increases in the 80 - 90 regime. The so-called "normal" winter anomaly could have been caused by a superposition of the annual mesopause temperature increase and some irregular increase with a time scale of the order of a month. The "anomalous" part of winter anomaly could be attributed to an additional temperature oscillation with a period of about 15 days and some shorter periods (possibly higher harmonics) superimposed on it.

The above hypothesis does not tell the detailed physical reason for the enhancement of electron densities, yet. A number of different mechanisms, which were discussed in the literature, may be active at the same time. Of course temperature enhancements as shown in Fig. 16 will strongly influence the neutral and ion chemistry in the regime around 80 km, with for instance the consequence of ion cluster density being decreased and electron density being increased. At the same time, however, increased atmospheric turbulence may be expected due to the steep temperature gradients above 90 km As the temperature variations have rather long periods and their vertical structure resembles a standing wave, there might be sufficient time for build up of turbulence. As a consequence increased vertical transport of O and NO into the 80 - 90 km regime is expected. Density increases of these species in this regime would directly or indirectly produce enhanced electron densities.

It will be a major objective of the ongoing data analysis to determine the relative importance of these competing mechanisms. Contributions are expected especially from the evaluation of wind and turbulence measurements, and of the mass spectrometric and photometric determination of various atmospheric and ionospheric species.

ACKNOWLEDGEMENT

The winter anomaly campaign was conceived by U. von Zahn, Bonn. It was supported by various institutions, especially by the Bundesministerium für Forschung und Technologie, Bonn, through DFVLR/BPT, Köln, and by the Instituto Nacional de Tècnica Aerospacial (INTA), Madrid. Geophysical alert and forecast data were supplied by K. Labitzke, Freie Universität Berlin; P. Simon, Observatoire de Paris, Meudon; and H. Heckman and H. Sargent, Space Environment Center, NOAA, Boulder and are gratefully acknowledged.

REFERENCES

Becker, M., F. Arnold, J. Bäte, M. Friedrich, V. Friedrich, D. Krankowsky, A. Loidl, D. Offermann, D.G. Papanikas, H. Schwentek, and E. Thrane, 1977, to be published

Beran, D., W. Bangert, F. Arnold, and D. Krankowsky, 1977, to be published

Bernard, R., M. Massebeuf, H.G. Muller, S.P. Kingsley, G. Rose, H.U. Widdel, H. Schwentek, M. Friedrich, J.M. Satrustegui, and J.M. Cisneros, 1977, to be published

Dahl, A.F., 1977, Forschungsbericht des Bundesministeriums für Wissenschaft und Technologie, to be published

Friedrich, M., G. Rose, K. Spenner, K. Torkar, and H.U. Widdel, 1977, to be published

Gregory, J.B., 1965, J.Atmos.Sci., 22, 18

Krankowsky, D., F. Arnold, V. Friedrich, and D. Offermann, 1977, to be published

Labitzke, K., 1972, J.Atmos.Sci., 29, 1395

Meira, L.G., Jr., 1971, J.Geophys.Res., 76, 202

Offermann, D., 1977, submitted to Z. Geophys.

Offermann, D., J. Barnett, M. Becker, J.M. Cisneros, K. Petzoldt, G. Rose, and J. Satrustegui, 1977, to be published

Reid, G.C., 1970, J.Geophys.Res., 75, 2551

Rose, G., H.U. Widdel, and J.M. Cisneros, 1977, Tech.Rep.,

MPAE-W-46-77-52, Max-Planck-Institut für Aeronomie, Lindau, Germany
Schwentek, H., 1971, J.Atmos.Terr.Phys., 33, 1647.
Sechrist, C.F.,Jr., 1975, Rev.Geophys.Space Phys. 13, 894
Spenner, K., W. Ott, and H.J. Bradenstein, 1977, submitted to Z.Geophys.
Thrane, E.V., F. Arnold, W. Bangert, D. Beran, M. Friedrich, B. Grandal, O. Hagen, D. Krankowsky, A. Loidl, H. Schwentek, and F. Ugletveit, 1977, to be published

DISCUSSION

Thrane: Based on measurements of X-rays (Dr. Loidl MPI Lindau) Lyman-α, particles (NDRE) and NO (DFVLR), we have estimated the ion production versus height in the range 50-100 km. The dominating production sources are for January 21. 1976, X-rays of 38-41 Å (90-100 km), Lyman-α ionization of NO (65-90 km), energetic electrons 60-65 km. Thus particle precipitation is not responsible for the winter anomaly in the case.

Friedrich: The temperature data from the highest PMR-channel (80-90 km) substantiate the extrapolation above the temperature node at 42 km. Whereas the temperatures shown from the 35 km range were in anticorrelation with the A-3 absorption, the 80-90 km data are in very good agreement with the observed absorptions.

Schlegel: I have looked at some of the wind data obtained with the chaff-method on the 21. Jan. They showed a high wind shear in the 80-90 km range. From these wind shears and the high temperature gradients measured simultaneously in the same height range, a Richardson number can be calculated. This Richardson number is quite low indicating strong turbulence in this height region at that time. Thus the turbulence may have considerable influence on winter anomalous conditions.

PARTIAL REFLECTION STUDIES IN THE QUIET POLAR D-REGION WITH RESPECT TO MESOSPHERIC TURBULENCE PARAMETERS

K. Schlegel, A. Brekke* and A. Haug*

Max-Planck-Institut für Aeronomie,
Lindau, West Germany

1. INTRODUCTION

In the past years the partial reflection (PR) method has been used to derive electron density profiles and (in the case of complex amplitude detection) simultaneously electron neutral collision frequency profiles in the ionospheric D-region [1]. In the following it is shown that PR measurements are also very useful in determining dynamical properties of the neutral atmosphere in the height region between 60 and 90 km. The measurements which should be interpreted in this way, have been obtained with the PR-installation at Tromsø [2] between April and October 1975. From this time period only those days have been selected which are quiet with respect to magnetic and particle activity. Therefore, the results should be valid to a certain extent, also for the D-region and mesosphere of lower latitudes.

The main operating parameters of the Tromsø PR-sounder are summarized in Table 1.

* Address: University of Tromsø, Tromsø, Norway

B. Grandal and J. A. Holtet (eds.), Dynamical and Chemical Coupling, 253-259.

Table 1. The PR-installation at Tromsø

Location	69.58 N, 19.22 E
Magnetic dip	77.5^{o}
Transmitter	
Frequency	2.75 MHz
Pulse repetition rate	50 Hz
Pulse width	10-100 μs, normally 30 μs
Power (peak)	100 kW
Receiver	
Bandwidth	20, 50, 100 kHz
Minimum detectable signal	-125 dBm
Sampling rate (height resol.)	150 kHz (1 km)
Antenna, 16 crossed dipols	
Beamwidth	17°
One way gain	17 dB
Data processing and control	
HP 2100, 16 k, Tape Unit, Display, TTY, Line Printer	

2. TEMPORAL FLUCTUATIONS OF PR AMPLITUDES

The basis of the following analysis are the temporal fluctuations of the PR amplitudes. Figure 1 shows a typical example. The temporal fluctuations of the PR amplitude at 4 different height levels are displayed in this figure in a logarithmic scale. It is obvious that the amplitude fluctuations become more rapid with increasing height. A suitable quantity to describe the fluctuation period is the correlation time of the amplitude signal. It is defined here as the time lag $\tau_{1/2}$ for which the autocorrelation function

$$R(\tau) = \langle A(t) \cdot A(t+\tau) \rangle$$

has decreased to 1/2 of its value at zero time lag. (A(t) being the envelope of the PR amplitude).

This correlation time is shown in Figure 2 as a function of height for different days in spring and summer. It decreases generally with increasing heights and exhibits greater values in spring than in summer. Extremely slow fluctuations are measured at about 60 km in spring.

An explanation of these temporal fluctuations may be the following: The PR's originate from irregularities in the D-region electron density. One can interpret these irregularities as eddied caused by turbulence in the mesosphere and they may represent the upper part of

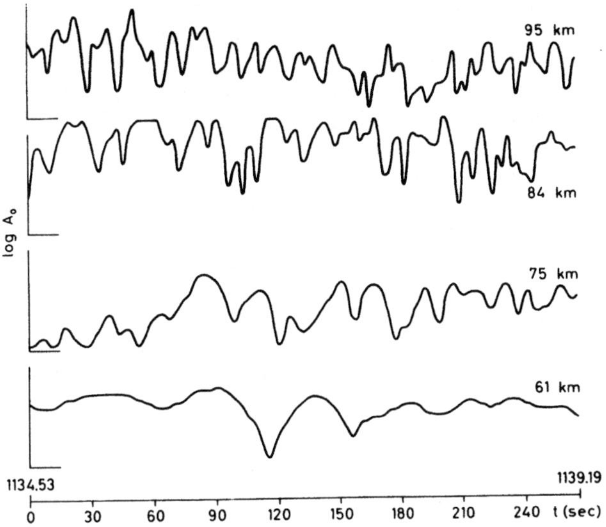

Fig. 1. Temporal fluctuations of the ordinary PR amplitude at 4 different height levels (29. April 1975, 1134.53 - 1139.19 UT).

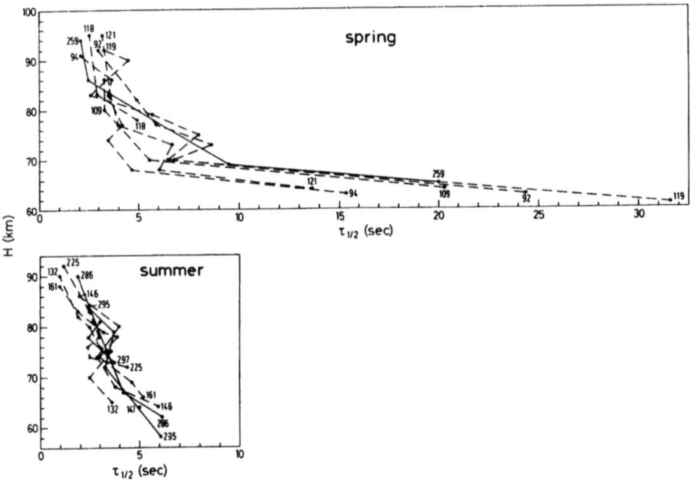

Fig. 2. Autocorrelation time profiles for different days in spring and summer. (The numbers at the endpoints of the curves indicate the number of the day in 1975).

the turbulence spectrum. The eddies move with different velocities causing a doppler shift of the partially reflected radio wave. Since there are several eddies in the scattering volume which is determined by the pulse width and the antenna beam width, the received PR signal exhibits a sum of different doppler shifts, corresponding to the different velocities of the eddies. The frequency spectrum of the PR signal thus has a finite width around the transmitted frequency. Since the width of the frequency spectrum and the correlation time are inversely proportional, a relation between the root mean square value of the eddy velocity and the correlation time $\tau_{1/2}$ can be established.

The full theory of this process has been developed by Booker, Ratcliffe and Shinn [3]. From their theory follows

$$v_o = \frac{\lambda_o \sqrt{\ln 2}}{4\pi \cdot \tau_{1/2}}$$

Here λ_o is the transmitter wavelength and v_o is the root mean square value of the random velocity of the eddies

$$v_o = \sqrt{\langle v_r^2 \rangle}$$

This quantity v_o is very useful in estimating turbulence parameters since it represents the so called characteristic velocity of the turbulence.

3. ESTIMATES OF TURBULENCE PARAMETERS

One of the most important parameters of turbulent flows is the energy dissipation by turbulence. It can be estimated from v_o by dimensional analysis which leads to

$$\varepsilon = \frac{v_o^3}{L}$$

(e.g. [4]). This expression is based on the assumption of isotropic turbulence being described by the Kolmogorov -5/3 law, where L is the characteristic scale of the turbulence. In this analysis L was set equal to λ_o, the wavelength of the partially reflected radio wave, as a first approximation. Calculating v_o from the $\tau_{1/2}$ values of Figure 2, the energy dissipation ε can be estimated. The results are shown in Figure 3.

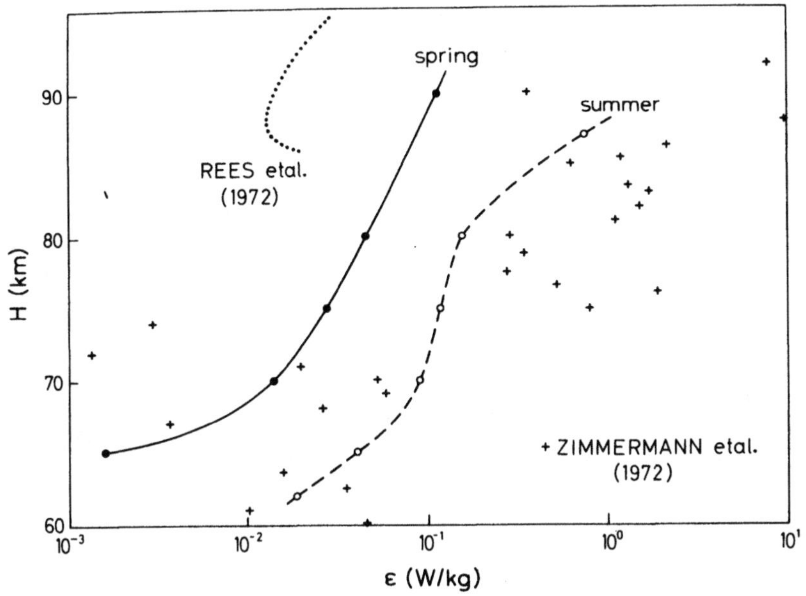

Fig. 3. Estimates of the turbulent energy dissipation ε as a function of height for spring and summer, respectively, and comparison with the results of other authors.

The ε values increase with increasing height and they are higher in summer than in spring. For comparison the results of two other authors are added in Figure 3. Zimmerman et al. [5] have estimated their values from in situ wind and temperature measurements at Ft. Churchill, and those from Rees et al. [6] have been obtained with the chemical release technique. Our ε values are in good agreement with those of Zimmerman et al. at heights below 75 km and they are well between the values of Rees et al. and Zimmerman et al. above this altitude. An explanation for the different energy dissipation in summer and in spring may be the following: It is well known that the mesospheric wind pattern changes its direction from W-E to E-W during spring times. The wind velocities are higher and therewith the driving forces of the turbulence are stronger, yielding a higher energy dissipation.

Besides the energy dissipation the coefficient of turbulent eddy diffusion can be estimated with the help of v_o. On the same theoretical basis as stated above [4] one obtains the expression

$$K_L = v_o \cdot L$$

For $L = \lambda_o$, the wavelength of the sounding wave, K_L represents the local diffusion coefficient due to eddies of dimension λ_o. Figure 4 shows the results for spring and summer, respectively. For comparison eddy diffusion coefficient estimates from other authors are displayed together with our results. Measurements are scarce in that height region, Roper [7] estimated his values from chemical release observations. The values of Johnson and Gottlieb [8] are based on a heat flux model and the curves labelled K(102) and K(112) are obtained from a transport model by Keneshea and Zimmerman [9], the number in the brackets indicating the height of the mesopause. Again our values obtained with the PR method are in satisfactory agreement with the other estimates.

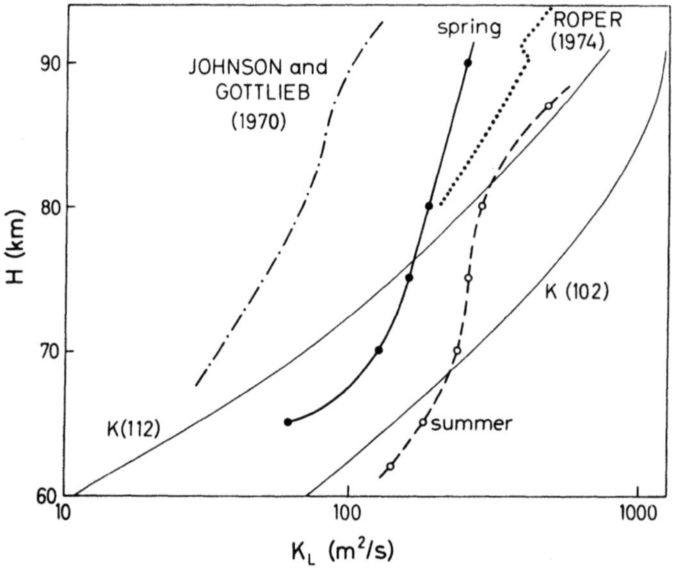

Fig. 4. Estimates of the eddy diffusion coefficient K_L as a function of height for spring and summer, respectively, and comparison with the results of other authors.

These examples for a new way of PR-data analysis show again that the PR method is a powerful tool for continuous remote sensing of the D-region and mesosphere.

REFERENCES

1. W. Dieminger and K. Schlegel, p. 164 in Manual on Ionospheric Absorption Measurements, ed. K. Rawer, World Data Center A, Boulder, 1976.
2. A. Haug, K. Bjørnå, A. Brekke, O. Holt and E.V. Thrane, Rept. No. 29-76, The Auroral Observatory, Tromsø, 1976.
3. H.G. Booker, J.A. Ratcliffe and D.H. Shinn, Phil. Trans. R. Soc. Lond. A242, 579, 1950.
4. J.E. Blamont, Planet. Space Sci. 10, 89, 1963.
5. S.P. Zimmerman, A.C. Faire and E.A. Murphy, p. 615 in Space Research XII, eds. S.A. Bowhill, J.D. Jaffe and M.J. Rycroft, Akademie Verlag, Berlin, 1972.
6. D. Rees, R.G. Roper, K.H. Lloyd and C.H. Low, Phil. Trans. R. Soc. Lond. A271, 631, 1972.
7. R.G. Roper, p. 642 in Proc. Int. Conf. on Structure, Composition and General Circulation of the Upper and Lower Atmospheres and Possible Anthropogenic Perturbations, (University of Melbourne, Australia), 1974.
8. F.S. Johnson and B. Gottlieb, Planet. Space Sci. 18, 1707, 1970.
9. T.J. Keneshea and S.P. Zimmerman, J. Atmos. Sci. 27, 831, 1970.

DISCUSSION

Thrane: One should realize that the lifetime of an electron against recombination is probably much longer than the lifetime of turbulent eddies, therefore the electrons can be used as tracers for atmospheric turbulence.

Reagan: How do you ascertain that electron precipitation which might result in large variations in N_e at altitudes of 75-100 km at high latitudes were not causing the variations you observed ?

Schlegel: We checked the riometer as well as the magnetometer to rule out particle and magnetic activity. In addition satellite data of proton and electron flux measurements were checked.

ODD NITROGEN IN THE MESOSPHERE AND THERMOSPHERE: ITS CHEMISTRY AND TRANSPORT*

D.W. Rusch

Department of Atmospheric and Oceanic Science, Space Physics Research Laboratory, University of Michigan, Ann Arbor, Michigan, USA

1. CHEMICAL PRODUCTION AND LOSS OF NITRIC OXIDE

Nitric oxide is produced in the thermosphere as an indirect result of an intricate series of ion recombination, ion-atom interchange, and atom-atom interchange reactions which initially result in the production of atomic nitrogen. Table 1 lists the relevant reactions and rate coefficients. The atomic nitrogen may be produced either in its ground (^4S) term or in the highly reactive metastable ^2D term. Nitric oxide is then produced in the atom-atom interchange reactions 1 and 2 of Table 1 (see e.g. Strobel et al. 1970; Norton and Barth 1970; Rusch 1973).

However, nitric oxide is chemically destroyed in its reaction with atomic nitrogen, reactions 3 and 4 of Table 1. Molecular nitrogen is reformed and the cycle is completed. The nitric oxide density, therefore, is not determined solely by the absolute production of atomic nitrogen but by the relative efficiency of the production of $N(^2D)$ with respect to $N(^4S)$ and the relative rates of reactions 1 through 4.

An important additional loss process for nitric oxide is predissociation in the δ-bands (Cieslik and Nicolet 1973), reaction 5. The responsible solar flux is shortward of 1908Å and the atmospheric penetration in this wavelength region is controlled largely by molecular oxygen absorption in the Schumann-Runge band system.

*This work has been supported by the National Aeronautics and Space Administration under Contract NAS5-23006.

B. Grandal and J. A. Holtet (eds.), Dynamical and Chemical Coupling, 261-266.

Table 1

Adopted Reactions and Rate Coefficients for Odd Nitrogen Chemistry

	Reaction	Rate*	Reference
1.	$N(^4S) + O_2 \rightarrow NO + O$	$2.4 \times 10^{-11} \exp(-3975/Tn)$	Wilson (1967)
2.	$N(^2D) + O_2 \rightarrow NO + O$	6×10^{-12}	Lin and Kaufman (1971)
3.	$N(^4S) + NO \rightarrow N_2 + O$	$1.5 \times 10^{-12} \sqrt{T}$	Phillips and Schiff (1962)
4.	$N(^2D) + NO \rightarrow N_2 + O$	7×10^{-11}	Black et al. (1969)
5.	$NO + h\nu \rightarrow N(^4S) + O$	$1.36 \times 10^{-5} sec^{-1}$	Cieslik and Nicolet (1973)
6.	$NO^+ + e \rightarrow N(^4S, ^2D) + O$	$5 \times 10^{-7} (T_e/300)^{-1.0}$ for $T_e > 1000K$	
		$4.2 \times 10^{-7} (T_e/300)^{-0.85}$ $T_e < 1000K$	
			Torr et al. (1976)
7.	$N_2^+ + O \rightarrow NO^+ + N(^2D, ^4S)$		
		$1.4 \times 10^{-10} (T/300)^{-0.44} [1 - 0.07(T/300)^{0.21}]$, $T < 1500K$	
		$5.2 \times 10^{-11} (T/300)^{0.2} [1 - 0.07(T/300)^{0.21}]$, $T > 1500K$	
			McFarland et al. (1974)
8.	$N_2^+ + e \rightarrow N(^4S) + N(^2D)$	$1.8 \times 10^{-7} (T/T_e)^{-0.4}$	Orsini et al. (1976)
9.	$e + N_2 \rightarrow N(^2D) + N(^4S) + e$		
10.	$N_2 + h\nu \rightarrow N(^4S) + N(^2D)$		Strobel et al. (1976)
11.	$N(^2D) \rightarrow N(^4S) + h\nu$	$1.06 \times 10^{-5} sec^{-1}$	Garstang (1956)
12.	$N(^2D) + O \rightarrow N(^4S) + O$	4.3×10^{-13}	Frederick and Rusch (1977)
13.	$N(^2D) + e \rightarrow N(^4S) + e$	$3.6 \times 10^{-10} (T_e/300)^{1/2}$	Frederick and Rusch (1977)

* Rates are in $cm^3 sec^{-1}$ unless otherwise noted

Although the major reactions governing the production and loss of nitric oxide were known for some time, (Strobel et al. 1970; Norton and Barth 1970) the detailed chemistry controlling the production and loss of atomic nitrogen atoms has only recently been uncovered (Rusch et al. 1975; Rusch et al. 1976; Frederick and Rusch 1977). The primary data in this analysis were the emission rate profiles of 5200Å photons (reaction 11) taken by the Visible Airglow Experiment (VAE) (Hays et al. 1973) on the Atmosphere Explorer (AE) satellite series.

Volume emission rate profiles derived from VAE measurements for three mid-latitude orbits are shown in Figure 1 (taken from Frederick and Rusch 1977). The volume emission rate for the brightest profile maximizes near 200 km at about 9 photons cm^{-3} sec^{-1}. $N(^2D)$ is produced by reactions 6 to 10 of Table 1 and lost through reaction 2 and reactions 11 to 13. Reaction 13, the quenching of $N(^2D)$ by atomic oxygen has been measured in the laboratory (Davenport et al. 1976). Their value of $1.8 \times 10^{-12} cm^3$ sec^{-1} has been found to be incompatible with the airglow observations of VAE which require a value for k_{12} of 4.3 x 10^{-13} cm^3sec^{-1} (Frederick and Rusch 1977).

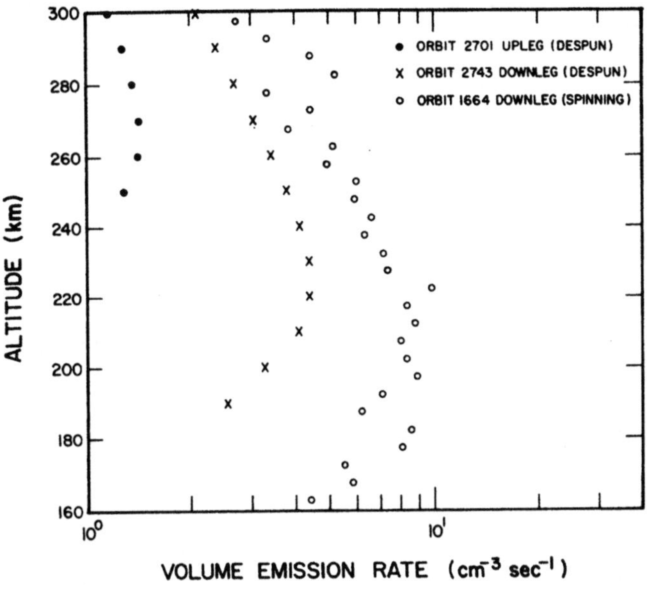

Fig. 1 Volume emission rates of 5200Å photons from AE-C data for solar zenith angles of 33° (000) 82° (xxx) and 95° (...).

The results of Rusch et al. (1975) and Frederick and Rusch (1977) may be summarized as follows: a) $N(^2D)$ is efficiently quenched by atomic oxygen; and, in the upper thermosphere, b) the reactions capable of producing $N(^2D)$ must do so with near unit efficiency, c) ground state atomic nitrogen is the most abundant odd nitrogen species; and d) the quenching of $N(^2D)$ by atomic oxygen is the largest source of $N(^4S)$.

3. THE EFFECT OF TEMPERATURE ON THE DISTRIBUTION OF ODD NITROGEN

An important variable not previously dealt with quantitatively is the temperature distribution below 150 km and its effect on the odd nitrogen chemistry. To illustrate this problem we have adopted two temperature profiles which differ by each other in their low altitude gradient. The profiles have equal temperatures at 120 km and T_∞ = 800°K. The profile with the smaller gradient is that of Jacchia (1971). The profile with the larger gradient is hotter than the first by 150°K at 130 km and is similar to profiles deduced by Donahue and Carignan (1976).

The calculated odd nitrogen distributions are shown in Figure

2 (taken from Rusch et al. 1976). The temperature profile with the larger low altitude gradient results in a factor of 3 increase in the nitric oxide densities and a corresponding low altitude decrease in $N(^4S)$ densities. In both cases the major source of NO is reaction 2. The major sink of NO is reaction 3. In the large temperature gradient case reaction 1 becomes competitive with reaction 3 as a sink for $N(^4S)$ and accounts for the increased NO concentrations.

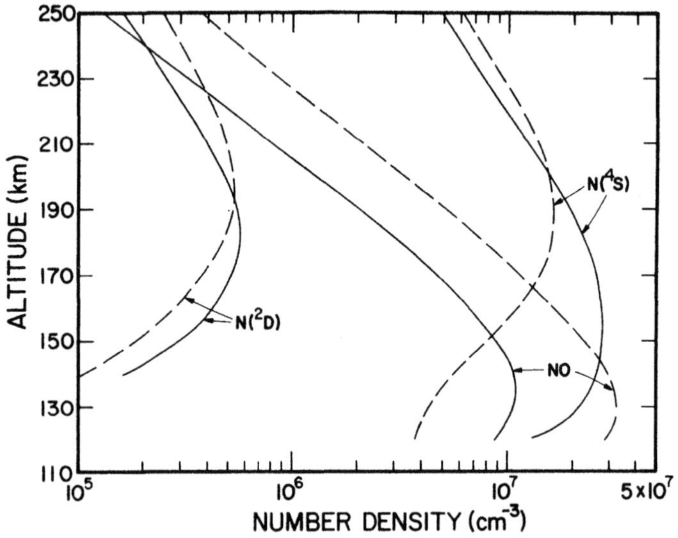

Fig. 2 The profiles of the calculated $N(^2D)$, $N(^4S)$ and NO densities. The dashed lines are for the high gradient temperature model and the solid lines are for the low gradient model.

4. THE LATITUDINAL DISTRIBUTION OF NITRIC OXIDE

Satellite measurements have shown that nitric oxide concentrations at high latitudes are highly variable in both time and space with average concentrations being 3 - 4 times higher than at mid-latitudes (Rusch and Barth 1975). Also increased nitric oxide densities are not confined to the latitudes of the auroral oval indicating efficient horizontal transport.

Recent theoretical calculations (Hyman et al. 1976; Gerard and Barth 1976; Roble and Reese 1977) have concluded that enhancements of 10 to 20 times the average mid-latitude concentrations have been generated during simulated auroral events.

In addition, recent satellite data from the ultraviolet nitric oxide experiment (UVNO) (Barth et al. 1973) on AE have shown latitude gradients of nearly a factor of 3 from equator to pole (Cravens and Stewart 1976). The UVNO data are shown in Figure 3 where zonally averaged counts are plotted against latitude for the NO density at 105 km (1000 counts implies 4 x 10^7 NO molecules cm^{-3}). The data are from an equinox period with spring in the southern hemisphere. The bars are a measure of nitric oxide variability. Calculations by Cravens et al. (1976) using a 1-D photo-chemical diffusion model for NO fail to predict the measured variation of NO with latitude. Their model did not include auroral sources of nitric oxide.

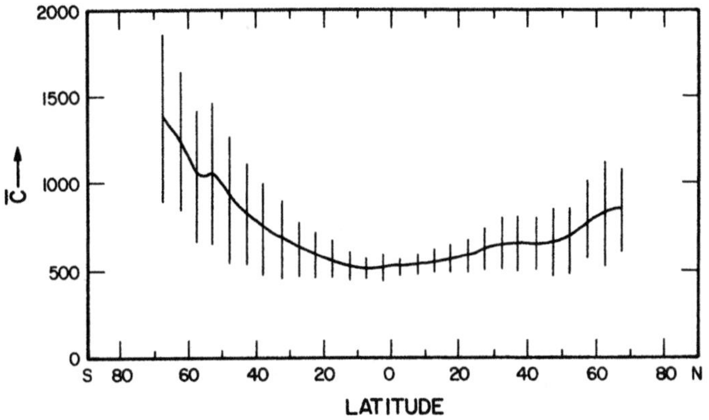

Fig. 3 The measured variation of nitric oxide density as a function of latitude. Calculations by Cravens et al. (1976) using a 1-D photo-chemical diffusion model for NO fail to predict the measured variation of NO with latitude. Their model did not include auroral sources of nitric oxide.

5. CONCLUSION

The nitric oxide density in the lower thermosphere and mesosphere is highly variable in both time and space. Future theoretical models should attempt to explain the fate of aurorally produced nitric oxide both as a source of mesospheric and possibly stratospheric NO and as a source of mid-latitude thermospheric NO.

REFERENCES

Barth, C.A., Rusch, D.W., and Stewart, A.I. 1973, Radio Science 8, 379.
Black, F., Slanger, T.G., St. John, G.A., and Young, R.A. 1968, J. Chem. Phys., 51, 116.
Cravens, T., and Stewart, A.I. 1977, J. Geophys. Res., 82.
Cravens, T., Stewart, A.I., and Gerard, J-C. 1977, J. Geophys. Res., 82.
Cieslik, S., and Nicolet, M. 1973, Planet. Space Sci., 21,925.

Davenport, J.E., Slanger, T.G., and Black, G.B. 1976, J. Geophys. Res., 81, 12.
Donahue, T.M., and Carignan, G.R. 1975, J. Geophys. Res., 80, 4565.
Frederick, J.E., and Rusch, D.W. 1977, J. Geophys. Res., 82.
Garstang, R.H. 1956, p. 324 in The Airglow and Aurorae (Eds. E.B. Armstrong and A. Dalgarno). Pergamon Press, New York.
Gerard, J-C, and Barth, C.A. 1977, J. Geophys. Res., 82, 674.
Hays, P.B., Carignan, G.R. Kennedy, B.C., Shepherd, G.G., and Walker, J.C.G. 1973, Radio Science, 8, 369.
Hyman, E., Strickland, D.J., Julienne, P.S., and Strobel, D.F. 1976, J. Geophys. Res., 81, 4765.
Jacchia, L.G. 1971, Spec. Rep. 332, Smithson. Astrophys. Observ., Cambridge, Mass.
Lin, C.-L., and Kaufman, F. 1971, J. Chem. Phys., 55, 3760.
McFarland, M., Albritton, D.L., Fehsenfeld, F.G., Ferguson, E.E., and Schmeltekopf, A.L. 1974, J. Geophys. Res., 79, 2925.
Norton, R.B., and Barth, C.A. 1970, J. Geophys. Res., 75, 3903.
Orsini, N., Torr, D.G., Hoffman, J.H., Hanson, W.B., and Walker, J.C.G. 1976, EOS, Trans. AGU, 57, 975.
Phillips, L.F., and Schiff, H.I. 1962, J. Chem. Phys., 36, 1509.
Roble, R.G., and Rees, M.H. 1977, Planet. Space Sci., 25.
Rusch, D.W. 1973, J. Geophys. Res., 78, 5676.
Rusch, D.W., and Barth, C.A. 1975, J. Geophys. Res., 3719.
Rusch, D.W., Roble, R.G., and Stewart, A.I. 1976, EOS, Trans. AGU, 57, 977.
Rusch, D.W., Stewart, A.I., Hays, P.B., and Hoffman, J.H. 1975, J. Geophys. Res., 81, 2300.
Rusch, D.W., Stewart, A.I., Hays, P.B., and Hoffman, J.H. 1976, J. Geophys. Res., 80, 295.
Strobel, D.F., Hunten, D.M., and McElroy, M.B. 1970, J. Geophys. Res., 75, 4307.
Strobel, D.F., Oran, E.S., and Feldman, P.D. 1976, J. Geophys. Res., 81, 3745.
Torr, D.G., Torr, M.R., Walker, J.C.G., Brace, L.H., Brinton, M. 1976, Geophys. Res. Lett., 3, 209.
Wilson, W.E. 1967, J. Chem. Phys., 46, 2017.

LOWER IONOSPHERE DISTURBANCES INDUCED BY HIGH ALTITUDE ION CLOUD RELEASES

T.B. Jones

Department of Physics, University of Leicester,
Leicester, England

The release of ion clouds at F-region heights produces marked local disturbances in the ambient thermosphere. In addition, it has been suggested that such releases can also disturb the lower ionosphere at E-region heights of about 100 km.

Shortly after release, the ion cloud polarizes and an electrostatic image is formed in the highly conducting E-layer below the cloud. Electron transport then occurs between the cloud and its image, in directions parallel to the geomagnetic field lines. The E-region electron density is enhanced at the point where the field line associated with the down-flowing electrons intersects the E-region and, similarly, a depletion is produced where there is an upward flow of electrons. A simple schematic representation of this situation is given in Fig. 1. Detailed calculations of the E-region disturbances produced during the coupling of the cloud and its image have been presented in the literature (Lloyd and Haerendel 1973; Perkins et al. 1973; Doles et al. 1976).

Attempts to observe cloud-induced E-region disturbances have usually been made by optical methods (Stoffregen 1970). The disturbances which have been reported from such observations appear to occur a few seconds after the release. This suggests that they may be produced by the precipitation of high energy electrons which can be triggered by the ion cloud. The formation of the electrostatic image type of disturbance takes considerably longer (~ 100 s) and no optical effects have been reported at these long delay times.

B. Grandal and J. A. Holtet (eds.), Dynamical and Chemical Coupling, 267-272. All Rights Reserved.

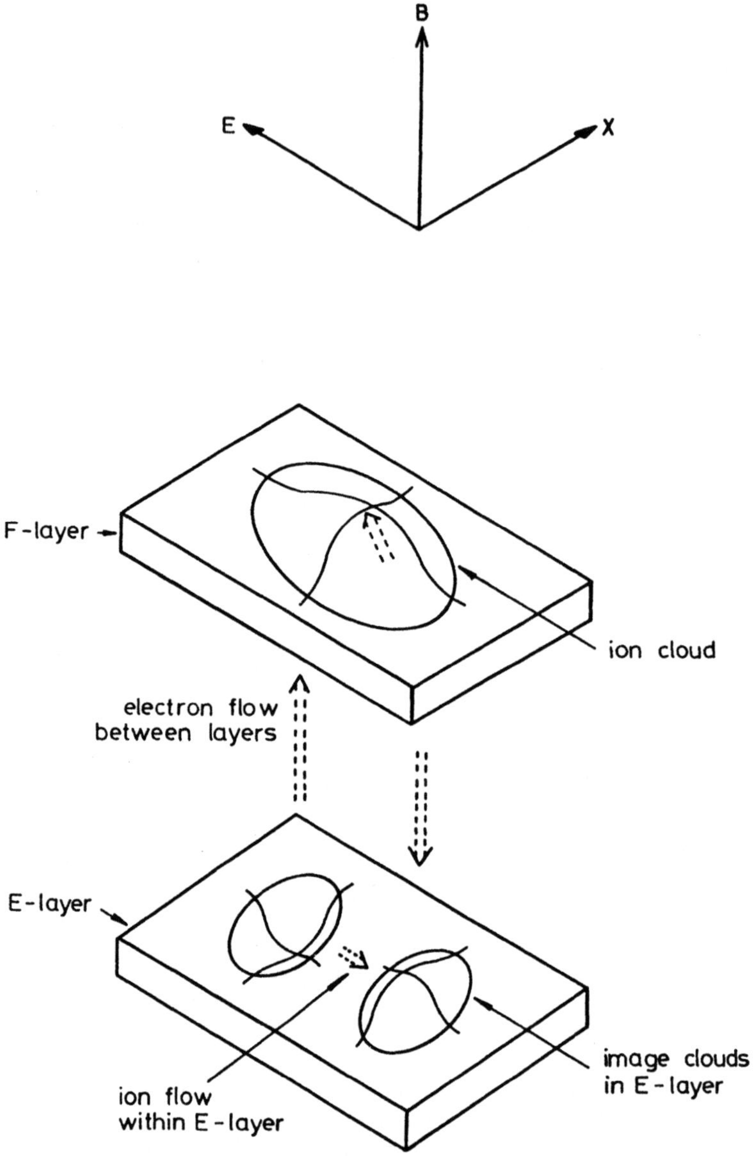

Figure 1: Schematic representation of the F-region cloud and its image in the E-region.

S21/3 TRIGGER 1977

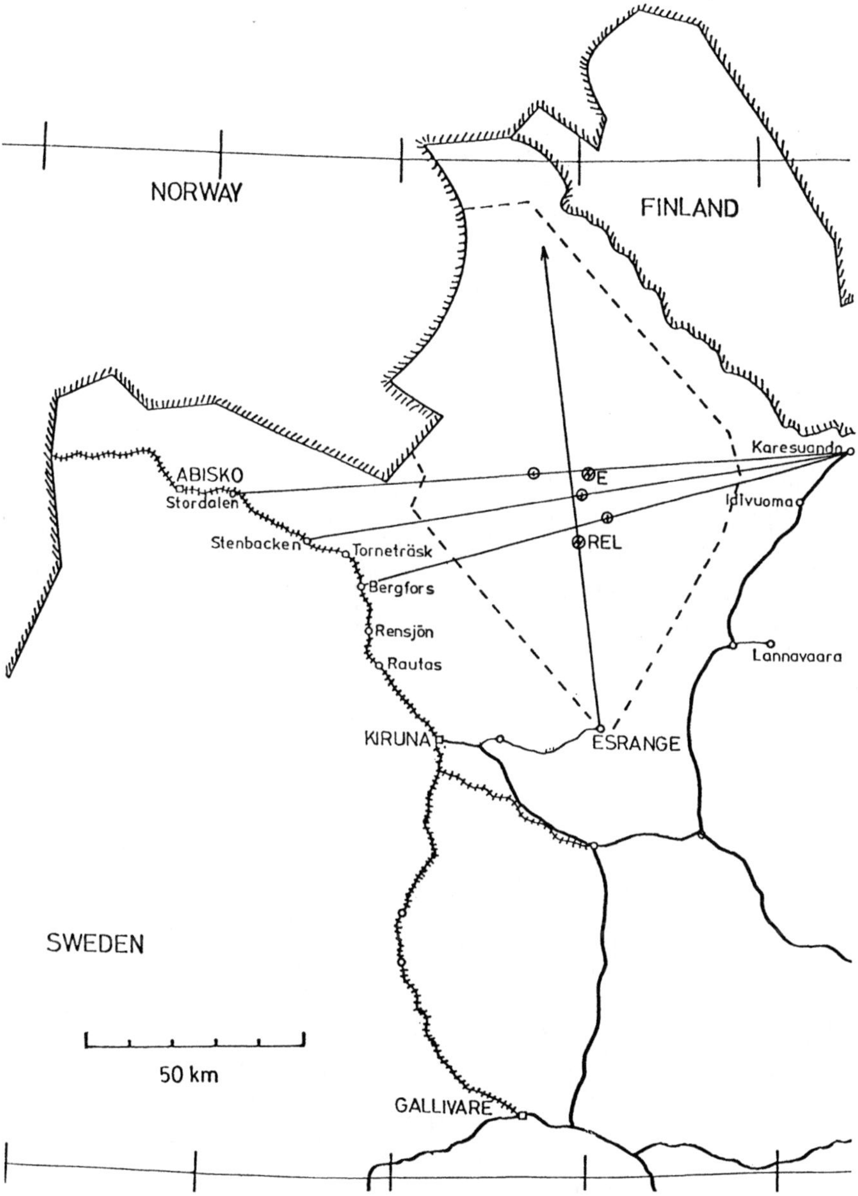

Figure 2: Locations of transmitting and receiving sites used at Esrange for radio wave phase experiments.

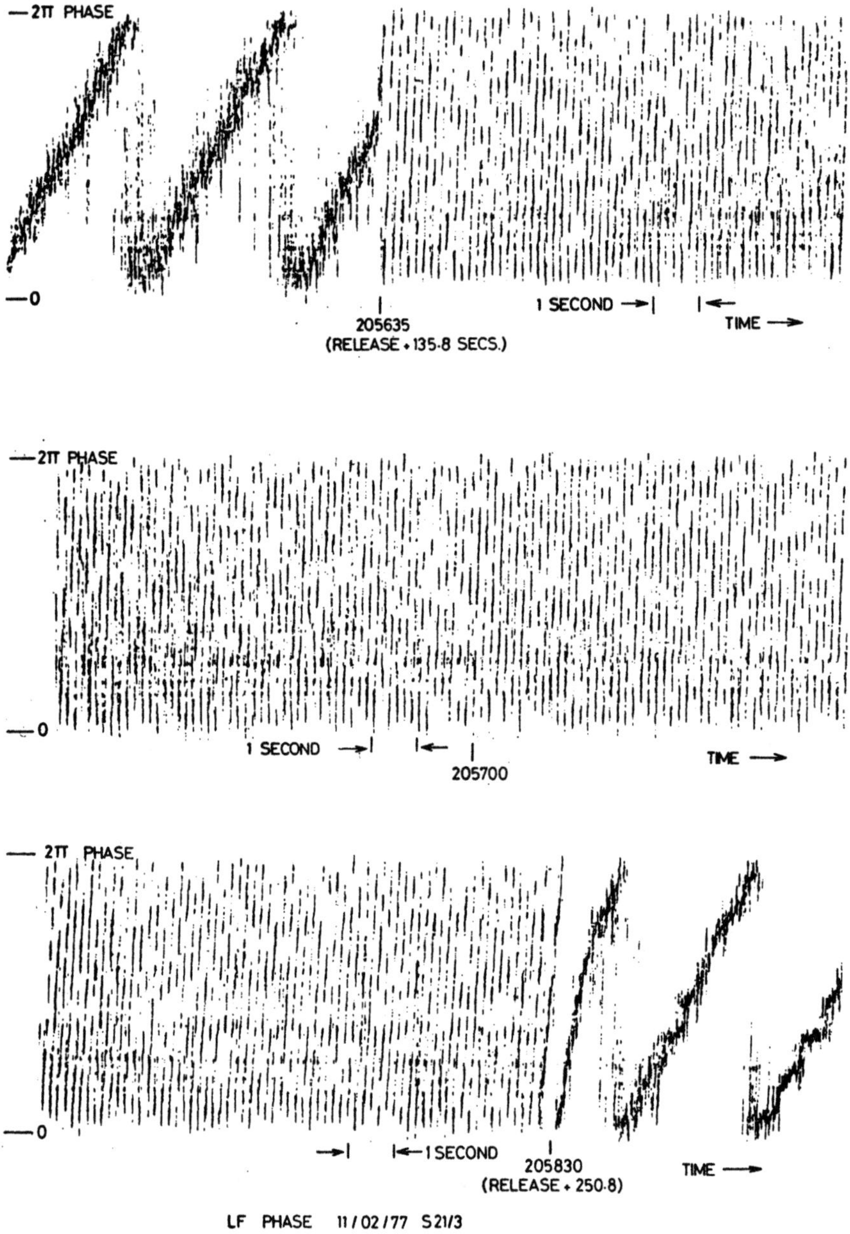

Figure 3: Phase disturbances recorded following cesium release 11-2-77 (1.486 MHz transmission).

An experiment is described in which a radio wave technique is used to search for the E-region disturbances (Jones and Spracklen 1976). The phase of a radio wave reflected from the disturbed region was monitored since this parameter is particularly sensitive to small changes in the electron density-height profile near the reflection point. Three transmitters radiating at 1.4779, 1.4809 and 1.4860 MHz respectively were deployed at the locations shown in Fig. 2. The receivers were located at Karasuando. Thus three monitoring points, spaced by about 10 km, are defined in the E-region. The position, at a height of 110 km, of the field line through the caesium cloud at release is also indicated. The ion cloud was released at 205419 UT at a height of 165 km. At 205635 UT a phase disturbance was noted on the Stenbacken to Karasuando (1.486 MHz) transmission and the recording is reproduced in Fig. 3. This was followed at 205655 UT by a marked amplitude increase which is clearly evident in the record reproduced in Fig. 4. No disturbances were noted on the other two propagation paths, 1.4779 and 1.4809 MHz.

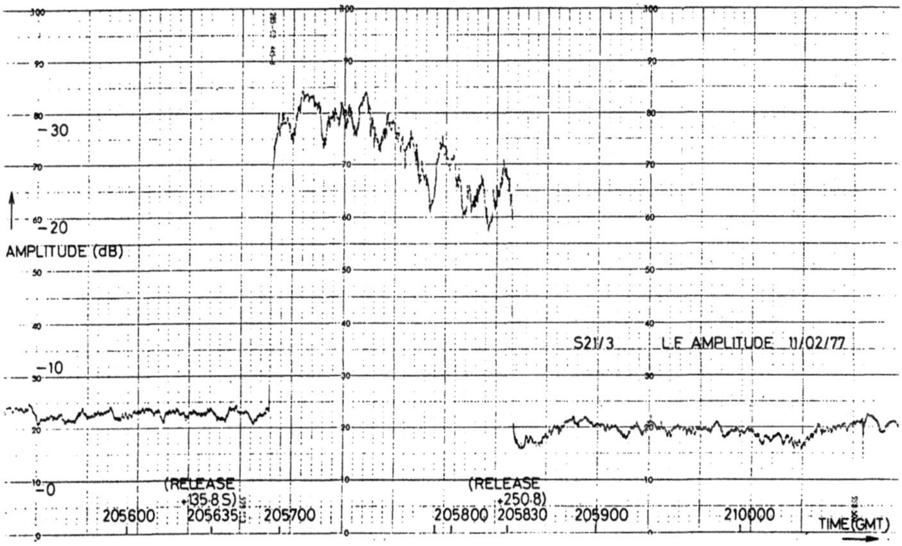

Figure 4: Amplitude increase following cesium release 11-2-77 (1.486 MHz transmission).

The time of onset and the duration of the event are comparable with those predicted theoretically for E-region image-induced disturbances. The initial form of the phase disturbance indicates that signals are being received over two propagation paths, i.e. a new reflection point has been produced in the E-region. The sudden amplitude increase some 20 sec later could be due to the steepening of the electron density gradients in the disturbance, thus producing a higher reflection coefficient which allows the second path signal to dominate that received over the original (undisturbed) propagation path.

CONCLUSIONS

From the results of the radio propagation experiment described the following features of the E-region disturbance produced by the ion cloud have been established.

1. There is a marked E-region disturbance some 130 s after cloud release.

2. The disturbance is very localised, probably less than 15 km in extent.

3. There are marked electron density gradient changes which produce a second reflection point, thus providing a second propagation path between the transmitter and the receiver.

4. The very fast phase changes could be interpreted in terms of a down-going plasma wave or a moving stream of electrons.

5. The delay time of the disturbance after cloud release and its duration are comparable to the corresponding times calculated by theoretical modelling of the cloud and its image.

ACKNOWLEDGEMENT

This work was undertaken at the invitation of the Ionospheric Institute at Uppsala, Sweden, and the assistance of Dr. G. Holmgren was greatly appreciated. The contributions of Dr. C.T. Spracklen and Mr. W. Smith are also acknowledged.

REFERENCES

Doles, J.H., Zabusky, N.J., and Perkins, F.W. 1976, J. Geophys. Res., 81, 5987.
Jones, T.B., and Spracklen, C.T. 1976, AGARD-CP-192, 16-1.
Perkins, F.W., Zabusky, N.J., and Doles, J.H. 1973, J. Geophys. Res., 78, 687.
Lloyd, K.H., and Haerendel, G. 1973, J. Geophys. Res., 78, 7389.
Stoffregen, W. 1970, J. Atmosph. Terr. Phys., 32, 171.

PART IV: AURORAL ZONE PHENOMENA

AURORAL STRUCTURE AND DYNAMICS

G.G. Shepherd

Centre for Research in Experimental Space Science,
York University, Toronto, Canada.

1. INTRODUCTION

The study of auroral structure has a long and honourable history (Egeland et al. 1973). Scientifically, it may be considered to have begun with Stormer's (1955) measurements of auroral height. This made it possible, for the first time, to deduce the location of the aurora from its visual position in the sky. Prior to that time it was not uncommon for observers to imagine that they saw auroral forms passing in front of nearby mountains, or illuminating clouds from below. Auroral dynamics dominates the impression on a visual observer but its scientific study still remains a challenge.

Auroral observations necessarily involve a 'probing' type of measurement. Ground-based all sky cameras and scanning photometers use optical probing to determine the horizontal distribution of aurora in the vicinity of the station. Auroral radars do the same, but with range resolution. With the first rocket launch into auroral forms, McIlwain (1960) probed the vertical distribution of the aurora. Many different instruments have been used as probes in this way. Since that time, separated ground-based photometers have been used to infer the vertical distribution of emission rate in the aurora.

More recently, optical measurements from satellites have added a new perspective (Anger et al. 1973; Pike and Whalen 1974). Viewing from this greater distance removes the limitations of a single station and of clouds, and provides "instantaneous" pictures of the auroral oval. Still more recently, optical photometers on the Atmospheric Explorer Satellite (Hays et al. 1973) have done a different type of probing, where the spacecraft passes right through the auroral form. The use of image reconstruction

B. Grandal and J. A. Holtet (eds.), Dynamical and Chemical Coupling, 275-290.

techniques will allow satellites to do both horizontal and vertical probing, revealing all of the available auroral information.

Particle measurements from both polar orbiting satellites and geostationary satellites can be usefully compared with the optical data. In this paper, a summary is given of our present stages of knowledge, derived from measurements of the type described.

2. GLOBAL AURORAL STRUCTURE

The pattern of OI 5577Å and N_2^+ 3914Å auroras in the auroral oval is well shown by Lui et al. (1975). Their work clearly demonstrates the continuity of the auroral oval, and reinforces the validity of the concept as put forward originally by Feldstein (1964). The ISIS-II results of Lui et al. (1975) show continuity in the sense of the presence of emission at all local times, for a photo taken during an "instant" of universal time. However, the character of the emission changes markedly around the oval, leaving us with the question of just what is that is continuous. Examination of their Figures 1a, 2a and 6a makes these changes in character apparent. Lui et al. (1975) suggest the continuous nature should be interpreted in "a topological manner only".

The dayside auroras are associated with the magnetospheric cleft (Heikkila et al. 1972). Dayside auroral images from the DMSP 8531 and 10533 satellites have been presented by Snyder and Akasofu (1976), exhibiting remarkable structures in the aurora near noon. During quiet times the evening auroral arc grows smoothly out of the equatorward boundary of the diffuse noon emission but during disturbed times a fan-like structure of arcs radiates from a "point" near noon. Some of these arcs enter deeply within the polar cap and become polar cap auroras; they tend to preserve their straightness for considerable distances, and then they turn into features parallel to the oval. These linear features are particularly pronounced on the morningside, but the same linearity is sometimes seen in the evening.

The nightside auroras have been identified with the plasma sheet (Winningham et al. 1975). One normally sees a separation into two distinct regions, the equatorward diffuse aurora and the poleward discrete aurora, as shown by Lui and Anger (1973), their Figures 2a and 6a. The equatorward boundary of the diffuse aurora may be shown to identify with the inner edge of the central plasma sheet (CPS) while the discrete auroras correspond with the boundary plasma sheet (BPS) (Winningham et al. 1975).

In the evening sector, the continuity of the oval has been demonstrated by Pike and Whalen (1974), who had earlier provided the term "continuous aurora", which for this region coincides with what has been called the diffuse aurora. Several features are evident in these pre-DMSP photographs. The aurora seen as "diffuse" from ISIS-II usually appears as multiple parallel arcs, with the equatorward one most intense, and sometimes the sole feature. The differences between the two arise partly from higher resolution on the Air Force satellites (3.2 vs 9.0 km), partly from lesser tropospheric scattering (albedo) of the near infrared emissions, and partly from the thresholding effects used in the data displays.

The morning sector exhibits perhaps the most variability. It is the home of widespread diffuse, patchy and pulsating auroras. The magnetospheric interpretation of evening and morning auroras is less clear than for the nightside and dayside auroras, because its three-dimensional aspects are less well developed. Eastman et al. (1976) have pointed out that auroral field lines in these regions must go out to the vicinity of the magnetospause, that is, into the boundary layer. We shall consider that this defines the "topology" of the oval in the interpretation that follows.

Inside the oval one finds polar cap auroras that tend to be sun aligned (Berkey et al. (1976). These must originate deep in the tail of the magnetosphere, or possibly on open field lines. The next section considers the spectral structure of the aurora, and this gives further insight into the differing characteristics around the oval. Then the structure of particle precipitation is examined, and following that is a discussion of auroral dynamics.

3. SPECTRAL STRUCTURE OF AURORA

Figure 1 shows a polar projection of 6300Å contours for a quiet period on December 14, 1971, obtained with the ISIS-II Red Line Photometer (RLP) (Shepherd et al. 1973). This is a polar projection onto a plane tangent to the earth's north geographic pole. The circular boundary is the equator and the radial line is the Greenwich meridian.° Latitude marks are shown at 10° intervals on it. The 6300A emission rates have been corrected to zenithal equivalent and contoured at levels shown in the upper right corner; the contour labels have the final zero deleted. The hatched slightly curved line is the foot of the field line through the spacecraft, traced down to the 250 km level, the same level as assumed for the emission. The hatch marks on this track are spacecraft rotations; they are numbered from pass turn-on. S_F denotes the position of the sun for the first spin and S_L for the last spin. Figure 2 shows an invariant latitude-longitude grid that can be overlaid on the geographic projection of Fig. 1 and other figures in this paper.

The equatorward boundary of the oval pattern is very smooth, very symmetrical and generally well defined. The nightside intensities are low, but quite distinct, at about 350R. On the dayside, the emission is considerably brighter; a large area is enclosed by the 500R contour, and the innermost contours are for 1400R. The morningside part of the dayside contours are much broader in latitude range than on the afternoon side of noon, where the contours tend to narrow and close off. On the morningside they stop in a more abrupt and well defined manner, although

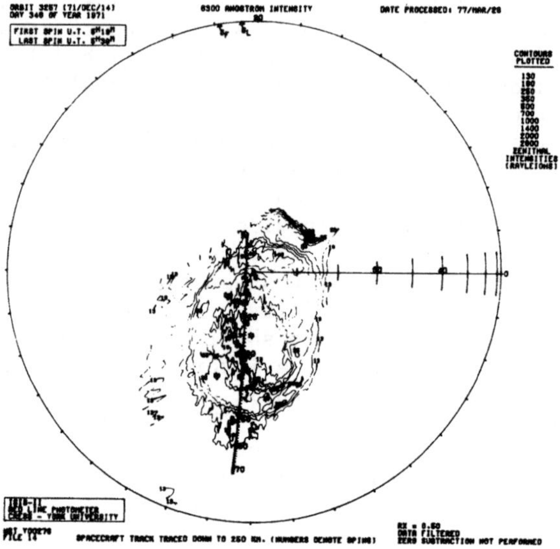

Fig. 1 Projection of 6300Å emission rate contours onto a plane tangent to the north geographic pole. Latitude marks are shown on the radial line which is the Greenwich meridian. ISIS-II pass date is Dec. 14, 1971, 05:19 U.T., a quiet period.

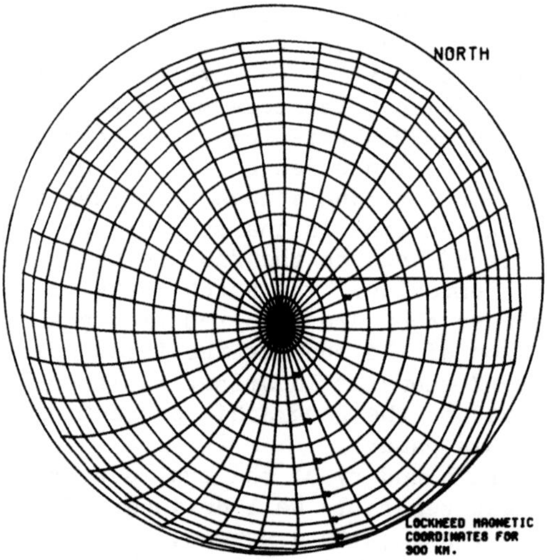

Fig. 2 Invariant coordinate overlay that can be placed on Fig. 1, and other ISIS-II figures in this paper. The outer circle is the geographic equator. Intervals are 5° in latitude, 20°-90° and 10° in longitude.

the intensity does not go to zero. A poleward oval boundary is also defined, though it is not as smooth or as symmetrical as the equatorward boundary. Inside the oval there are patchy regions that are sun aligned, of about 130R intensity. Examination of similar contours for the 5577A emission show a nightside emission of about 700R, confined to the region near spin 45; this is the diffuse aurora with I(6300/I(5577) ≈ 0.5. Going into the evening sector this ratio rises to unity and this is also so in the morning. On the dayside the 5577Å emission is lost in sunlight, for this orbit.

Figure 3 shows the dramatic change that occurred between these two successive orbits, owing to a substorm that originated near midnight. The nightside emission intensifies and the brightening extends as a smooth continuous form back towards the daytime hours. The intensity minimum between evening and noon has advanced from about 19 hrs. back to about 16 hrs. The dayside contours have narrowed a little but the intensity is just the same.

Figure 4 shows selected 6300Å contours from this pass, overlaid on the corresponding auroral scanning photometer photo. The intensity ratio changes around the oval are clearly seen. These ratios can give good indications of the energies of the precipitating electrons as described by Mende and Eather (1975). Figure 5 shows a 6300Å polar plot for a dawn-dusk Tromsϕ-Alaska pass that shows very clearly the asymmetry between dawn and dusk. The 5577Å contours follow the dusk 6300Å emission closely, but the 5577Å contours on the dawnside are displaced to lower latitude.

The region of hydrogen emission has been recently reanalysed by Montbriand (1976). The northern boundary of the hydrogen region is quite close to the diffuse auroral region at quiet times, while the equatorward boundary is in the region of the diffuse electron aurora for disturbed times. The optical emissions we are concerned with here are dominantly produced by electrons.

4. THE STRUCTURE OF PARTICLE PRECIPITATION

The pattern of particle precipitation has been investigated in great detail and many characteristics have become well known. See for example the results of Heikkila and Winningham (1971), Winningham and Heikkila (1974) and Winningham et al. (1975). Deehr et al. (1976) have compared ISIS-II particle observations with airborne photometric measurements and they note the response of I(6300/I(5577) to the electron characteristic energy.

By comparing the ISIS-II soft particle spectrometer data directly with the optical data measured on the same spacecraft we are able to relate the optical patterns to the particle features. The cartwheel configuration was used for the analysis and portions of each spin were selected so that the time delay between the direct observation of particles and the observation of emission at the foot of the field line was never more than

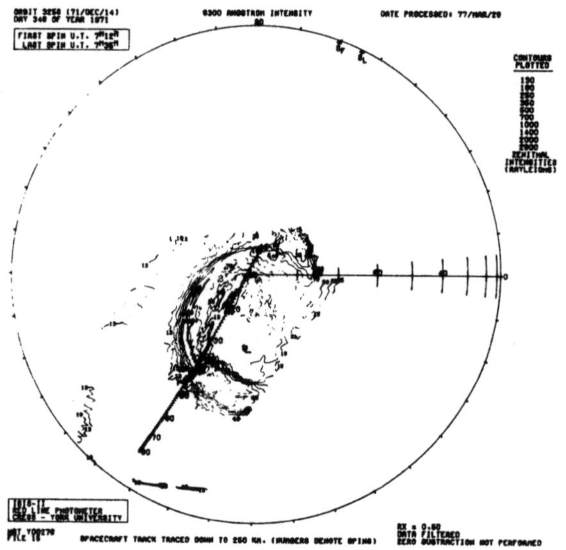

Fig. 3 Same as Fig. 1 but one orbit later, at 07:12, following a small substorm. The S marks near the top border the sun's location during the pass, and the curved hatched line is the spacecraft's ionospheric footprint.

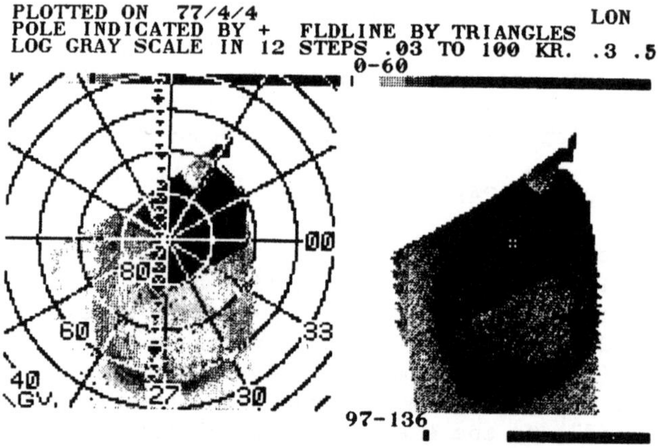

Fig. 4 An auroral scanning photometer photo for the same pass as Fig. 1, in identical format and size. The 6300Å contours may thus be directly overlaid, though contrary to the text, that is not done in the figure.

one-half spin period. Figure 6 shows the 6300Å auroral emission along the spacecraft track, as a function of time, using a logarithmic scale, along with the energy fluxes integrated over four energy bands: 5-60 eV, 60-300 eV, 300-1000 eV and 1-15 keV. Beginning on the right, at low invariant latitude and moving to the left one finds in the 300-1000 eV energy channel (the second from the top) at about 60.5° invariant some trapped or quasi-trapped fluxes. The pitch angle of the detector is plotted at the top and the downward pointing tooth of the sawtooth corresponds to downward going particles. There is clearly a deep loss cone corresponding to quasi-trapping that continues to about 64° where the loss cone suddenly disappears and the fluxes become isotropic. These isotropic fluxes rise smoothly to a maximum and fall, disappearing at about 68°. Extending poleward of this are highly structured fluxes corresponding to discrete aurora, that are absent in the 1-15 keV channel. This smooth broad region corresponds to the diffuse aurora, and the central plasma sheet. For the lower energy channels the pattern is similar but there is a clear shift to the right, to lower latitude. Thus the plasma sheet edges occur at larger radial distance for larger energies. For the three lowest energy channels one sees at the northern boundary a dramatic change to a highly structured type of precipitation. This is the discrete aurora, which is seen to have narrow 6300Å features that are as intense as in the diffuse region. This is consistent with the soft precipitation seen in the particle data. These data were taken close to local midnight.

The structure of the dayside cleft emission has been described by Shepherd et al. (1976).

5. VERTICAL AURORAL STRUCTURE

Vertical structures in aurora have been measured with rocket probes. In principle, the vertical distribution of luminosity is equivalent to a knowledge of the electron energy spectrum, which can be obtained by other means. But in fact there are a number of other important considerations that must be introduced in detailed analyses. This is illustrated in Fig. 7 obtained by Deans and Shepherd (1977), which is the overhead emission rate measured from a sounding rocket for the (0-0) 2POS N_2 band at 3371Å and the (1-10) VK N_2 band at 3425Å. The ratio between the two is clearly altitude dependent and this was interpreted here as quenching of the N_2 $A^3\Sigma_u^+$ state, largely by atomic oxygen. The atomic oxygen profile deduced from these and some other flight data is shown by Deans and Shepherd (1977) and gives concentrations of 1.7×10^{11} cm^{-3} at 100 km and 2×10^9 at 200 km. A certain fraction of these quenching collisions should result in energy transfer from the N_2A state to the O^1S state, leading to 5577Å emissions. Deans and Shepherd (1977) concluded that a substantial fraction of the 5577Å emission observed could have been produced in this way. Atmospheric temperature profiles can

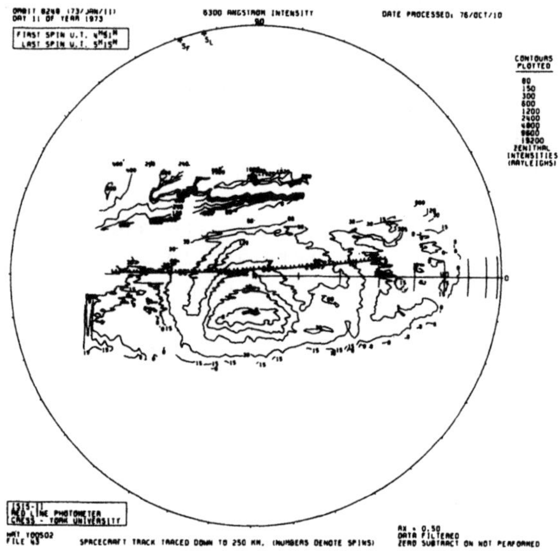

Fig. 5 Same as Fig. 1 but for Jan. 11, 1973, showing a clear asymmetry about noon in the cleft-associated 6300Å emission, which cannot be a limb viewing effect in this dawn-dusk orbit.

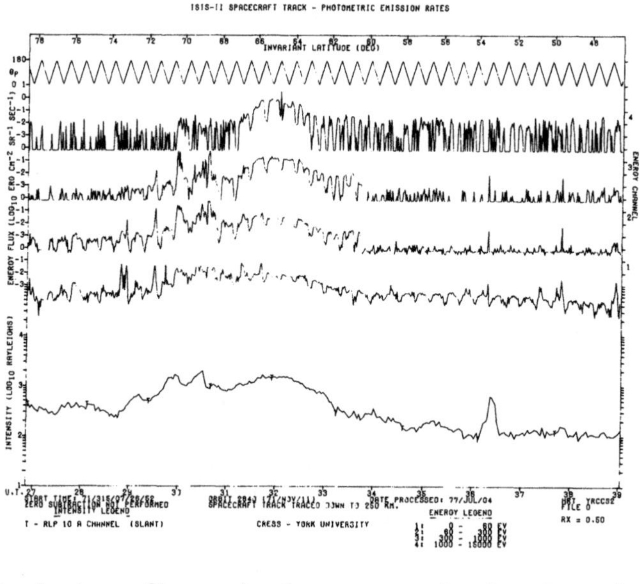

Fig. 6 Soft electron fluxes in four energy bands, from the top down 0-60 eV, 60-300 eV, 300-1000 eV, 1-15 keV, with the 6300Å emission rate plotted below. Pitch angle and invariant latitude are shown at the top.

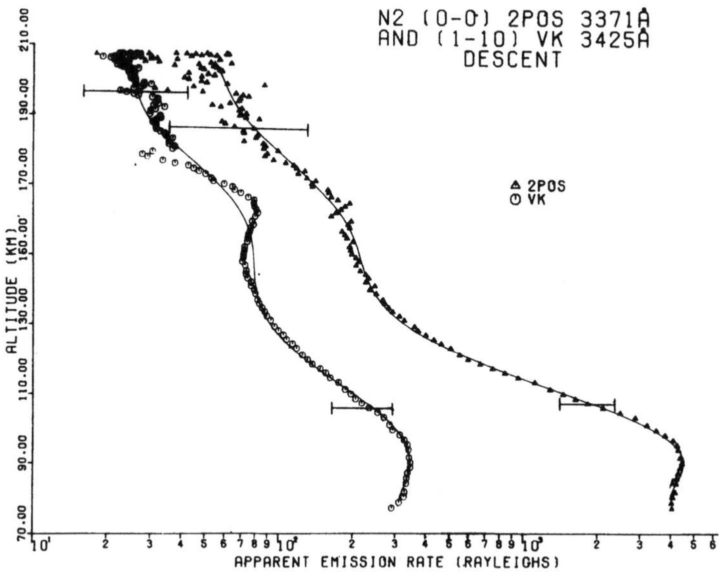

Fig. 7 The overhead (apparent) emission rate measured from a rocket launched at Fort Churchill, for the N_2 2POS and N_2 Vegard-Kaplan emissions. Note the change in ratio between the two (Deans and Shepherd, 1976).

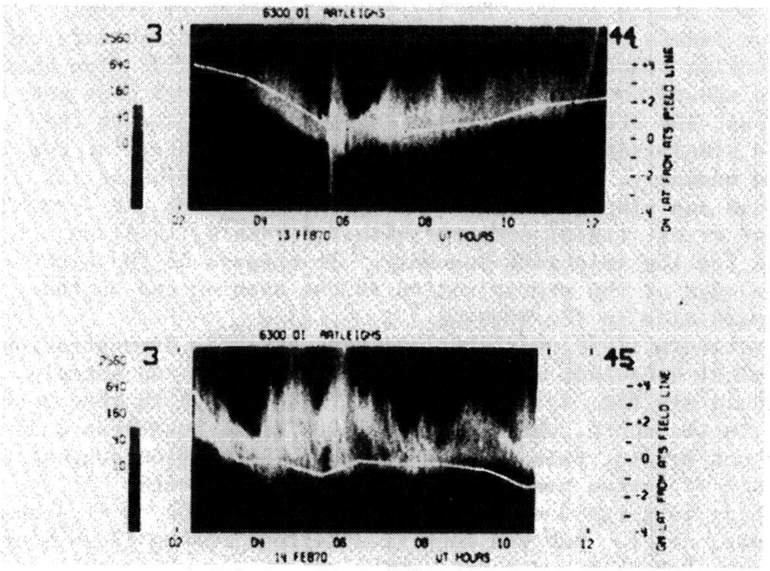

Fig. 8 Two "keograms" in the 6300A emission, obtained from Gillam, Manitoba, compared with the injection boundary calculated from McIlwain (1974), (Eather et al. 1976).

also be derived from auroral emissions.

Thus in general, to utilize the vertical emission profiles one must know the characteristics of the atmosphere and ionosphere involved, and understand the production mechanisms involved for these emissions. One might also expect that local electric field or other processes would alter the particle characteristics within the ionosphere in a way that would cause emission rate profiles to depart from those predicted from the particle characteristics measured at high altitude. Measurements of sufficient accuracy to detect such effects have not yet been made.

6. SUBSTORM DYNAMICS

Existing techniques all have their limitations, but in the study of substorm dynamics ground based optical scanners have much to offer. In a recent paper, Eather et al. (1976) present optical data obtained simultaneously with the ATS 5 particle data, located on the field line passing near the ground station at Gillam, Manitoba. The photoelectric instrumentation has a threshold for N_2^+ 4278Å and Hβ emission of 2R, so that subvisual auroras are included. Eather et al. find that essentially continuous aurora is detected, even at very quiet times, over a 2° latitude range at about 70° invariant. The location shows a diurnal variation and a response to magnetic activity. The equatorward boundary of this precipitation is identified with the inner edge of the plasma sheet. During moderate activity plasma injection events occur in a region having an inner boundary called the injection boundary (McIlwain 1974), which extends from about 19 hours around to 02 hours. The injection boundary does not extend past the prexisting plasma sheet. In weak events the enhanced precipitation resulting from injection occurs at the poleward boundary, but in more active cases it begins at the inner edge and propagates poleward. Figure 8 shows the "keograms" of Eather et al. (1976) compared with McIlwain's (1974) empirical equation for the injection boundary. It appears on the high latitude edge of the precipitation in the evening and on the equatorward side in the morning.

Creutzberg (1976) has obtained some striking demonstrations of dynamical behaviour in aurora, also at Gillam. An example is shown in his Fig. 4 in which the diffuse aurora is seen very clearly in the south, while discrete arcs drift equatorward down to, but not beyond, this diffuse aurora. This motion suggests the inward diffusion associated with injection events. At substorm onset there is a rapid northward movement; followed by a further equatorward drift. But the localized diffuse region is obliterated by the substorm, with emission covering the whole sky. After the substorm it begins to reform again. Creutzberg (unpublished data) has also obtained scanning photometer patterns of dayside aurora. Figure 9 shows his results obtained at Cape Parry, N.W.T. for a quiet period just following noon. In Fig. 9(a) one

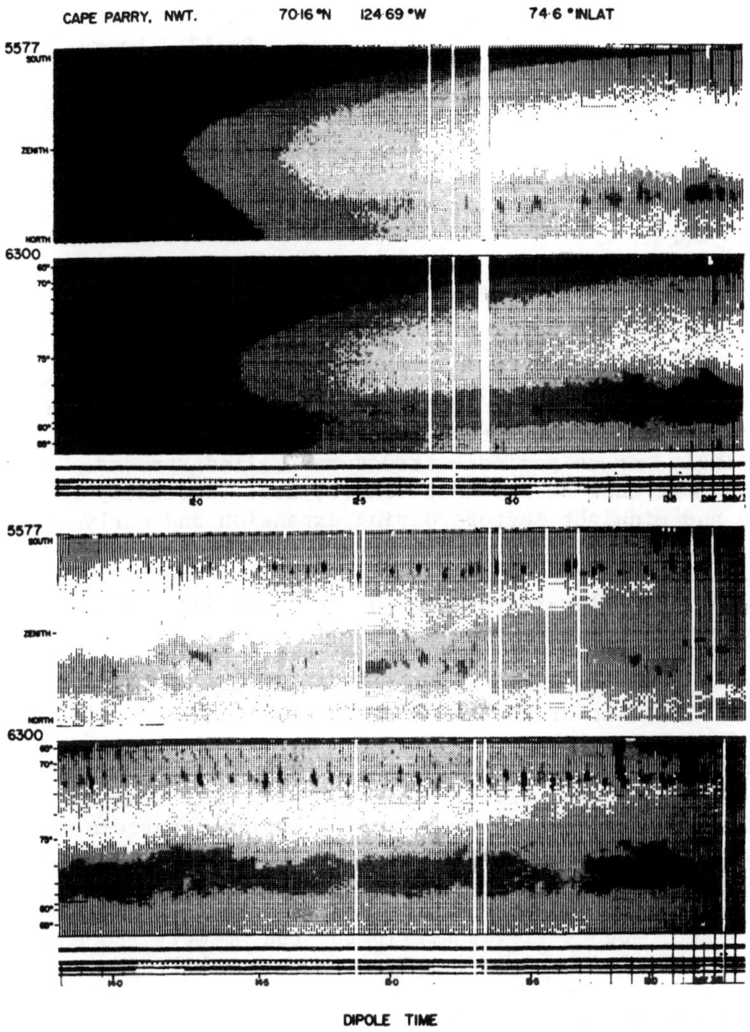

Fig. 9 (a) Scanning photometer patterns of the post-noon dayside cleft emission, obtained for the 5577A (above) and 6300Å (below) emissions from Cape Parry. The intense region at the left is twilight; the cleft emission is seen to the right.

Fig. 9 (b) Continuation of Fig. 9(a) in time. Invariant Latitude is shown at the left. The cleft emission which grows out of twilight at 12.5 hrs. is little changed at 16.2 hrs. The 5577Å emission shows larger dynamical changes.

sees the 6300Å emission emerging out of twilight at about 12.5 hrs. magnetic, at about 78.5° invariant. By 13.7 hours it has drifted very slowly down to 77.5° invariant. The continuation is shown in Fig. 9(b) where the location remains essentially constant from 13.7 to 16.2 hours. In the 5577Å trace one can see "spots" of emission appearing and disappearing so there is some dynamical behaviour. During active periods these become poleward drifting features. Vorobjev et al. (1975) have also noticed these drifting arcs.

7. MEDIUM SCALE STRUCTURE AND DYNAMICS

The small scale structure is what gives the visual aurora its distinctive character, but ideas about these processes are not well developed. Davis and Hallinan (1976) described auroral spirals, which have diameters about 20 and 1300 km. They can occur singly, or as periodic structures called streets. They occur on more than 30% of DMSP images, most frequently in the evening and midnight sectors during expansion and early recovery phases of substorms. Thus they develop in the same region in which plasma injection occurs. Under active conditions they occur near the poleward boundary of the expanded oval, possibly well poleward of the normal location.

A spiral can wind up (or unwind) three revolutions in roughly one minute, and apparent tangential velocities of 10 km/sec are involved. The authors believe the spirals to be associated with field aligned currents.

The dynamical behaviour of the morningside is characterized by pulsations. Oguti (1976) has shown data that demonstrate the recurrence of auroral patterns, which require a pattern memory in the magnetosphere. The period of recurrence is of the order of 2-5 sec and the on/off behaviour is restricted to domains which can be independent of nearby domains. The scattering time is less than 0.7 sec. Oguti (1976) attributes the "memory" to thermal plasma which initiates the ion cyclotron instability.

8. MAJOR GEOMAGNETIC STORMS

Major geomagnetic storms, as opposed to substorms, cause dramatic changes in the precipitation pattern that persist for several days. Examples of the 6300Å distribution during the storm of December 14-21, 1971 are shown in Figure 10, taken on December 18, just after the peak of the storm, four days following a solar outburst that generated a large proton flux in interplanetary space (Bunn and Shepherd, manuscript in preparation). One can see an enormous fraction of the polar cap is filled with 6300Å emission, and thus with soft particle precipitation. The dayside emission is particularly extensive, reaching nearly to the magnetic pole. The nightside region is similarly expanded so that only a

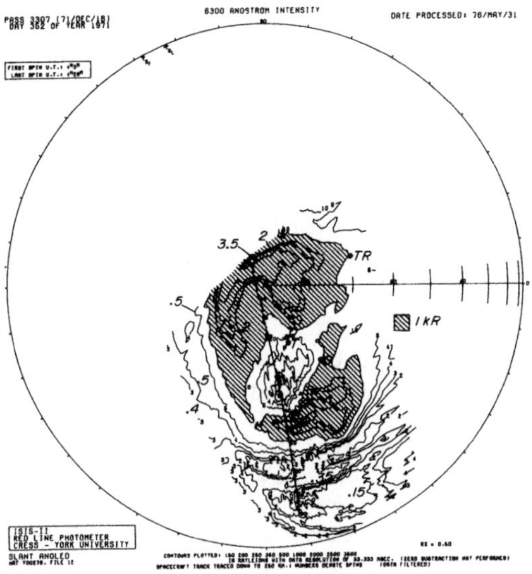

Fig. 10 A 6300Å polar plot (as in Fig. 1) for a major geomagnetic storm, Dec. 18, 1971, 04:03 U.T. The region of emission rate greater than 1 kR is shaded.

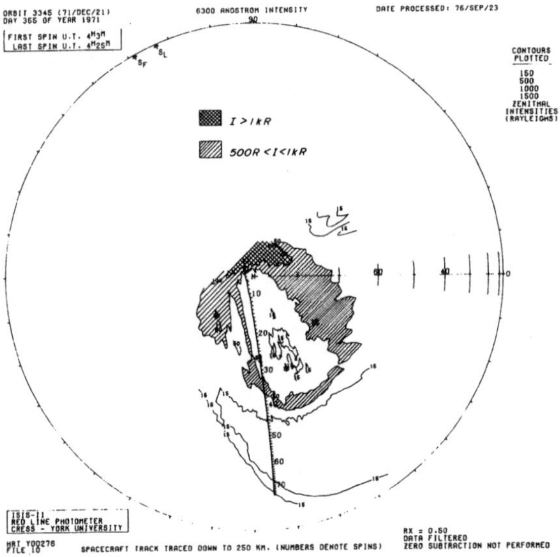

Fig. 11 A 6300Å polar plot late in recovery of the storm of Fig. 10, obtained Dec. 21, 1971, 04:03 U.T. Two regions are shaded, showing an intense dayside region and a sun aligned polar cap feature.

small portion of the polar cap is precipitation-free, just on the nightside of the magnetic pole. Note also a SAR arc at lower latitudes, near spin 65, marked .35 kR. Figure 11 was from an orbit three days later, showing a recovery pattern of unusual character. The day before, the polar cap was full of sun aligned patchy emission. Here a structure extends from dayside to nightside, right across the polar cap. Patchy emission is also evident in the middle of the polar cap. It seems that such storms cause highly unusual magnetospheric configurations, such as that proposed by Foster and Burrows (1976), for this and another event.

9. DISCUSSION

The global pattern of the auroral oval has a smooth and well defined equatorward boundary. The poleward edge is less well defined and sun aligned forms are formed within it. Within the oval one finds a diffuse auroral belt, comprising a smooth distribution of isotropic E > 1 keV precipitation that has drifted subsequent to injection. Poleward of this is a region of highly structured fresh E < 1 keV flux. These features are equatorward drifting arcs within the diffuse region. During major storms the precipitation pattern is dramatically changed.

10. ACKNOWLEDGEMENTS

Dr. F.W. Thirkettle created the software that produced all of the RLP data displays. Drs. J.D. Winningham, C.D. Anger and F. Creutzberg all generously supplied unpublished data. This research is supported by the National Research Council of Canada.

REFERENCES

Anger, C.D., Lui, A.T.Y. and Akasofu, S.-I. 1973, J. Geophys. Res., 78, 3020.

Berkey, F.T., Cogger, L.L., Ismail, S.,and Kamide, V. 1976, Geophys. Res. Lett., 3, 145.

Creutzberg, F. 1976, Magnetospheric Particles and Fields, Ed. B.M. McCormac, D. Reidel, Dordrecht, Netherlands.

Davis, T.N.,and Hallinan, T.J. 1976, J. Geophys. Res., 81, 3953.

Deans, A.J., and Shepherd, G.G. 1977, Planet Space Sci. (In press).

Deehr, C.S., Winningham, J.D., Yasukara, F., and Akasofu, S.-I. 1976, J. Geophys. Res., 81, 5527.

Eastman, T.E., Hones, E.W., Bame, S.J. and Asbridge, J.R. 1976, Geophys. Res. Lett., 3, 685.

Eather, R.H., Mende, S.B., and Judge, R.J.R. 1976, J. Geophys. Res., 81, 2805.

Egeland, A., Holter, Ø., and Omholt, A. 1973, Cosmical Geophysics, Universitetsforlaget, Oslo-Bergen-Tromsö.

Feldstein, Y.I. 1964, Tellus 16, 252.

Foster, J.C., and Burrows, J.R. 1976, J. Geophys. Res., 81, 6016.

Hays, P.B., Caringnan, G., Kennedy, B.C., Shepherd, G.G., and Walker, J.C.G. 1973, Radio Science, 8, 369.

Heikkila, W.J., and Winningham, J.D. 1971, J. Geophys. Res., 76, 227.

Heikkila, W.J., Winningham, J.D., Eather, R.H., and Akasofu, S.-I. 1972, J. Geophys. Res., 77, 4100.

Lui, A.T.Y., Anger, C.D., Venkatesan, D., Sawchuk, W., and Akasofu, S.-I. 1975, J. Geophys. Res., 80, 1795.

Lui, A.T.Y., and Anger, C.D. 1973, Planet. Space Sci., 21, 799.

McIlwain, C.E. 1960, J. Geophys. Res., 65 , 2727.

McIlwain, C.E. 1974, Magnetospheric Physics, Ed. B.M. McCormac, D. Reidel, Dordrecht, Netherlands.

Mende, S.B. and Eather, R.H. 1975, J. Geophys. Res., 80, 3211.

Montbriand, L.E. 1976, Canad. J. Phys., 54, 2310.

Oguti, T. 1976, J. Geophys. Res., 81, 1782.

Pike, C.P. and Whalen, J.A. 1974, J. Geophys. Res., 79, 985.

Shepherd, G.G., Fancott, T., McNally, J., and Kerr, H.S. 1973, Appl. Opt.,12, 1767.

Shepherd, G.G., Whitteker, J.H., Winningham, J.D., Hoffman, J.H., Maier, E.J., Brace, L.H., Burrows, J.R., and Cogger, L.L. 1976, J. Geophys. Res., 81, 6092.

Snyder, A.L., and Akasofu, S.-I. 1976, J. Geophys. Res., 81, 1799.

Størmer, C. 1955, The Polar Aurora, Clarendon Press, Oxford.

Vorobjev, V.G., Gustafsson, G., Starkov, G.V., Feldstein, Y.I., and Shevnina, N.F., Planet. Space Sci., 23, 269.

Winningham, J.D. and Heikkila, W.J. 1974, J. Geophys. Res., 79, 949.

Winningham, J.D., Yasuhara, F., Akasofu, S.-I., and Heikkila, W.J. 1975, J. Geophys. Res., 80, 3148.

DISCUSSION

Stadsnes: In the examples shown of ISIS 2 measurements of aurora and low energy electrons it appeared that the electrons causing the discrete aurora mainly have energies less than 1 keV. Rocket measurements often show peaks in the electron spectrum at a few keV in auroral arcs. How does this fit into your picture ?

Shepherd: The optical photometers average over a large area, about 60 km across, and so the high I(6300)/I(5577) ratio there suggests a widespread region of soft precipitation (< 1 keV) within which the hard precipitation you describe must exist as narrow features. I cannot explain why the hard precipitation is not apparent in the accompanying particle data, but I do know that on other occasions inverted V's are seen by this instrument in the soft precipitation region.

RECENT ADVANCES IN THE USE OF RADAR AURORAL BACKSCATTER TO MEASURE IONOSPHERIC ELECTRIC FIELDS

Raymond A. Greenwald

Max-Planck-Institut für Aeronomie
3411 Katlenburg-Lindau 3, F.R.G.

ABSTRACT. An improved understanding of the spatial and temporal structure of the ionospheric electric field is important for a better understanding of magnetospheric process as they relate to the auroral zone ionosphere and to ionospheric dynamics and chemistry. One as yet unexploited technique for measuring ionospheric electric fields is radar auroral backscatter. This technique remains unexploited because, at the present time, its applicability is still questionable. However, several recent studies have provided further evidence for its usefulness, either by explaining away some of the old arguments or by new comparative measurements with more accepted techniques. In this paper we briefly review the theory of the plasma instabilities that produce radar auroral irregularities, we compare the predictions of theory with spectral observations, and we survey the recent studies. Finally, we present some initial observations obtained with the new Scandinavian Twin Auroral Radar Experiment - STARE - that demonstrate the spatial and temporal structure of radar auroral irregularity motions. Presumeably, this structure is also present in the associated auroral zone electric fields.

INTRODUCTION

One of the most important physical quantities in the study of the auroral zone ionosphere and its relationship with the magnetosphere is the ionospheric electric field. If one assumes that the auroral zone magnetic field lines are electrically equipotential, then one

B. Grandal and J. A. Holtet (eds.), Dynamical and Chemical Coupling, 291-312.

can use the ionospheric electric field topology to study plasma convection patterns in the magnetosphere, particularity in the ring current and plasma sheet regions. By studying the variations of this topology throughout a magnetospheric substorm, it may be possible to obtain a better understanding of the changes that occur in magnetospheric convection during these events.

While the possibilities of studying magnetospheric convection in the ionosphere seem exciting, it has been argued that electrical potential drops can develop along magnetic field lines as a result of plasma instabilities [Kindel and Kennel, 1971; Block, 1972]. Under these circumstances ionospheric electric field measurements could not be used to measure magnetospheric plasma convection, however, through coordinated ground-based and satellite measurements, it may be possible to determine the times and locations at which these potential drops occur.

Ionospheric electric field measurements are also important for studies of ionospheric energy transport and chemistry. For example Joule heating is one of the principle means by which energy is deposited in the auroral zone ionospheric, and Schunk et al., [1975] have pointed out that appreciable Joule heating can modify aeronomic reactions and cause ionic composition changes in the ionosphere. It would be important to determine the spatial and temporal extent of these effects.

The standard techniques for measuring ionospheric electric fields have been electrostatic probes on rockets and balloons [Mozer and Bruston, 1967; Mozer and Serlin, 1969], barium releases [Haerendel, 1971] and incoherent scatter radars [Banks et al., 1973]. They are characterized in Table 1. One sees from the table that the rocket-borne techniques, probes and barium releases, are of the shortest duration; however, they provide the best temporal and spatial resolution. These techniques have produced most of our knowledge of the smaller scale spatial and temporal structure associated with auroral zone electric fields.

TABLE 1: <u>TECHNIQUES TO MEASURE AURORAL ZONE IONOSPHERIC ELECTRIC FIELDS</u>

Techniques	Duration of Measurements	Temporal Resolution (T)	Spatial Coverage	Spatial Resolution
Electrostatic probes (balloons)	several days	$\simeq$ 1 sec	several hundred km square	several hundred km square
Electrostatic probes (rockets)	$\doteq$ 10 min	$<$ 1 sec	path of rocket	velocity rocket /T
Barium releases	1-2 hours max	seconds	several km square	several km square
Incoherent scatter	hours to days	$\simeq$ 5 min	variable dependent on radar operating mode	$\simeq$ 50 km diameter circle
Radar auroral scatter (?)	continuous	1-60 sec	300,000km^2	$\triangleq$ 30 km square

Electric field studies over a longer period of time, particularly diurnal variations, have been provided by incoherent scatter radar observations. To date, the only operational, auroral zone, incoherent scatter radar is in Chatanika, Alaska; however, in the near future the European Incoherent Scatter Radar - EISCAT - will be operational near Tromsö, Norway. Radar observations of this type have inherently good spatial resolution ($\simeq$ 1 sq km), however, with the Alaskan system it is necessary to combine measurements in different directions in order to determine vector electric fields and the spatial resolution is degraded [Banks et al., 1973]. The EISCAT system will avoid this problem through the use of multiple receiving antennas. The temporal resolution of incoherent scatter studies is limited by poor signal-to-noise ratios. Most of the reported electric field data from the Chatanika radar has a temporal resolution of 6 minutes.

Electric field measurements of still longer duration have been obtained with balloon-borne electrostatic probes. Typical measurements last for periods of several days. While the temporal resolution of these measurements is quite good, o(1sec), balloon-borne probes have the disadvantage that they are only sensitive to electric field structure that is at least several hundred kilometers in extent. Thus, while balloon measurements are good for studying the diurnal dependence of ionspheric electric fields or their dependence on magnetospheric variations, they are not particularly useful for studying the finer scale structure of auroral zone electric fields.

The final entry in Table 1 is radar auroral backscatter. This techniqes is based on the assumption that the net drift velocity of radar auroral irregularities is determined by the electron drift velocity which in the E-region is given by $\underline{E}\times\underline{B}/B^2$. At the present time this assumption is not completely verified; however, recent observations have given it strong support [Ecklund et al., 1977]. Should this assumption be proven valid, radar auroral backscatter would become a powerful technique for measuring ionospheric electric fields. It can have a number of advantages including 1) continuous observation, 2) a large area of coverage, and 3) good spatial and temporal resolution.

In this paper we briefly review the theory of the plasma instabilities that produce radar auroral irregularities, and we compare the predictions of this theory with Doppler spectral observations of the radar signals backscattered from the irregularities. Although some radar auroral spectral observations have been given other interpretations, we will attempt to show that they are consistent with theory. Noteably, we feel that the irregularities have a net drift velocity equal to the electron drift velocity and that the velocity observed at any angle Θ between the radar wave vector and the electron drift direction varies as cos Θ. We then examine some recent measurements by Ecklund et al., [1977] that further support these predictions. Finally, we present several examples of the irregularity drift patterns that have recently been observed with the new Scandinavian Twin Auroral Radar Experiment - STARE -. The patterns were derived using the cos Θ assumption. The observed irregularity drifts are in the same direction as the electron drift in the auroral electrojets that, on the basis of ground-based magnetometers, were flowing at that time.

LINEAR THEORY OF RADAR AURORAL IRREGULARITIES

Radar auroral irregularities have been shown to be located in the auroral electrojet current [Greenwald et al., 1975; Siren et al., 1977]. These irregularities and their counterparts in the equatorial electrojet have been explained in terms of the cross-field two stream [Buneman, 1963; Farley, 1963] and gradient drift [Maeda et al., 1963; Reid, 1968] plasma instabilities. Both of the instabilities are produced by electrons drifting through ions in the plane perpendicular to the earth's magnetic field. The unstable waves propagate in the plane perpendicular to the magnetic field or at angles that are nearly perpendicular. This accounts for the strong aspect angle dependence of the scattering cross section. For a backscatter radar, the radar wave vector must be nearly perpendicular to the earth's magnetic field. Historically, these instabilities have generally been treated as separate entities; however, more recently, the instabilities have been combined into a single dispersion relation [Rogister and D'Angelo, 1970; Sudan et al., 1973] that more accurately expresses the nature of the irregularities.

This relation is written as

$$V_{ph} = \gamma_k/k = \underset{\sim}{k} \cdot \underset{\sim}{V}_d/k\ (1+\psi) \tag{1}$$

$$\gamma_k = (\psi/(1+\psi))\ \{\Omega_e \omega_k/\nu_e kL + (\omega_k^2 - k^2C_s^2)\ /\ \nu_i\} \tag{2}$$

$$\psi = \nu_i\ \nu_e\ /\ \Omega_i\ \Omega_e$$

where V_{ph} is the phase velocity of the kth mode; V_d is the electron drift velocity; ν_e, ν_i, Ω_e, and Ω_i are respectively the electron and ion collision frequencies with neutrals and the gyrofrequencies; L is the electron density scale length in the direction of the ambient electric field; and γ_k is the growth rate of the kth mode. In the auroral E-region $\psi << 1$.

If the electron density gradient term is negligible, (2) reduces to the two-stream growth condition, $V_d > C_s$, whereas, if $V_d < C_s$ and an appropriately directed electron density gradient exists, then (2) is still gradient-drift unstable for sufficiently long wavelength (small k) irregularities. It is important to note, however, that in both cases the predicted phase velocity of the unstable waves is equal to the component of the electron drift velocity in the direction of propagation.

One other point should be noted from the linear analysis. For a given electron drift velocity, assumed greater than the ion acoustic velocity, and a given electron density scale length, irregularities of a given k can only be generated over a certain range of angles about the electron drift direction. For larger angles, the instability is damped. In particular, one should never observe irregularities perpendicular to the electron drift.

COMPARISON OF THEORY AND EQUATORIAL IRREGULARITY OBSERVATIONS

At this point it is worthwhile comparing the linear instability with radar observations of irregularities in the equatorial electrojet. The comparison is meaningful since most instability theories have been developed to explain the equatorial observations and most auroral studies have been analyzed in terms of the equatorial results. In particular, many auroral radar researchers have tried to relate radar auroral spectra with specific types of equatorial spectra [e.g. Abel and Newell, 1969; Tsunoda, 1975].

The equatorial irregularity spectral observations are summarized in Table II. Type 1 spectra have been associated with the two-stream instability while type 2 spectra are related with the gradient drift instability. In this table we have noted which of the various spectral features are explained by present linear and non-linear plasma theories. For the present discussion we have not compared the widths of the two types of spectra with theory, one sees that the linear theory described in the previous section explains several of the spectral features; however, it does not explain the presence of type 1 or 2 irregularities propagating at large angles to the electron drift, the presence of short wavelength type 2 irregularities, or why the type 1 spectral peak is only observed at the ion acoustic velocity. With a non-linear analysis, such as presented by Sudan et al., [1973] several more features can be explained, but it is still not possible to explain why the type 1 spectral peak is only observed at the acoustic velocity. One can only propose that the electrons must also be drifting at the treshold velocity for the instability [Rogister, 1973]. This velocity is approximately equal to V_d when $\psi \ll 1$.

TABLE II: CHARACTERISTICS OF EQUATORIAL ELECTROJET SPECTRA

Spectral Characteristic	Explained by: Linear Theory	Explained by: Non-Linear Theory
Type 1 spectra		
Not observed when $V_d \ll C_s$	X	X
When observed, type 1 spectra are see at nearly all angles relative to the electrojet current direction		X
Spectral peak occurs at or very near the ion acoustic velocity		
Spectral width narrow relative to Doppler shift	-	-
Wavelength of the irregularities	X	X
Type 2 spectra		
Wavelength of the irregularities		X
Observed when $V_d < C_s$	X	X
Type 2 spectra seen at nearly all angles relative to electrojet current direction		X
Doppler shift varies as the cosine of the angle between the current direction and the radar wave vector	X	X
Spectral width broader than that of type 1 spectra	-	-

COMPARISON OF RADAR AURORAL SPECTRA AND THEORY

In Table III the spectral characteristics of radar auroral irregularities have been summarized. Since radar auroral irregularities have been observed over a wider frequency range than equatorial irregularities, it is more difficult to obtain a consensus of observations; however, the characteristics given are more or less generally accepted. The important points to note are the large mean Doppler velocities that have been observed in radar auroral spectra and the curious angular variation of the mean Doppler shift.

Over the past 12 years many observations have been made of radar auroral backscatter spectra, and mean Doppler velocities in the range from 500 to 1000 m/sec have often been observed [e.g. Leadabrand et al., 1965; Ecklund et al., 1973]. Greenwald and Ecklund [1975] have even reported mean Doppler velocities approaching 2000 m/sec. These Doppler velocities are well above any exected acoustic velocity, and they seem to indicate that the radar auroral two-streamlike spectra do not have Doppler shifts that are limited to the acoustic velocity.

TABLE III : CHARACTERISTICS OF RADAR AURORA SPECTRA

1. $V_d \simeq o(C_s)$

2. Irregularities observed at nearly all angles relative to the current direction.

3. Near 90° to the current direction the irregularity spectra are broad.

4. Near 0° and 180° to the current direction the irregularity spectra are narrow.

5. Mean Doppler velocities well in excess of the expected ion acoustic velocity have been observed.

6. Doppler shift varies with angle between radar wave vector and current direction. Near 90° the variation is more rapid than sinusoidal and away from 90° the variation tends to flatten.

One example of the azimuth variation of the mean Doppler velocity, after Tsunoda [1976] is shown in Figure 1. It can be seen that the mean Doppler velocity changes quite rapidly for those angles that are approximately normal to the mean irregularity drift direction and remains approximately constant in those directions that are more aligned with the irregularity flow. The standard interpretation has been that type 2 irregularities are present in the region with changing Doppler velocities, while type 1 or two-stream irregularities dominate in the approximately constant velocity region. Large Doppler velocities in excess of the expected acoustic velocity have generally been interpreted as due to neutral winds.

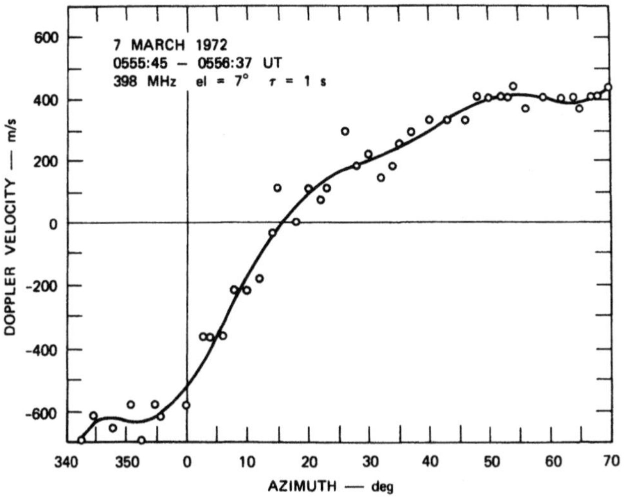

Fig. 1. Plot of the azimuthal variation of the mean Doppler shift for afternoon radar aurora. The Doppler variation is more rapid than sinusoidal about the null Doppler azimuth and is relatively constant at larger angles. (Courtesy of R. Tsunoda).

Recently, Brekke et al.,[1976]have attemped to explain the rapid variation of the mean Doppler velocity as being due to the inherent curvature of the auroral oval. In their model they note that the auroral electrojets tend to follow the auroral oval and hence a given current element would follow a curvlinear rather than straight path. The effect is shown schematically in Figure 2. In this figure we have drawn straight line and curved current elements in which irregularities are assumed to be flowing toward the left at a velocity V_d. For the straight line case the observed Doppler velocity would vary with the expected sinusoidal dependence. In contrast, for the curved line case, not only is the orientation of the radar wave vector being varied, but also the orientation of V_d, at the point of intersection of the radar wave vector and the current element. The net result is a more rapid than sinusoidal variation in the Doppler velocity with the observed Doppler velocity approaching V_d for angles as small as 30°. An additional by-product of this model is an extended region of relatively uniform Doppler velocity as the angle between the radar wave vector and V_d go through 0° or 180°.

Alignment of the radar wave vector with V_d could occur at Θ angles as small as 40°.

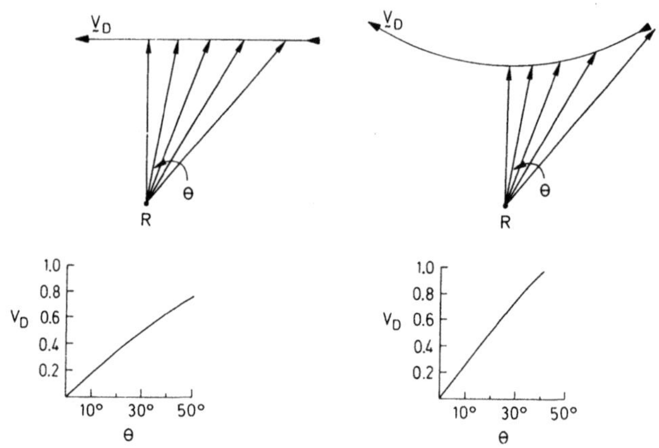

Fig. 2. Plots of the expected angular variation of the mean Doppler shift for irregularities flowing in straight-line and curvilinear current systems. For the curvilinear case one sees that the Doppler variation is more rapid than the sine of the azimuth angle.

In addition to auroral electrojet curvature, the earth's curvature will also produce a more rapid than sinusoidal variation of the mean Doppler velocity. This effect also results from a change in orientation of V_d with azimuth, and it occurs even if the electrojet current is flowing along a latitude contour. Earth curvature effects reinforce the effects of electrojet curvature and produce an even more rapid azimuth variation. Unfortunately, at the present time the exact magnitude of earth curvature effects have not been calculated. A third source of unusual azimuthal Doppler variations is smaller scale structure in the currents. Brekke et al., [1976] have noted this possibility and Greenwald and Ecklund [1975] have stressed it strongly. The latter authors have produced Doppler velocity maps as a function of range and antenna azimuth. One of these maps is shown in Figure 3. In this example one sees a relatively uniform azimuthal Doppler variation at the nearer ranges, whereas the Doppler contours at the further ranges are relatively far apart to the west and relatively close together to the east. This type of Doppler structure is quite common and it undergoes large temporal variations. Only a small percentage of the Doppler maps observed by Greenwald and Ecklund[1975] were described as having uniform Doppler variations.

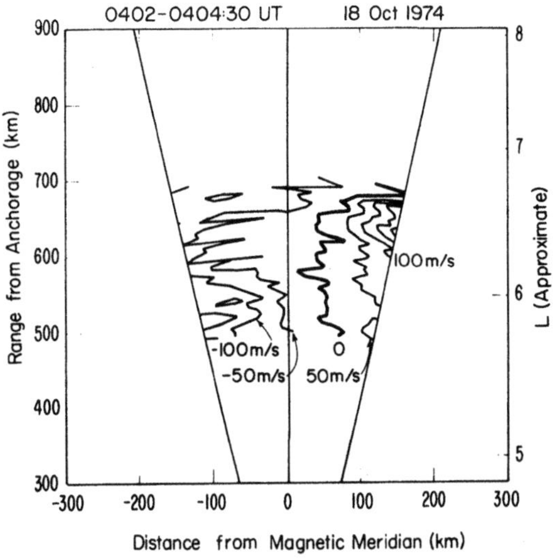

Fig. 3. Mean Doppler velocity contours as observed from Anchorage, Alaska, for 0402-0404:30 UT on 18 October 1974. Contours are printed for every 50 m/sec with the heavier contour representing zero mean Doppler. Reprinted from Greenwald and Ecklund [1975].

In conclusion, it appears that the effects of large scale auroral electrojet curvature and smaller structure, render it unlikely that azimuthal scans with a single auroral radar can be used to determine the angular variation of radar auroral Doppler velocities.

DRIFT COMPARISONS BETWEEN AURORAL AND INCOHERENT SCHATTER RADARS

As a result of the inherent difficulties associated with determining irregularity drift velocities from azimuth scanning techniques, other methods have been attempted. One of the most successful has been a recent comparison of E-region irregularity drifts and F-region plasma drifts by Ecklund et al., [1977]. For this comparison an auroral radar was temporarily operated from Aniak in southwestern Alaska. This radar had a 5° wide horizontal beamwidth that was directed in a line passing to the west of the Chatanika incoherent scatter radar. The Chatanika radar was directed to the west and intersected in the F-region a magnetic field line that passed through the auroral radar E-region scattering volume. The geometry of the experiment was so chosen that both radars would observe the same radial Doppler velocity, except for sign, if 1) the ionspheric electric field were directed toward magnetic north or south, 2) radar auroral irregularities have a net drift velocity equal to the electron drift velocity and 3) the observed radar auroral Doppler velocity at any angle Θ between the radar wave vector k and the electron drift direction varies as cos Θ.

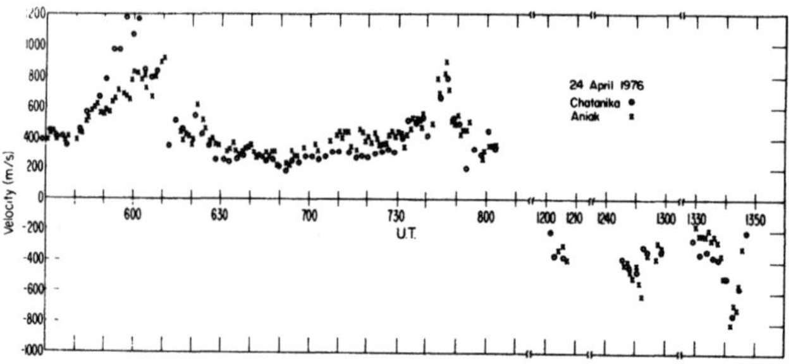

Fig. 4. Comparison of F-region plasma drifts as measured with the Chatanika incoherent scatter radar and E-region radar auroral irregularity drifts as measured with a 50 MHz pulsed radar located at Aniak, Alaska. The two drifts should be equal if the ionospheric electric field is directed toward magnetic north or south, and if the irregularities drift at the electron drift velocity.

The results of this experiment, shown in Fig. 4, demonstrates the excellent agreement that has been observed between the two techniques over a extended time period. Those periods when the two data sets are not in agreement have been interpreted by the authors as times when the electric fields have rotated from a northward or southward direction. While this was only a preliminary measurement it presents strong evidence that, at least at lower Vhf frequencies, irregularity Doppler measurements can be used to measure auroral zone electric fields.

IRREGULARITY DRIFT MEASUREMENTS WITH STARE

Another approch for studying irregularity drift motions is to analyze a given scattering volume from two or more different directions. This approach is used in the new Scandinavian Twin Auroral Radar Experiment - STARE - [Greenwald, 1976]. STARE was developed to 1) avoid the inherent difficulties associated with azimuth scanning radars, 2) study the longitudinal and latitudinal structure of radar auroral irregularity drift patterns, and 3) provide information on the ionospheric electric field.

The experiment consists of two nearly identical auroral radars located near Trondheim, Norway, and Sauvamaki, Finland with frequencies of 140 MHz and 143.8 MHz, respectively. In the present operating mode each radar is programmed, through an on-line computer, to determine the backscatter intensity and mean Doppler velocity from fifty different ranges in eight different directions. The range interval studied extends from 495 km to 1235 km with a resolution of 15 km. The eight beams cover an azimuth sector of slightly less than 29°. Each beam is separated by 3.6^{o} and has a half-power beamwith of 3.2^{o}. All eight directions are probed simultaneously.

The temporal resolution of the measurements is selectable and has typically been in the range from 2-60 sec.

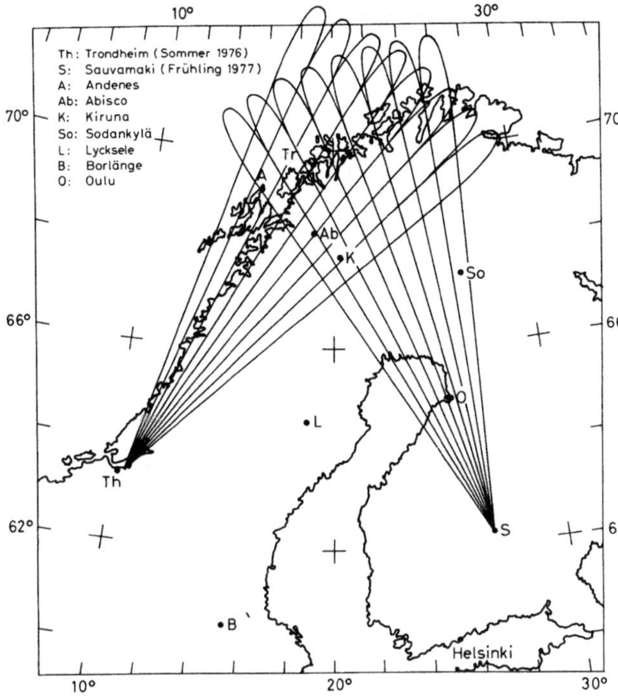

Fig. 5. Map of northern Scandinavia showing the eight commonly used beams from each of the STARE radars.

One sees that there is a large region covering most of northern Scandinavia where the two sets of beams overlap. Two distict Doppler velocity measurements can be obtained for any E-region volume element from which backscatter is observed by both radars. Assuming that the measured Doppler velocities vary as the cosine of the angle between the electron drift direction and the radar wave vectors, one can derive the mean irregularity drift velocity within the colume element. By additionally assuming that the mean drift velocity is equal to the electron drift velocity, as indicated by both the Ecklund et al., [1977] experiments and plasma theory, one can derive a two dimensional map of the ionspheric electric field.

As an example of the type of results that are obtained with this technique, we have examined the radar auroral motions associated with a weak magnetic substorm that took place between 1925 UT and 2015 UT on 2 February 1977 (Day 33). This example, which is one of the first that has been obtained with the STARE system, is particularly interesting because it involves the formation and subsequent decay of an Harang discontinuity in the pre-midnight sector of the auroral zone. The Tromsö magnetogram for this period is shown in Fig. 6. It indicates a weak positive magnetic bay in the H-component prior to 1927 UT and after 2014 UT. The magnetogram is predominantly negative in the intervening period.

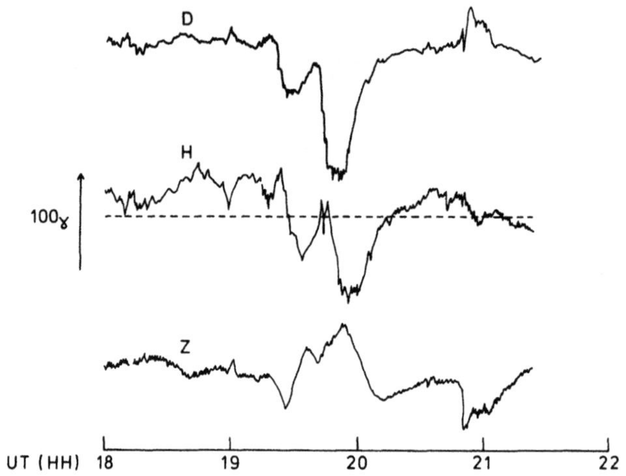

Fig. 6. Tromsö magnetogram for 2 February 1977 from 1800 UT to 2200 UT.

During the entire substorm the STARE radars were operating with a 20 sec integration time. From this data base we have selected the records for 1902 UT, prior to the onset of the discontinuity; 1927 UT, just after the formation of the discontinuity; 1953 and 2001, during the maximum phase of the substorm; and 2017, after the storm. The 1902 data is shown in Figure 7. At this time ΔH at Tromsö was approximately + 30 γ. This magnetic deflection is produced by a relatively weak eastward auroral electrojet and the observed backscatter is from the irregularities that are embedded in this current. The backscatter intensity observed by the two STARE radars is shown to the left of the figure. Here we have plotted digits representing the signal-to-noise ratio (SNR) in dB divided by 4 and truncated. For example, a 3 implies an SNR between 12 dB and 16 DB. SNRs of less than 4 dB have not been plotted. For this and the subsequent examples, the data have been plotted in a geomagnetic coordinate system. One can see that the poleward and equatorward edges of the scattering region are inclined at about 25° to a geomagnetic latitude contour. This alignment is more closely related to a geographic or corrected geomagnetic contour, being about half away between the two.

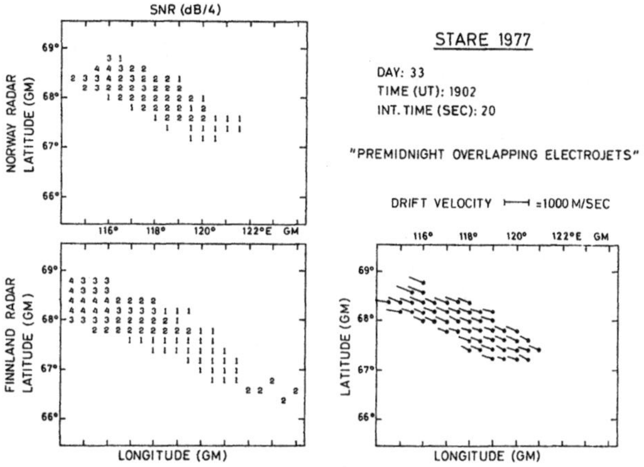

Fig. 7. Plot of the backscattered signal intensities observed by the Norwegian and Finnish STARE radars and the associated irregularity drift velocities for 1902 UT on 2 February 1977. The intensities are given by digits representing the SNR/4 and truncated. In the velocity plots, the dots represent the location of the velocity measurement and the length and direction of the lines represent the speed and direction of the irregularity drift. This plot was obtained prior to the onset of the substorm.

All of the SNRs have been corrected for range dependence, aspect angle dependence, and antenna beamwidth dependence. One sees that the irregularities observed from the two widely separated stations are in essentially the same region and they have nearly the same intensity distribution. In subsequent examples one will note that the similarity is not nearly as good.

The drift velocities associated with these irregularities are plotted to the right in this figure. The dots represent the location of the measurements and the length and direction of the lines indicate the magnitude and direction of the drift velocities. A value is plotted for a given location only if both STARE radars observed more than a 3 dB SNR. One can see that the velocity vectors are approximately aligned with the poleward or equatorward edge of the scattering region. This drift direction will be referred to as westward in the subsequent discussion. The drift speeds are between 400 and 700 m/sec at this time.

If we now assume that the observed irregularity drift velocities are equal to the local electron drift velocity, we can derive the electric field topology at this time. In the figure this is accomplished by rotating all of the velocity vectors 90° clockwise and setting 1000 m/sec = 50 mV/m. One observes generally northward electric fields with magnitudes ranging from 20 to 35 mV/m. In those regions where irregularities are not observed, one can only say that the electric field is probably below some critical value necessary for the onset of the instability.

The SNR and drift velocity maps shortly after the formation of the Harang discontinuity are shown in Figure 8. At this time, a poleward westward auroral electrojet has intensified and the radar auroral irregularties associated with it are observed on both STARE radars. At the same time, the eastward electrojet associated scatter has moved equatorward and occupies a broader latitudinal extent. One sees that the drift velocities in the poleward scattering region are also aligned with the structures; however, in this case the drift is toward the east. Some of these drift speeds are of the order of 800 m/sec, while others are appreciably lower. The exact reason why the Norwegian radar sees westward electrojet associated scatter over a much larger region than the Finnish radar is currently being studied.

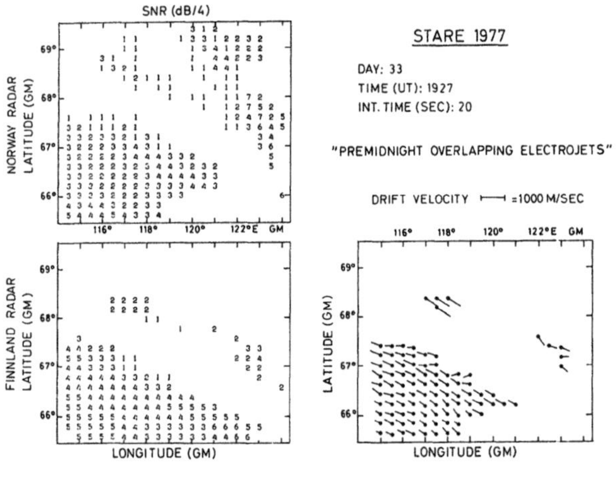

Fig. 8 Plots of SNRs and drifts velocities for 1927 UT on 2 February 1977. This plot was obtained shortly after the onset of the Harang discontinuity.

Figure 9 illustrate the SNR and velocity data for 1953 UT near the maximum phase of the substorm. One sees a latitudinally confined scattering region associated with the eastward electrojet and a latitudinally extended scattering region associated with the poleward westward electrojet. Between the two regions of scatter, one observes a region with little or no scatter. This region is presumeably filled with active visual auroras; however, we have no optical data to support this conjecture. One sees more structure in the westward electrojet-associated irregularity drift velocities and speeds of more than 1400 m/sec corresponding to southward electric fields of 70 mV/m are observed.

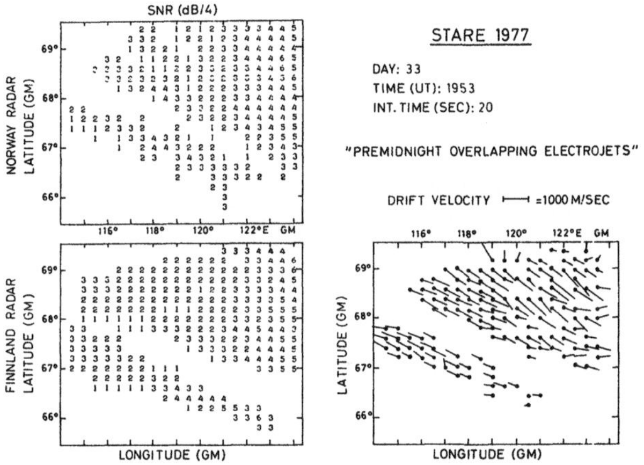

Fig. 9. Plots of SNRs and drift velocities for 1953 UT on 2 February 1977. This plot was obtained near the maximum phase of the substorm and it shows a well developed poleward westward electrojet.

The SNR and drift velocity data observed later in the maximum phase of the substorm, 2001 UT, are shown in Figure 10. One can at this time see considerable differences in the SNR data from the two stations. The counterstreaming electrojet-associated irregularity regions still exist and the westward electrojet associated irregularity regions still indicate large velocities. However, the velocity patterns now show considerable structure and they appear to be breaking up. This break-up is particularly apparent in the lower right hand corner of the velocity plot.

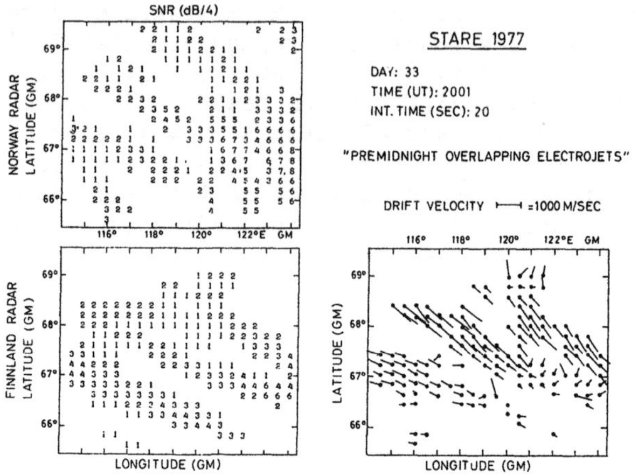

Fig. 10. Plot of the SNRs and drift velocities for 2001 UT on 2 February 1977. This plot was obtained late in the substorm maximum phase when the velocity patterns start to break-up.

Finally in Figure 11, we see the SNR and velocity distributions for 2017 UT at the end of the substorm. The region of westward electrojet associated scatter has virtually disappeared and the eastward electrojet associated scatter has moved poleward. The irregularities are drifting in a westward direction at speeds of 400 - 600 m/sec. Essentially, the configuration is quite similar to that of 1902 UT, prior to the start of the substorm.

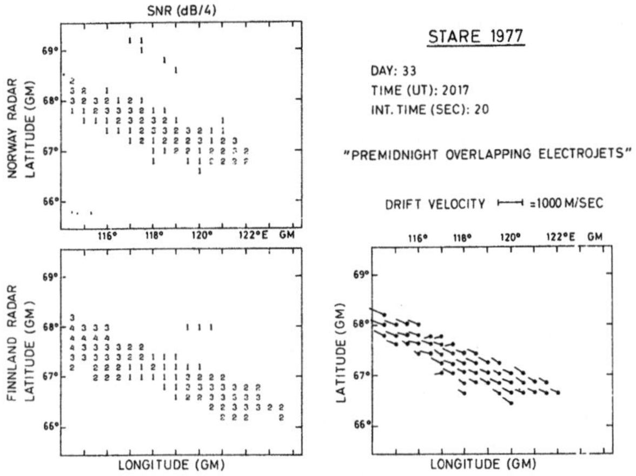

Fig. 11. Plot of the SNRs and drift velocities for the 2001 UT on 2 February 1977. This plot was obtained at the end of the substorm.

SUMMARY

In this paper we have briefly reviewed the theory of electrojet-associated, electron density irregularities and compared it with spectral observations at the equator and in the auroral zone. We have seen that theory predicts both two-stream and gradient-drift irregularities to have a net drift velocity equal to the electron drift velocity; however, the equatorial observations indicate that only gradient drift irregularities display this characteristic. Equatorial two-stream irregularities appear to propagate at the ion acoustic velocity. Since some characteristics of radar auroral spectra are similar to equatorial two-stream spectra, it has been conjectured that auroral electrojet two-stream irregularities also propagate at the ion acoustic velocity. However, there is a growing amount of evidence that this conjecture may not be correct. This evidence includes 1) the observations, by many observers, of Doppler velocities considerably higher than any expected acoustic velocity, 2) the recent explanation by Brekke et al. [1976] of the radar auroral azimuthal Doppler variation in terms of electrojet curvature, and 3) the recent measurements by Ecklund et al. [1977] indicating that radar auroral irregularities have a net drift equal to the electron drift velocity.

We have also presented some initial STARE observations that were obtained during a weak magnetic substorm. During the substorm we observed radar auroral irregularities associated with a premidnight eastward auroral electrojet and a poleward located westward auroral electrojet that temporarily extended into the premidnight sector forming a Harang discontinuity. By assuming that the observed Doppler velocity varies as the cosine of the angle between the radar vector and the net irregularity drift direction, it was possible to calculate the net irregularity drift velocity for each spatial region where both STARE radars observe backscatter. The irregularity drift direction were generally aligned with the structure, being westward in the eastward electrojet and eastward in the westward electrojet. Generally higher speeds were measured for the westward electrojet associated scatter, where they were as high as 1400 m/sec. Under the additional assumption that the irregularity drift velocities are equal to the electron drift velocity, we obtained generally northward electric fields, between 20 and 35 mV/m, in association with the eastward electrojet and somewhat larger southward electric fields, approaching 70 mV/m, in association with the westward electrojet.

The drift patterns that one observes on the STARE velocity plots are quite similar to the expected electron drift pattern for a Harang discontinuity conditon. It is tempting to believe that the two are the same and that the STARE velocity plots could be used to study the spatial and temporal structure of ionospheric electric fields. At the present time additional confirmation of the results obtained by Ecklund et al. [1977] is of the utmost importance. Fortunately, a number of rockets containing electric field probes and barium releases were launched into the STARE scattering volume during the first few months of 1977. This data set is currently being evaluated and additional comparisons of electric fields and the drift velocity of radar auroral irregularities should be available in the near future.

REFERENCES

Abel, W.J. and R.E. Newell, Measurements of afternoon radio aurora at 1295 MHz, J. Geophys. Res., 74, 231, 1969.

Banks, P.M., J.R. Doupnik, and S.I. Akasofu, Electric field observations by incoherent scatter radar in the auroral zone, J. Geophys. Res., 78, 6607, 1973.

Block, L.P. Potential double layers in the ionosphere, Cosmic Electrodynamics, 3, 349, 1972.

Brekke, A., R.I. Tsunoda, and M.J. Baron, On auroral electrojet curvature and the interpretation of azimuth-scan, radar Doppler

data, Radio Science, 12, 141, 1977.
Buneman, O., Excitation of field aligned sound waves by electron streams, Phys. Rev. Letters, 10, 285, 1963.
Ecklund, W.L., B.B. Balsley, and R.A. Greenwald, Doppler spectra of diffuse radar aurora, J. Geophys. Res., 78, 4797, 1973.
Ecklund, W.L., B.B. Balsley, and D.A. Carter, A preliminary comparison of F-region plasma drifts and E-region irregularity drifts in the auroral zone, J. Geophys. Res., 82, 195, 1977.
Farley, D.J., A plasma instability resulting in field-aligned irregularities in the ionosphere, J. Geophys. Res., 63, 6083, 1963.
Greenwald, R.A., W.L. Ecklund and B.B. Balsley, Radar observations of auroral electrojet currents, J. Geophys. Res., 80, 3635, 1975.
Greenwald, R.A. and W.R. Ecklund, A new look at radar auroral motions, J. Geophys. Res., 80, 3642, 1975.
Greenwald, R.A., STARE - First report. Max-Planck-Institut für Aeronomie, MPAE-W-04-76-24, 1976.
Haerendel, G., Plasma drifts in the auroral ionosphere derived from barium releases, Rep 57, Max-Planck-Institut Munich, Oct. 1971.
Kindel, J.M. and C.F. Kennel, Topside current instabilities, J. Geophys. Res., 76, 3055, 1971.
Leadabrand, R.L., J.C. Schlobohm, and M.J. Baron, Simultaneous very high frequency and ultra high frequency observations of the aurora at Fraserbrugh, Scotland, J. Geophys. Res., 70, 4235, 1965.
Maeda, K., J. Tsunoda, and K. Maeda, Theoretical interpretation of the equatorial sporadic-E layers, Phys. Res. Letters, 11, 406, 1963.
Mozer, F.S. and P. Bruston, Electric field measurements in the auroral ionosphere, J. Geophys. Res., 72, 1109, 1967.
Mozer, F.S. and R. Serlin, Magnetospheric electric field measurements with Balloons, J. Geophys. Res., 74, 4739, 1969.
Reid, G.C., The formation of small scale irregularities in the ionosphere, J. Geophys. Res., 73, 1627, 1968.
Rogister, A. and N. D'Angelo, Type II irregularities in the equatorial electrojet, J. Geophys. Res., 75, 3879, 1970.
Schunk, R.W., W.J. Raitt, and P.M. Banks, Effect of electric fields on the daytime high latitude E and F regions, J. Geophys. Res., 80, 3121, 1975.
Siren, J.C., J.R. Doupnik, and W.L. Ecklund, A comparison of auroral currents measured by the Chatanika radar with 50 MHz backscatter observed from Anchorage, Submitted to J. Geophys. Res., 1977.
Sudan, R.N., J. Akinrimisi, and D.T. Farley, Generation of small scale irregularities in the equatorial electronjet. J. Geophys. Res., 78, 240, 1973.
Tsunoda, R., Doppler velocity maps of diffuse radar aurora, J. Geophys. Res., 81, 425, 1976.

DISCUSSION

Stockflet Jörgensen: Is the general direction of the observed drift velocity more or less parallel to the auroral oval ?

Greenwald: The irregularity drift velocities are aligned with radar auroral forms and these appear to be more or less aligned with the auroral oval. One must be somewhat careful, however, since the auroral oval is a statistical quantity. For any given time period, both electrojet currents and auroral forms can have alignments in other directions.

Stockflet Jörgensen: Is it a necessary condition to have an instability for getting radar reflections or could an electron density gradient for example also produce a radar eccho ?

Greenwald: It is theoretically possible that electron density gradients could produce radar auroral backscatter. However, the scattering cross-sections are so high, that I think it is unlikely that an electron density gradient alone could produce the necessary scatter. One must remember that the radar wavelength is only 1 meter and the electron density scale length must approach this dimension in order to get appreciable partial reflections.

Shepherd: You did not identify the two scattering regions associated with the westward and eastward electrojets with the discrete and diffuse aurora, respectively. But that is, I believe, the association that was made by Rostoker et al.

Greenwald: The equatorward scattering region that I have associated with the eastward electrojet appears to be in a region of diffuse aurora. The position of the poleward scatter relative to the aurora is not so clear. I am not sure that westward electrojets are necessarily associated with discrete aurora and I do remember Kisabeth and Rostoker noting that westward electrojets can flow in regions where there were not any bright visual forms.

AURORAL EFFECTS ON NEUTRAL DYNAMICS

Asgeir Brekke

The Auroral Observatory, University of Tromsø
Tromsø, Norway

1. INTRODUCTION

Before a discussion about possible auroral effects on the neutral motion is started it can be illuminating to compare the available energy of an auroral substorm with the energy of the neutral air in which the substorm takes place.

Let us for simplicity assume that the neutral air density at 80 km altitude is 10^{-5} kg m^{-3} and that the scale height all the way to the exosphere is on the average H_o = 10 km, then the potential energy density of a column of air above 80 km with unit cross section is

$$E = g \int_{80}^{\infty} \rho_{80} \exp(-\frac{z-80}{H_o})(z-80)dz = H_o^2 g \rho_{80} \qquad (1.1)$$

where ρ_{80} is the air density at 80 km, z is the altitude in km, and g is the acceleration of gravity (10 m/s^2). By inserting the adopted values: E = 10^4 Joule/m^2. For strong auroral substorms it can be expected [1] that an effect of P_a = 1 watt/m^2 is released. This means that during a strong substorm lasting for about 3 hours the same amount of energy is dumped into the atmosphere above 80 km as is present in terms of potential energy of the neutral air above this level.

It is therefore conceivable that an auroral substorm may cause major effects in the neutral atmosphere dynamics at and above auroral levels. Two of these effects will be discussed in the following order.

B. Grandal and J. A. Holtet (eds.), Dynamical and Chemical Coupling, 313-336.

1) The acceleration of the neutral atmosphere by ion drag.

2) The joule heating caused by strong electrojets at auroral altitude.

We will not treat the energy released by the precipitating particles and the effect this might have on the neutrals. The talk will be partly of the review character and partly based on some new results.

2. ION-DRAG EFFECTS

2.1 Altitude variations of the atmosphere

In this report we will treat the upper atmosphere as the region between 80 and 500 km.

In this altitude region we also find the major part of the ionosphere, but the electron density hardly exceeds more than 10^{12} particles per m^3 at any height. Since the neutral air density varies from 10^{20} per m^3 at 80 km to 10^{14} per m^3 at 500 km, the degree of ionization will hardly be larger than 10^{-3}.

As the density of the two gases, the neutral atmosphere and the ionospheric plasma, changes strongly and partly independently by altitude, the degree of interaction will be highly variable from one altitude to another. While the motion of the neutral gas at 80 km will dominate the motion of the ionospheric plasma at this level, the opposite can be true at higher levels where the ion motion may at times strongly oppose the neutral air motion.

2.2 The mobility equation of the neutral air

To investigate the relative importance of the different forces acting on the neutral air mass, the momentum equation for the neutrals will be used

$$\frac{d\underline{u}}{dt} = - 2\underline{\Omega} \times \underline{u} + g - \frac{1}{\rho} \nabla \underline{p} - V\psi + \frac{\mu}{\rho} \nabla^2 \underline{u} - \nu_{ni}(\underline{u} - \underline{V}_i) \qquad (2.1)$$

where the symbols are: $\underline{u}$ the neutral air velocity, $\underline{\Omega}$ the angular velocity of the earth, ρ the neutral gas density, $\underline{p}$ the neutral pressure, ψ a potential function due to tideraising forces, μ the coefficient of viscocity, ν_{ni} the neutral ion collision frequency or more precisely the rate of transfer of momentum, $\underline{V}_i$ the velocity of the ion gas.

The five terms on the right hand side arise because of 1) Coriolis force, 2) acceleration of gravity including the centripetal acceleration due to a rotating coordinate frame, 3) pressure gradient force, 4) scalar potential due to tide raising forces, 5) viscous force, and 6) ion drag term due to collisions between neutrals and ions.

In Fig. 1 are shown the different terms in Eq. (2.1) as function of altitude between 120 and 500 km for 45°N.

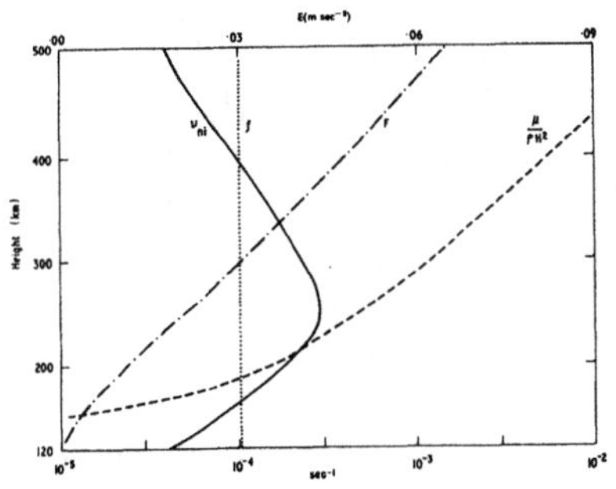

Fig. 1. The different forms in the neutral mobility equation for midday at sunspot minimum. The upper scale applies to the pressure gradient F. The lower scale applies to the ion-drag parameter ν_{ni}, the Coriolis parameter f, and the normalized kinematic viscosity $\mu/\rho H^2$. According to [2].

According to this model [2] the most important forces acting on the neutrals above say 150 km are the viscosity, the pressure gradients and the ion drag. Below 150 km the Coriolis force is important. The model however assumes an ion density at 150 km equal to 1.7×10^{11} m^{-3}. In the auroral zone the ion density in the E-region can attain values at least a factor of 10 larger than this. and thereby leave the ion-drag term of dominant importance.

2.3 The thermospheric global scale wind models

Several attempts have been made to develope global scale wind models at thermospheric levels. From drag measurements on satellites it has been found that there is a significant difference in the air drag on the satellite between day and night and such observations have been used to deduce the global temperature distribution at the termopause. Fig. 2 shows such a model according to [3]. A nonuniform global temperature is evident, a maximum of 1300 K occurs at equator at about 1400 hours local time, and a corresponding minimum of 1000 K occurs at equator at 0400 hours local time. The model however is symmetric with respect to equator.

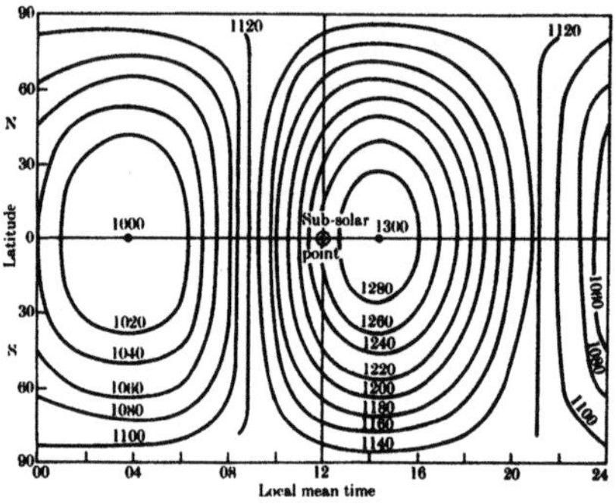

Fig. 2. The global temperature distribution at equinox. According to [3].

These type of data have been used [4,5] to infer the global neutral wind system at thermospheric levels. It is usually assumed that the driving force is the pressure gradients due to the temperature assymetry, and by using reasonable models for the viscosity coefficients and ion densities a global neutral wind pattern as shown in Fig. 3 can be derived. The wind is found to blow mainly across the isobars antiparallel to the pressure gradient, thus indicating that the ion-drag is of considerable importance.

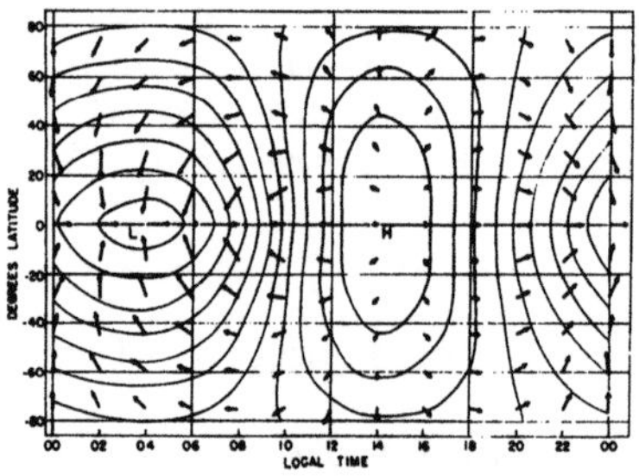

Fig. 3. The global wind field at 300 km altitude plotted on a background of isobars. According to [5].

For all other terms neglected except the ion-drag and the pressure, Eq. (2.1) reduces in a steady state and for a static ionosphere ($\underline{V}_i = 0$) to:

$$u = \frac{-1}{\rho \nu_{ni}} \nabla p \qquad (2.2)$$

This method has recently been improved [6,7], but the wind-pattern at high latitudes are not drastically different from the one already shown in Fig. 3, and as indicated by Roble (this issue), a more realistic model must take both joule heating and momentum sources into account.

2.4 Observations of the neutral wind at thermospheric levels

At the upper thermosphere in the F-region, neutral wind observations have been obtained by Barium releases from rockets. A synthesis of such measurements according to [8] is shown in Fig. 4 together with the corresponding drift of the F-region ionization when the interplanetary magnetic field is directed away from the sun (to the left). For a reversal of the magnetic field polarity the smaller convection cell will appear in the evening sector and the larger cell will be situated in the morning sector (to the right).

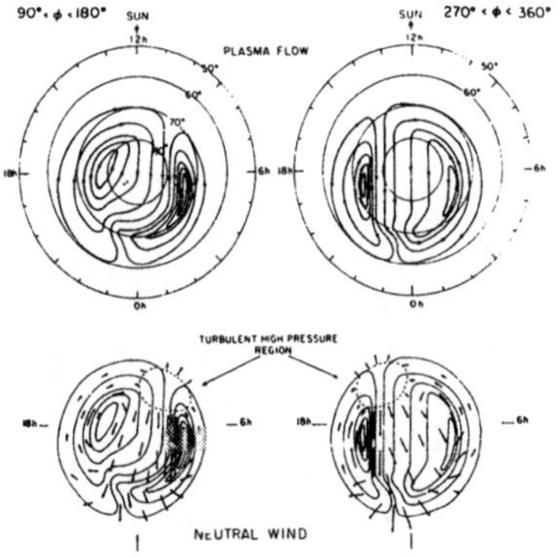

Fig. 4. Idealized polar patterns of the convective plasma flow for "away" and "toward" sectors of the interplanetary magnetic field (top panel). Idealized vector representation of neutral winds above 180 km (lower panel). According to [8].

According to this picture, the thermospheric neutral wind pattern between 200 and 400 km of altitude is very similar to the F-region ionization drift, thus indicating that the ion-drag in the neutral mobility-equation (Eq. 2.1)) is not at all a drag on the neutral motion but a dominant driving force in the polar region.

2.5 The ion-drag term in the neutral mobility equation

Since Heppner's observations so clearly indicate that the neutral air at upper thermospheric levels in the polar region closely follows the ions, it can be of interest to examine further the ion-drag term in a situation where the ions are moving with a velocity different from zero. Neglecting all other terms in Eq. (2.1) except the inertia term and the ion-drag term the equation reduces to

$$\frac{d\underline{u}}{dt} = - \nu_{ni}(\underline{u} - \underline{V}_i) \tag{2.3}$$

The neutral velocity will then approach $\underline{V}_i$ in a time scale equal to

$$\tau = \frac{1}{\nu_{ni}} \tag{2.4}$$

The collision frequency is given by

$$\nu_{ni} = K_{ni} \cdot N_i \tag{2.5}$$

where the collision rate $K_{ni} = 3.75 \times 10^{-16}$ m^3/s is a reasonable approximation in the thermosphere. For ion densities of the order of $N_i \approx 10^{12}$ m^{-3} the time scale will be $\tau \approx$ 2600 s or approximately 45 minutes. For ion densities of the order of 10^{10} m^{-3} however the time constant will be more like 75 hours. It is therefore conceivable that for reasonable high ion densities say above 10^{11} m^{-3} that the ionosphere during a substorm actually impose enough momentum on the neutral air in the thermosphere that the neutral wind will be strongly modified according to the ion motion.

2.6 The ion mobility equation and the ionospheric convection

In periods of large convection electric fields the ions will be driven by large velocities antisunward across the polar cap and sunward in the auroral oval. Enhanced electric fields can last for a few hours and extend across sizable areas illustrated by Greenwald (this issue). With reference to what is said above concerning the neutral air response time to the ion motion, substorm related electric fields will be of particular importance to the neutral motion.

In order to understand the neutral motion in the auroral and polar cap region, the motion of the ions have to be considered simultaneously with the motion of the neutral gas.

A simplified ion-mobility equation perpendicular to the earth's magnetic field $\underline{B}$ can be written as,

$$\frac{d\underline{V}_{i_\perp}}{dt} = \frac{e}{m_i}(\underline{E}_\perp + \underline{V}_{i_\perp} \times \underline{B}) - \nu_{in}(\underline{V}_{i_\perp} - \underline{u}_\perp) \qquad (2.6)$$

where pressure gradients and gravity are neglected.

m_i is the ion mass

e is the ion charge

$\underline{E}_\perp$ is the electric field perpendicular to $\underline{B}$

$\underline{B}$ is the magnetic field assumed vertical on earth.

ν_{in} is the ion-neutral collision frequency and related to ν_{ni} by

$$\frac{\nu_{ni}}{N_i} = \frac{\nu_{in}}{N_n} = K_{ni} \qquad (2.7)$$

where N_n is the density of the neutral gas.

Since the neutral air density varies so strongly in the region of interest between 80 and 500 km the ion velocity vector will also change by altitude for a constant E-field and neutral velocity. We will here be concerned with the ion motion perpendicular to $\underline{B}$ only.

In the most simple case when the ion motion is stationary and $\underline{u} = 0$ the ion velocity will be given by:

$$\underline{V}_{i_\perp} = \frac{\Omega_i^2}{\Omega_i^2 + \nu_{in}^2}\left[\frac{\nu_{in}}{\Omega_i}\frac{\underline{E}_\perp}{B} + \frac{\underline{E}_\perp \times \underline{B}}{B^2}\right] \qquad (2.8)$$

where Ω_i is the ion gyro frequency $\Omega_i \approx 175\ s^{-1}$.

In Fig. 5 are shown how the two components of $\underline{V}_i$ will vary by altitude for a certain E-field. For an imposed E-field the ion-velocity will be given by the full $\underline{E}_\perp \times \underline{B}/B^2$ drift in the F-region while at about 120 km it will be reduced to $\sqrt{\frac{1}{2}}$ times this according to the present model. At 80 km the ion drift will only be a negligible fraction of the full $\underline{E}_\perp \times \underline{B}/B^2$ drift and directed parallel to $\underline{E}_\perp$.

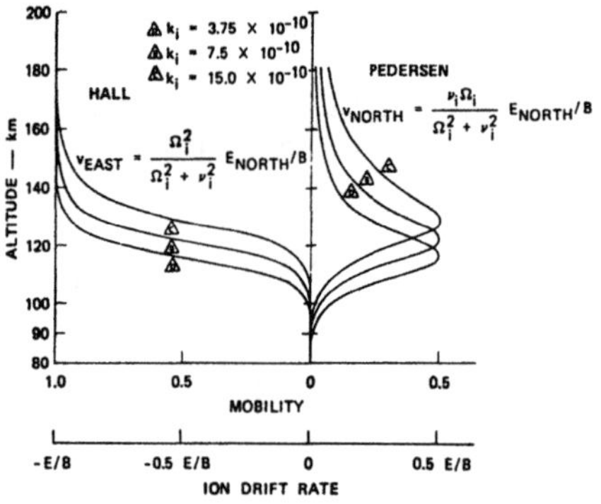

Fig. 5. Relative ion drifts for an applied electric field. According to [9].

2.7 The effects of enhanced convection electric field on the neutral gas motion

The ionmobility equation and a simplified mobility equation for the neutral gas where only the inertia term, the ion-drag term and the viscosity term are included have been solved simultaneously by [10]. Starting out with $\underline{V}_i = \underline{u} = 0$ at time t = 0 when an electric field of 40 mv/m is turned on, the ion and neutral velocities can be solved for different altitudes and times. The time variation of the neutral wind and ion velocities at different altitudes are shown in Fig. 6 for the component along $\underline{E}_\perp \times \underline{B}$.

The neutral gas at 300 km will be driven in the $\underline{E}_\perp \times \underline{B}$ direction almost instantaneously. After 1 hour for instance the angle between the neutral wind and the E-field will be 89.92° and 16.70° at 300 and 117 km, respectively.

In the auroral zone one knows that nor the electric field neither the ion density are constant during an auroral substorm but varying on time scales of minutes or less. The picture will therefore not be as simple as indicated where a constant electron density model is assumed.

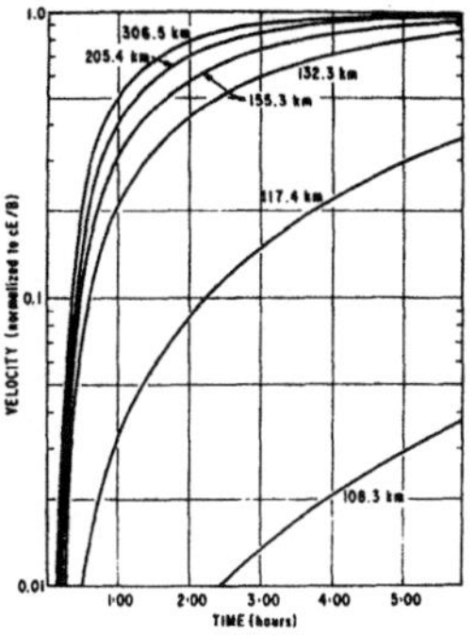

Fig. 6. The $\underline{E}_\perp \times \underline{B}$ component of the neutral wind arising from a quasi-step function electric field. According to [10].

By retaining the Coriolis force in the neutral mobility equation (Eq. (2.1)) [11] have recently reanalysed this model. Their results shown in Fig. 7 are compared with the results of [10]. The main results of their reanalysis are that the neutral velocity in the $\underline{E}_\perp \times \underline{B}$ direction is strongly reduced at 150 km and that it does not reach a steady state at 125 km and 115 km within a time period of 24 hours. On the other hand the neutral velocity component at 150 km along the E-field is strongly enhanced when the Coriolis force is included. At 125 and 115 km however this component is reduced and does not reach a steady state. It is also evident that the Coriolis force has effects on the ion velocity due to the mutual coupling between the neutrals and the ions.

From this analysis it appears rather complicated to infer anything about the ion momentum transfer on the neutrals from simultaneous and instantaneous measurements of ion and neutral velocities at one place in space, because long time constants and long distance effects are involved. The momentaneous neutral wind measured at one place in space may have been accelerated by the ions long distances away several hours ago. This fact should in particular be kept in mind when analysing rocket measurements of the neutral wind in terms of acceleration by ions.

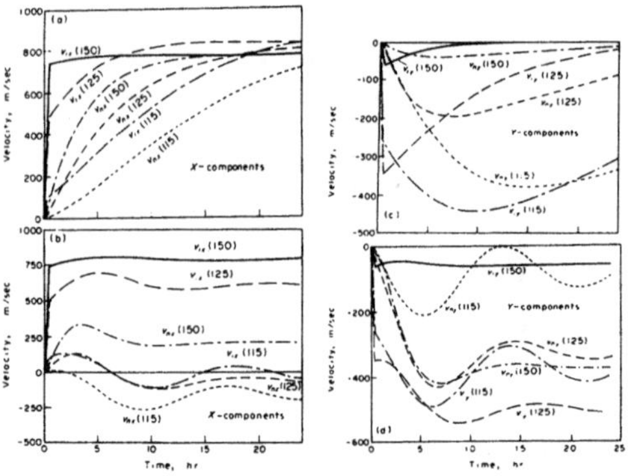

Fig. 7. Velocity components as function of time and altitudes 115, 125 and 150 km from model calculations: For Coriolis force omitted (upper panel). For Coriolis force included (lower panel). According to [12].

2.8 Neutral wind measurements in the E-region and below

The neutral wind between 80 and 170 km has been measured by rocket born chemical releases in the Auroral ionosphere [13]. Highly variable wind velocities from day to day and also by altitude are found. One such profile is shown in Fig. 8 observed at the recovery of a negative magnetic bay of 100 γ. The neutral wind is directed in the southeastward direction above 110 km, and in

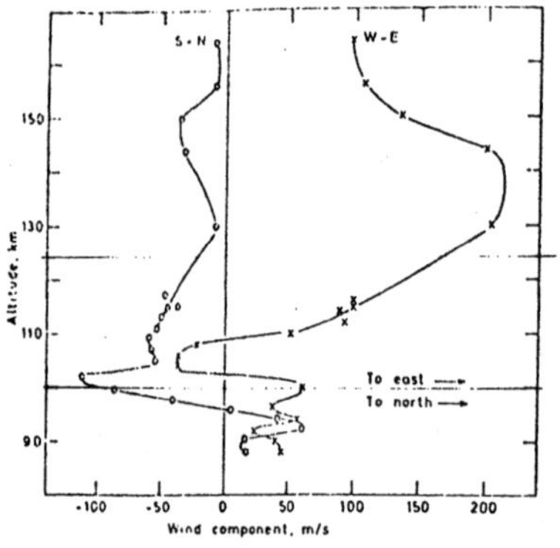

Fig. 8. Neutral wind components derived in the auroral E-region. According to [13].

particular the eastward velocity is strong in this region. Below 110 km the wind velocity is highly irregular and characterized by several shears.

An interesting relationship between the neutral wind measurements at the E-region level and magnetic field perturbations preceeding the observations by from one to two hours was observed by [13]. For positive magnetic bays the wind is toward west and for negative bays it is toward east. These measurements can be explained by assuming that the magnetic bays are mainly caused by Hall-currents and that the driving electric field which will be directed in the meridional plane will drive the ions in the zonal direction and thereby accelerate the neutrals in the east-west direction.

Height averaged neutral winds in the auroral E-region have been derived by an incoherent scatter radar at Chatanika, Alaska [14]. During daytime the wind velocities appear to be in accord with the models derived by assuming the natural pressure gradient as the driving force. At nighttime however, large deviations from this picture occur as can be seen in Fig. 9, where northward velocities are observed before midnight, and rapid turning toward south is seen after midnight. There is however a tendency for the E-region wind to blow westward in the late evening and eastward in the morning related to the switch from north to south in the meridional electric field observed simultaneously, indicating some influence of the ion motion.

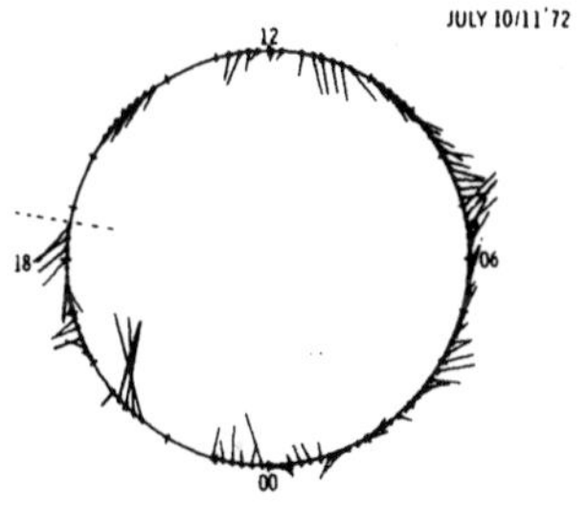

Fig. 9. Height averaged Auroral zone E-region neutral wind velocities. According to [14].

Recently [9] have obtained simultaneously altitude profiles of the auroral zone E-region neutral winds and ion drifts by a high resolution incoherent scatter radar at Chatanika, Alaska. The radar being capable of measuring ion drifts by 6 km resolution in the altitude region from 80 to 200 km can be used to derive neutral wind data with the same height resolution.

From the simple ion mobility equation used earlier (Eq. (2.6)) the neutral wind velocity at a particular altitude will in a steady state be given by

$$\underline{u} = \underline{V}_{i_\perp} - \frac{\Omega_i}{\nu_{in}} (\underline{E}_\perp + \underline{V}_{i_\perp} \times \underline{B}/B) \tag{2.9}$$

Because the ions are drifting in the true $\underline{E}_\perp \times \underline{B}$ direction above the 200 km level, ion velocity measurements made in this region will give direct information on the electric field which can be substituted in Eq. (2.9). At altitudes say below 150 km the obtained ion velocities can then be used to derive $\underline{u}$ directly from Eq. (2.9).

One notices however in Eq. (2.9) the critical dependence of the neutral wind estimates on the ν_{in}-profile. Indeed, with increasing height Ω_i/ν_{in} ultimately becomes much larger than unity. Hence any error in the measured value of $\underline{E}_\perp$ and/or $\underline{V}_{i_\perp}$ is amplified at higher altitudes.

This is illustrated in Fig. 10 where measured ion velocity profiles in the height region between 90 and 140 km are shown together with neutral wind profiles derived for different values of ν_{in}. For larger values of ν_{in} the derived neutral wind follows more closely the ion velocity as can be expected from Eq. (2.9). For small values of ν_{in} unrealistic values of the neutral wind velocities are derived particularly at the uppermost heights.

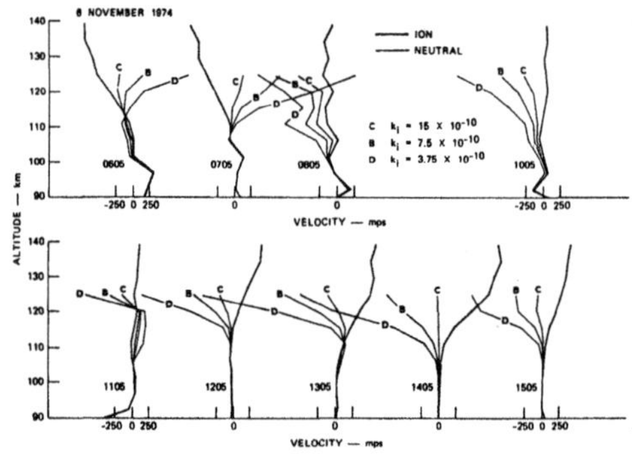

Fig. 10. Measured eastward ion drift velocities and the corresponding neutral-wind estimates derived for different collision-rate parameters. According to [9].

In Fig. 11 are presented neutral wind velocities at three different altitudes in the auroral E-region derived according to the scheme explained above. The most striking feature is the strong northward velocities at all heights in the morning hours. Strong equatorward motion is also evident before midnight in particular at 125 km.

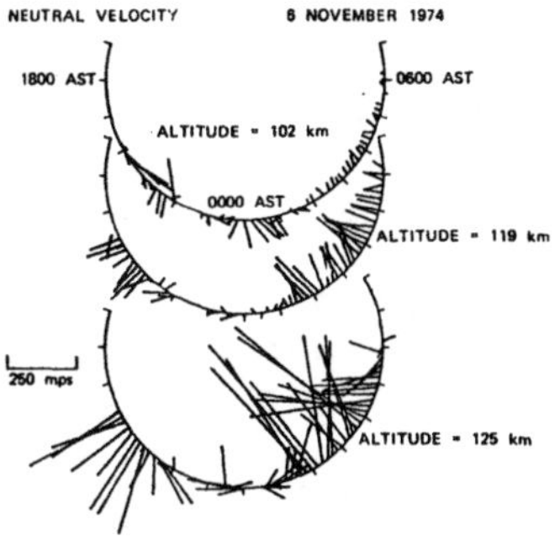

Fig. 11. Neutral wind estimates at three different altitudes. According to [9].

The strong northward motion after midnight can neither be explained by the global pressure asymmetry nor by ion drag. The global pressure gradients will drive the neutrals equatorward in the night section and the electric field being directed mainly southward in this period is equivalent to a southeastward ion motion. Therefore the effect must be due to local sources caused by the very special dynamical situation in the auroral ionosphere.

Recently [15,16] used the following steady-state ion mobility equation

$$2\underline{\Omega} \times \underline{V}_{i_\perp} \doteq \frac{e}{m_i}(\underline{E}_\perp + \underline{V}_{i_\perp} \times \underline{B}) + \nu_{in}(\underline{u} - \underline{V}_{i_\perp}) \qquad (2.10)$$

where the Coriolis force is retained, and solved this simultaneously with Eq. (1.1) where the gravity, the pressure gradients and tidal forces are neglected, thus reducing to:

$$\frac{d\underline{u}}{dt} = 2\underline{\Omega} \quad \underline{u} + \frac{\mu}{\rho} \nabla^2 \underline{u} - \nu_{ni}(\underline{u} - \underline{V}_{i_\perp}) \qquad (2.11)$$

By assuming vertical $\underline{B}$-field, a high latitude electric field model in agreement with the data supplied by [17] and simplified models for the latitudinal variation of the ion density at day and night above 60° of latitude [16] obtained neutral wind patterns in the auroral oval and polar cap at different altitudes. Poleward the auroral oval just after midnight he finds northward velocities above 100 km in good agreement with the data obtained by [9].

From the data presented it is evident that the neutral wind pattern is rather complicated at high latitudes and that the relative position between the auroral oval and the site of observation is of extreme importance for understanding the observed wind pattern. The auroral oval which by itself is very dynamic and responding to the general state of the magnetosphere can sweep across the site of observation within minutes and leave behind a region of confusion.

To increase our knowledge of the neutral wind in the auroral thermosphere a network of observations are needed, which are capable of giving longtime as well as good height resolution measurements. More model analysis like the one of Roble (this issue) are highly wanted, but in order to give meaningful comparison with data other boundary conditions at lower altitudes than is used must be introduced. As this short review indicate, it is just not true that the neutral air is stationary at 120 km, particularly not in the auroral zone. Not only must an auroral heat source be included in the models but also a momentum source due to the ion convection, and both these sources must be time dependent in order to give useful models.

3. JOULE HEATING EFFECTS OF THE HIGH LATITUDE NEUTRAL ATMOSPHERE

3.1 Observations indicating heat sources in the auroral and polar cap region

Early investigations on the drag of artificial satellites established the fact that the temperature and density of the upper atmosphere increase during magnetic storms [18] and that the atmospheric perturbation is enhanced at high latitudes [19].

Neutral wind velocities by chemical rocket releases in the F-region above Kiruna that opposed the thermal wind pattern derivable from the global temperature distribution have been observed by [20]. It was therefore assumed by [20] that local heat sources in addition to ion-drag where present in the auroral zone resulting in pressure gradients opposing the global distribution at high latitudes. In Fig. 12 are shown the wind measurements by [20] superimposed on a global temperature distribution model according to [3]. Also indicated are the expected thermal wind pattern at high latitudes according to this model, and the local heat sources in the auroral zone needed to drive the observed neutral wind.

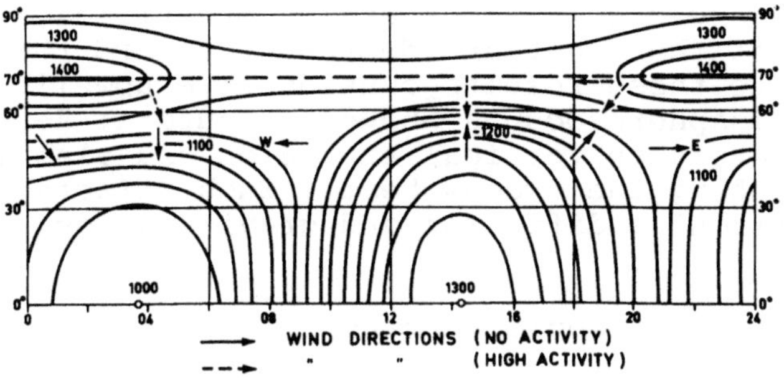

Fig. 12. Temperature distribution in the exosphere when an additional heating source at the night side of the auroral zone is introduced. The change of wind directions is indicated by arrows. According to [20].

Recent composition measurements with gas analyzers on the OGO 6 and ESRO 4 satellites have shown atmospheric perturbations of geomagnetic origin [21,22]. In Fig. 13 are shown the concentration of 4 different atmospheric species molecular Nitrogen, atomic Oxygen, Argon and Helium relative to the quiet day mean. The

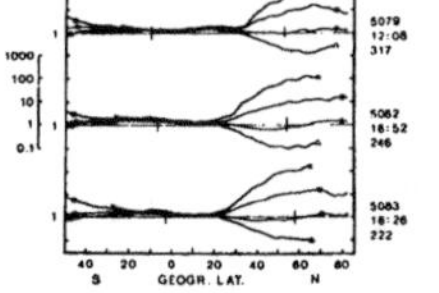

Fig. 13. The concentration of 4 different atmospheric species in subsequent satellite passes during a period of geomagnetic disturbance. According to [22].

observations which are obtained during a period of geomagnetic disturbance demonstrate large variations in the neutral atmospheric species above 300 km at high latitudes. While the heavy constituents like molecular Nitrogen and Argon increase in density during the disturbance, the light species like Helium decreases in density. Oxygen being somewhere in between as far as weight is concerned, increases in concentration at the beginning of the event and decreases toward the end. According to these results considerable upwelling of the neutral atmosphere occurs above the auroral E-region but other effects than vertical expansion are also involved in distributing the excess heat. Such effects are most likely horizontal diffusion and gravity waves.

3.2 The origin of Joule heating

According to conventional textbooks in electromagnetism the rate at which work is done on a plasma by an electric field in the presence of a current is given by

$$P = \int \underline{E} \cdot \underline{J} \, d\tau \tag{3.1}$$

where $d\tau$ is a volume element. Since we are only concerned with E-fields and currents perpendicular to the earth's magnetic field the ionospheric current of interest is given by

$$\underline{J}_\perp = N_i \, e(\underline{V}_{i_\perp} - \underline{V}_{e_\perp}) \tag{3.2}$$

where the electron velocity $\underline{V}_{e_\perp}$ is given by the $\underline{E}_\perp \times \underline{B}/B^2$ velocity.

By introducing Eq. (2.8), Eq. (3.2) can be written as

$$\underline{J}_\perp = \sigma_P \, \underline{E}'_\perp - \sigma_H(\underline{E}'_\perp \times \underline{B})/B \tag{3.3}$$

where $\underline{E}'_\perp = \underline{E}_\perp + \underline{u} \times \underline{B}$ and the Hall and Pedersen conductivities are given by

$$\sigma_H = \frac{N_i e}{B} \frac{\nu_{in}^2}{\Omega_i^2 + \nu_{in}^2} \tag{3.4}$$

$$\sigma_P = \frac{N_i e}{B} \frac{\Omega_i \, \nu_{in}}{\Omega_i^2 + \nu_{in}^2} \tag{3.5}$$

We notice that the electron neutral and the electron

ion collision frequencies are set equal to zero.

We consider the reference frame moving with the neutral wind, in this reference frame the measured electric field is given by

$$\underline{E}'_\perp = \underline{E}_\perp + \underline{u} \times \underline{B} \tag{3.6}$$

and therefore the energy dissipation rate per unit volume becomes particularly simple

$$q = \underline{J}_\perp \cdot \underline{E}'_\perp = \sigma_p \, \underline{E}'^{\,2}_\perp = \sigma_p (\underline{E}_\perp + \underline{u} \times \underline{B})^2 \tag{3.7}$$

In order to obtain experimental estimates of this quantity simultaneous measurements of the current, the electric field and the neutral wind is needed. Very few experimental arrangements except incoherent scatter radars are capable of conveying such information. From a radar measurement however the neutral wind must be derived from ion velocity measurements at different altitudes according to Eq. (2.8). The estimated neutral wind in Eq. (3.7) is therefore very strongly depending on the model neutral atmosphere through the ion-neutral collision frequency.

To interpret q further we write

$$q = \underline{J}_\perp \cdot \underline{E}'_\perp = \underline{J}_\perp \cdot \underline{E}_\perp + \underline{J}_\perp \cdot \underline{u} \times \underline{B} = \underline{J}_\perp \cdot \underline{E}_\perp - \underline{J}_\perp \times \underline{B} \cdot \underline{u} \tag{3.8}$$

The first term to the right is now equal to the energy rate at which work is done on the plasma in the earth fixed frame and the last term is the rate at which work is done by the Lorenz force on the neutrals. For $d\underline{V}_i/dt = 0$ in Eq. (2.6) and by inserting the expression for $\underline{J}_\perp$ from Eq. (3.3) one obtains

$$\underline{J}_\perp \cdot \underline{E}_\perp = e\, N_i \cdot \underline{V}_{i_\perp} \cdot \underline{E}_\perp = - N_i\, m_i\, \nu_{in} (\underline{u} \cdot \underline{V}_{i_\perp} - \underline{V}_{i_\perp}^{\,2}) \tag{3.9}$$

$$\underline{J}_\perp \times \underline{B} \cdot \underline{u} = e\, N_i (\underline{E}_\perp + \underline{V}_{i_\perp} \times \underline{B}) \cdot \underline{u} = N_i\, m_i\, \nu_{in} (\underline{u}^2 - \underline{V}_{i_\perp} \cdot \underline{u}) \tag{3.10}$$

And finally

$$q = \underline{J}_\perp \cdot \underline{E}'_\perp = N_i\, m_i\, \nu_{in} |\underline{u} - \underline{V}_{i_\perp}|^2 \tag{3.11}$$

and the energy dissipation rate in the reference frame moving with the neutral wind is equal to the total rate of change of kinetic energy to the extent that $dV_i/dt=0$.

According to [23] however we have for the rate of change of thermal energy for species α when external forces are neglected in an inviscid gas mixture

$$\frac{3}{2} N_\alpha k\dot{T}_\alpha = \frac{m_\alpha \nu_{\alpha\beta}}{m_\alpha + m_\beta}\left[k(T_\alpha - T_\beta) - m_\beta(\underline{U}_\beta - \underline{U}_\alpha)^2\right] \quad (3.12)$$

α and β may denote the ions and the neutrals, T is the temperature and k Boltzmann's constant. The first term to the right indicates that heat is transported from the hot species to the cooler and the second term indicate that heat is interchanged by collisions. The total rate of change of thermal energy in the mixture of ions and neutrals is given by

$$q_{th} = \frac{3}{2} N_i k\dot{T}_i + \frac{3}{2} N_n k\dot{T}_n = m_i N_i \nu_{in}|\underline{u} - \underline{V}_{i_\perp}|^2 \quad (3.13)$$

and equal to the expression for q in Eq. (3.11). The rate of change of kinetic energy in the reference frame moving with the neutral wind is therefore equal to the rate of change of the total thermal energy of the gas mixture, and we call this the Joule heating rate.

For $T_i = T_n$ we see that the heating is distributed between the ions and neutrals by the fractions m_i/m_i+m_n and m_n/m_i+m_n. In the E-region ionosphere we have $m_i = m_n$ so that

$$q_F = \frac{1}{2} q_{th} = \frac{1}{2} \underline{J} \cdot \underline{E}'_\perp \quad (3.14)$$

is the frictional heating rate of the neutral gas.

4.3 The procedure of deriving Joule heating estimates

To return to the procedure for deriving estimates of the Joule heating rate from observations of ion and electron velocities at different altitudes we notice from Eq. (2.9) when $d\underline{V}_i/dt = 0$ that

$$(\underline{U} - \underline{V}_{i_\perp}) \times \underline{B} = \frac{-e}{m_i \nu_{in}} (\underline{E}_\perp + \underline{V}_{i_\perp} \times \underline{B}) \times \underline{B}$$

and therefore

$$|\underline{U} - \underline{V}_i|^2 = \left(\frac{e}{m_i \nu_{in} B}\right)^2 |(\underline{E}_\perp + \underline{V}_{i_\perp} \times \underline{B}) \times \underline{B}|^2 =$$

$$= \left(\frac{e}{m_i \nu_{in}}\right)^2 |\underline{V}_{i_\perp} - \underline{V}_e|^2 \cdot B^2$$

Finally

$$q_F = \frac{1}{2} \frac{e^2 N_i B^2}{m_i \nu_{in}} |\underline{V}_{i_\perp} - \underline{V}_{e_\perp}|^2 = \frac{1}{2} \frac{\underline{J}_\perp{}^2 B^2}{N_i m_i \nu_{in}}$$

The estimated Joule heat entering into the neutrals therefore is inversely related to the ion neutral collision frequency. This is illustrated in Fig. 14 where the derived Joule heating rate is plotted versus altitude for 3 different ion-neutral collision models. By increasing the collision rate from 3.75×10^{-16} m^3 sec^{-1} to 15×10^{-16} m^3 sec^{-1} the heating rate decreases at all altitudes by a factor 4. The ion-neutral collision frequency in those altitudes may be uncertain by 100% and therefore the estimated heating rates are only indicative.

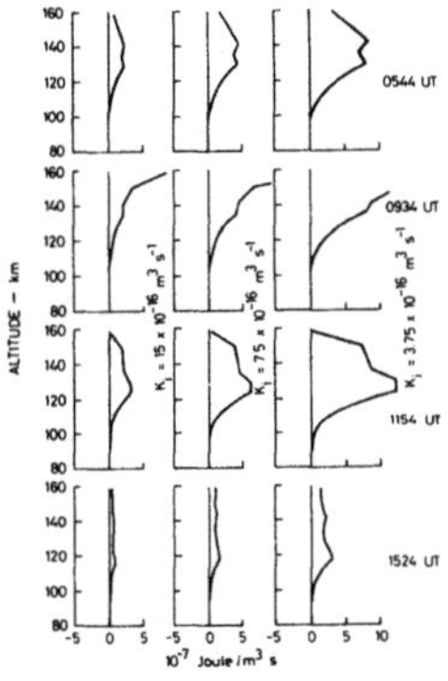

Fig. 14. Joule heating rate versus altitude for three different models of the collision frequency. According to [30].

The upper panel is obtained during a positive magnetic bay, and the heating rate maximizes between 130 and 150 km, the third panel from above is obtained during a negative bay and shows a maximum at about 125 km.

This difference in altitude is also reflected in the current densities during different phases of magnetic disturbances. The current is usually intensified at lower altitudes during periods of negative bays. In Fig. 15 the total integrated Joule heating rate is plotted versus time for a period of 10 hours when a positive as well as a negative bay occurred. The total heating rate during the positive bay between 05 and 07 UT is of the same order og magnitude as the heating rate during the negative bay between 11 and 16 UT, even though the current in the negative bay is stronger.

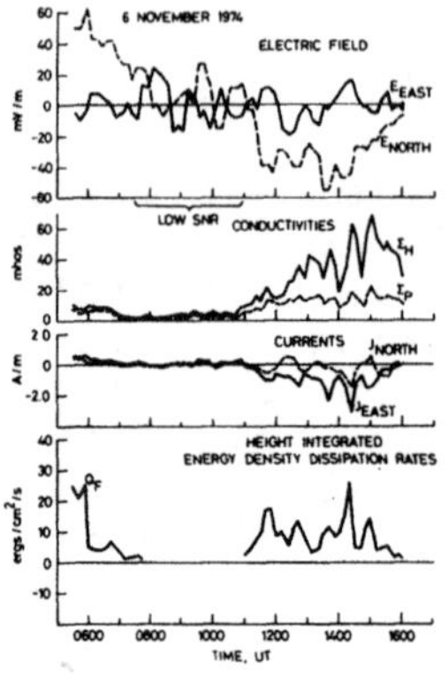

Fig. 15. Electric fields, height integrated conductivities, currents and Joule heating rates derived during a radar experiment conducted in the auroral zone. According to [30].

In Fig. 16 however are shown the Joule heating rate for an experiment where the largest heating occurred during a positive bay. The reason for this can be seen in the large difference in the electric field which was almost 50 mV/m in the positive bay and only 20 mV/m in the negative bay. The height integrated Pedersen conductivities being only a factor of 2 different in the two events. Since the Joule heating rate is proportional to the square of the E-field, and only linearly related to the Pedersen conductivity the Joule heating rate is much more sensitive to variations in the E-field.

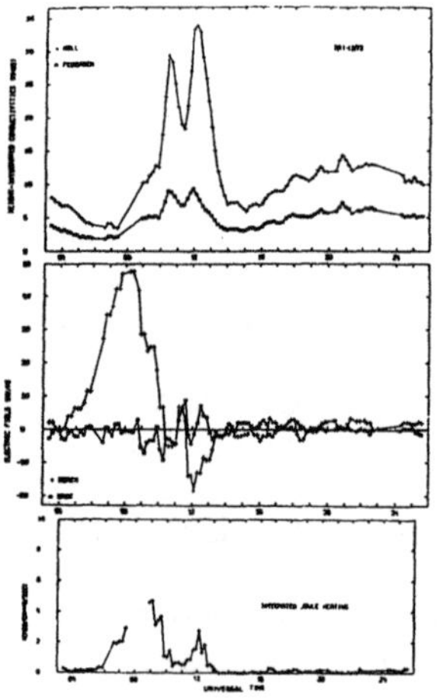

Fig. 16. Electric fields, height integrated conductivities and Joule heating rates derived for a 24 hour radar experiment in the auroral zone.

One has seen attempts in the past to derive estimates of the Joule heating rate by using magnetometer data. The data presented here clearly underlines the caution that must be maintained when such enterprises are undertaken, and attempts should be made to calibrate magnetograms with respect to Joule heating rates.

4. RECENT THEORETICAL DEVELOPMENTS CONCERNING THE EFFECTS OF ION-DRAG AND JOULE HEATING. CONCLUSION

We have seen experimental evidences for large dynamical variations in the auroral zone neutral atmosphere due to auroral disturbances like enhanced electric fields and particle precipitation. Arguments have been put forward on one hand that ion-drag is the major factor in modifying the neutral motion, while on the other Joule heating effects are also of major concern. Effects of the Corioli's force and the inertia have also been considered by some authors.

A fully consistent theoretical investigation of the complex of interaction appears next to unsolvable, but even so large progress has been made on this point the latest years. Important contributions to the understanding of this complex relationship have been made in addition to [24] by [25,26,27,11,15,16,28,29,10,12], and Roble (this issue).

Particularly the three dimensional analysis by [29] of the dynamical effects due both to a momentum source (ion-drag) and an energy source (Joule heating) caused by enhanced convection electric fields in the auroral zone, shows how these two effects will modify the thermal motion differently at different time zones and latitudinal regions.

It appears hazardeous from single measurements in space of the neutral air motion and related ion velocities to infer what source has caused modified neutral winds if any. A large global grid of observational data of the main dynamical parameters is needed before the present theories can be tested experimentally. Possibly careful calibrations of magnetograms with respect to Joule heating, may represent a potential source for global scale data in this respect.

Acknowledgement: I am greatly indebted to Drs. Charles L. Rino and Murray J. Baron for allowing me to use unpublished data from the Chatanika Incoherent Scatter Radar.

REFERENCES

1. D.M. Willis, The energetics of Sun-weather relationships: magnetospheric processes, J. Atmos. Terr. Phys., 38, 685-698, 1976.
2. H. Rishbeth, Thermospheric winds and the F-region: A review, J. Atmos. Terr. Phys., 34, 1-47, 1972.
3. L.G. Jacchia, The temperature above the thermopause, Space Research V, p. 1152, North-Holland, Amsterdam, 1965.
4. Geisler, Atmospheric winds in the middle latitude F-region, J. Atmos. Terr. Phys., 28, 703-320, 1966.
5. Geisler, A numerical study of the wind system in the middle thermosphere, J. Atmos. Terr. Phys., 29, 1469-1482, 1967.
6. P.W.Blum and I.Harris, Full nonlinear treatment of the global thermospheric wind system - I. Mathematical method and analysis of forces, J. Atmos. Terr. Phys., 37, 193-212, 1975a.
7. P.W. Blum and I.Harris, Full non-linear treatment of the global thermospheric wind system - II. Results and comparison with observations, J. Atmos. Terr. Phys., 37, 213-235, 1975b.
8. J. Heppner, High latitude winds related to solar-interplanetary conditions, Goddard Space Flight Center, Maryland, November 7-8, 1973.
9. C.L. Rino, A. Brekke and M.J. Baron, High-resolution auroral zone E-region neutral wind and current measurements by incoherent-scatter radar. Accepted by J. Geophys. Res., 1977.
10. J.A. Fedder and P.M. Banks, Convection Electric Fields and Polar Thermospheric Winds, J. Geophys. Res., 77, 2328-2340, 1972.
11. S.T. Wu and K.D. Cole, Transient Thermospheric Heating and Movement caused by an Auroral Electric Field, Planet. Space Sci., 24, 727-730, 1976.
12. R.H. Comfort, S.T. Wu and G.R. Swenson, An Analysis of Auroral E-region neutral winds based on incoherent scatter radar observations at Chatanika, Planet. Space Sci., 24, 541-560, 1976.
13. D. Rees, Ionospheric Winds in the Auroral Zone, J. Brit. Int. Soc., 24, 233-246, 1971.
14. A. Brekke, J.R. Doupnik and P.M. Banks, A Preliminary Study of the Neutral Wind in the Auroral E-region, J. Geophys. Res., 78, 8235, 1973.
15. H. Maeda, Nuetral winds and ion drifts in the polar ionosphere caused by convection electric fields, J. Atmos. Terr. Phys., 38, 197-205, 1976.

16. H. Maeda, Neutral winds and ion drifts in the polar ionosphere caused by convection electric fields - 2. Preprint, submitted to J. Atmos. Terr. Phys., 1977.
17. D.A. Gurnett, Critical Problems of Magnetospheric Physics, p. 123, IUCSTP Secretariat, Washington D.C., 1972.
18. L.G. Jacchia, Corpuscular radiation and the correlation of artificial satellites, Nature, 183, 1662, 1959.
19. L.G. Jacchia and J. Slowey, Atmospheric heating in the auroral zones: A preliminary analysis of the atmospheric drag of the Injun 3 satellite, J. Geophys. Res., 69, 905, 1964.
20. W. Stoffregen, Magnetosphere-Ionosphere Interactions, ed. K. Folkestad, Universitetsforlaget, Oslo, pp. 83-86, 1972.
21. G.W. Prølss and U. von Zahn, Large and small scale changes in the disturbed upper atmosphere, J. Atmos. Terr. Phys., 38, 655-659, 1976.
22. C.A. Reber, A.E. Hedin and S. Chandra, Equatorial phenomena in neutral thermospheric composition, J. Atmos. Terr. Phys., 35, 1223, 1973.
23. B.S. Tannenbaum, Plasma Physics, McGraw-Hill, New York, p. 112, 1967.
24. K.D. Cole, Joule Heating of the upper atmosphere. Austr. J. Phys., 15, 223, 1962.
25. A.D. Richmond and S. Matsushita, Thermospheric Response to a Magnetic Substorm, J. Geophys. Res., 80, 2839-2850, 1975.
26. M.G. Heaps, The Effects of Including the Coriolis Force of Joule Dissipation in the Upper Atmosphere, Planet. Space Sci., 22, 1031-1035, 1974.
27. M.G. Heaps and L.R. Megill, Circulation in the High-Latitude Thermosphere due to Electric Fields and Joule Heating, J. Geophys. Res., 80, 1829-1831, 1975.
28. P.B. Hays, R.A. Jones and M.H. Rees, Auroral heating and the composition of the neutral atmosphere, Planet. Space Sci., 21, 559-561, 1973.
29. J.M. Strauss and M. Schulz, Magnetospheric Convection and Upper Atmospheric Dynamics, J. Geophys. Res., 81, 5822-5832, 1976.
30. A. Brekke, C.L. Rino and M.J. Baron, Incoherent-Scatter Radar Measurements of the Auroral-Zone Energy Dissipation due to Ionospheric Currents. Preprint, Stanford Research Institute, Menlo Park, California, 1976.

DISCUSSION

Roble: By examining both Chatanika ion drift motions and Fabry-Perot interferometer measured red line winds at 250 km Dr.P.B. Hays, University of Michigan and I were able to estimate the importance of the pressure term in the horizontal momentum equations. We found a very large horizontal scale size for E x B momentum addition in the east-west direction. As a result, the pressure force in the east-west direction could not adjust sufficiently to balance this force with the net effect of winds moving with the E x B drift in the zonal direction. The north-south component of the E x B drift, however, had a smaller horizontal scale size and the derived pressure force effectively balanced the momentum addition and therefore large horizontal winds in the N-S direction did not develop. In the E-W direction the motion was approximately 50% E x B driven and 50% pressure driven.

During geomagnetic storms when Ap exceeds some value like 35 or 40, Dr. G. Hernandez from the Fritz Peak Observatory near Boulder measures strong westward winds at 250 km near sunset and at the same time he measures a decrease in the local temperature with time. The temperature decrease should drive eastward winds and the measured westward winds suggest that perhaps strong E x B drifts move equatorward to mid-latitudes during geomagnetic storms and affect the mid-latitude zonal winds. During quiet geomagnetic conditions the sunset winds are generally eastward or weakly westward.

LARGE-SCALE CHARACTERISTICS OF FIELD-ALIGNED CURRENTS DETERMINED FROM THE TRIAD MAGNETOMETER EXPERIMENT*

T. A. Potemra

The Johns Hopkins University
Applied Physics Laboratory
Laurel, Maryland 20810 U.S.A.

1. INTRODUCTION

Following earlier suggestions of Edmund Halley and Anders Celsius for the "magnetic behavior" of auroral phenomena, Kristian Birkeland discovered in his polar expeditions of 1902-3 that large-scale currents were associated with the aurora. He was also the first to suggest that these currents originated far from the earth and that they flowed into and away from the polar atmosphere along the geomagnetic field lines. The existence of such field-aligned currents was widely disputed because it was not possible to unambiguously distinguish between current systems that are field aligned *(Birkeland, 1908)* and those which are completely ionospheric *(Vestine and Chapman, 1938)* only from a study of surface magnetic field measurements. Furthermore, recent theoretical calculations by *Fukushima (1969, 1976)* predict that the effects of field-aligned currents may be impossible to detect with ground-based magnetometers for certain models of these current systems.

The presence of field-aligned currents has been confirmed in the last decade with particle and magnetic field observations acquired from a variety of rocket and satellite instruments (e.g., the reviews of *Arnoldy, 1974 and Anderson and Vondrak, 1975)*. The rocket-borne experiments have provided detailed observations of the fine structure of field-aligned currents (and their relationship to energetic particles and visual auroral arcs) which are

* This work supported by the National Science Foundation and the Office of Naval Research.

B. Grandal and J. A. Holtet (eds.), Dynamical and Chemical Coupling, 337-352.

confined to a limited range of space and time. The satellite experiments have provided observations of the large-scale characteristics of field-aligned currents over a much larger spatial range. The first satellite measurements of field-aligned currents were made by *Zmuda et al. (1966, 1967)* with a single axis magnetometer onboard the polar orbiting satellite 1963-38C at ~ 1100 km altitude. The magnetic disturbances observed by *Zmuda et al.* were initially interpreted as hydromagnetic waves, but it was soon realized that their latitudinal extent was too small for waves of the appropriate wavelength *(Cummings and Dessler, 1967)* and they were interpreted as being due to field-aligned, or Birkeland currents. Several observations of field-aligned currents have been reported during the last few years by virtue of their associated magnetic variations detected by satellite borne magnetometers. Among these have been the observations in the auroral and polar cap regions acquired with a two-axis magnetometer onboard the Azur satellite at altitudes between 400 and 3100 km *(Theile and Praetorius, 1973)*; the three-axis measurements by the IMP-4 and IMP-5 satellites in the magnetotail plasma sheet *(Fairfield, 1973)*; the three-axis measurements on the OGO-5 satellite at altitudes $> 4\ R_e$ *(Sugiura, 1975)*; and at synchronous altitude with the ATS-1 satellite *(Coleman and McPherron, 1970; Iijima, 1974)*.

The vector magnetometer onboard the low altitude (~ 800 km) polar orbiting TRIAD satellite has been used to determine for the first time the flow directions, spatial distribution, and intensities of field-aligned currents at all magnetic local times in the northern high-latitude region *(Armstrong and Zmuda, 1973; Zmuda and Armstrong, 1974a,b; Armstrong, 1974; Iijima and Potemra, 1976a,b; Sugiura and Potemra, 1976)*. The TRIAD satellite has to date obtained the only continuous three-axis, high-resolution magnetic measurements at relatively low altitudes (~ 800 km) in the north and south polar regions. The large-scale average properties of field-aligned currents determined from the TRIAD observations are reviewed here. These include the location, flow directions and intensities of field-aligned currents in the auroral and polar regions and their relationship to optical and radar auroral forms and geomagnetic and interplanetary magnetic field variations. The importance of these Birkeland currents to the coupling between the magnetosphere and the auroral atmosphere and ionosphere is emphasized by their total intensity, which ranges between 10^6 and 10^7 amperes, and by the energy which they dissipate in the upper atmosphere, which can exceed by a considerable factor the energy deposition associated with the dramatic visual auroral forms.

2. THE TRIAD MAGNETOMETER EXPERIMENT

The characteristics of the TRIAD magnetometer experiment have been described in detail by *Armstrong and Zmuda (1973)*, and the principal features are listed as follows. The Navy/APL TRIAD satellite was launched into a nearly circular polar orbit at 800 km altitude in November 1972. Each of the three orthogonal axes of the magnetometer on TRIAD are sampled at a rate of 2.25 samples/axis/sec to a resolution of 12 γ. At each observation time, theoretical vector magnetic field values are computed from the IGRF (International Geomagnetic Reference Field) 1965.0 geomagnetic field model extrapolated to the epoch of the observations and are subtracted from the values measured by each magnetometer. The vector difference ΔB is plotted against time, invariant latitude, and magnetic local time for analysis.

Data from the TRIAD satellite have been recorded on a real-time basis at ground stations located at College, Alaska (Λ = 64.6°), Resolute, N.W.T., Canada (Λ = 84.1°) and McMurdo, Antarctica (Λ = 79.7°). The data were received at College during TRIAD crossings of the auroral oval roughly between 180°E and 240°E geographic longitude. For these data, the three orthogonal axes of the TRIAD magnetometer are aligned along the following directions: (1) radially downward and nearly parallel to the main geomagnetic field, referred to in the following as the Z sensor, (2) transverse to the main geomagnetic field and approximately in the magnetic east-west direction, referred to as the A sensor, and (3) transverse to the main geomagnetic field and approximately in the magnetic north-south direction, referred to as the B sensor *(see Figure 2 of Armstrong and Zmuda (1973))*.

Two kinds of variations are observed in ΔB: gradual and slowly varying changes and distinct short-period variations with large amplitudes. The short-period variations in ΔB are observed mainly in the A sensor, which is predominantly transverse to the main geomagnetic field and aligned in the geomagnetic east-west direction for the College data. As reported earlier by *Zmuda and Armstrong (1974b)*, these disturbances are nearly always observed at high invariant latitudes, Λ = 60°-80°, and are statistically coincident with the visual auroral oval defined by *Feldstein (1966)*.

Figures 1a-1d show examples of the TRIAD magnetometer variations which were measured from College. In each figure the record from the A magnetometer sensor is plotted against invariant latitude (which increases from left to right in all cases). A positive eastward disturbance is plotted upward in all figures, and the reference scale of 400 γ is the same for all figures. The magnetic local time is shown near the principal disturbance in each plot. The data have been selected to represent typical variations often observed in the following local time sectors: evening to premidnight sector (1700-2200 MLT) in Figure 1a, midnight to early morning (2300-0600 MLT) in Figure 1b, late morning (0700-1000 MLT) in

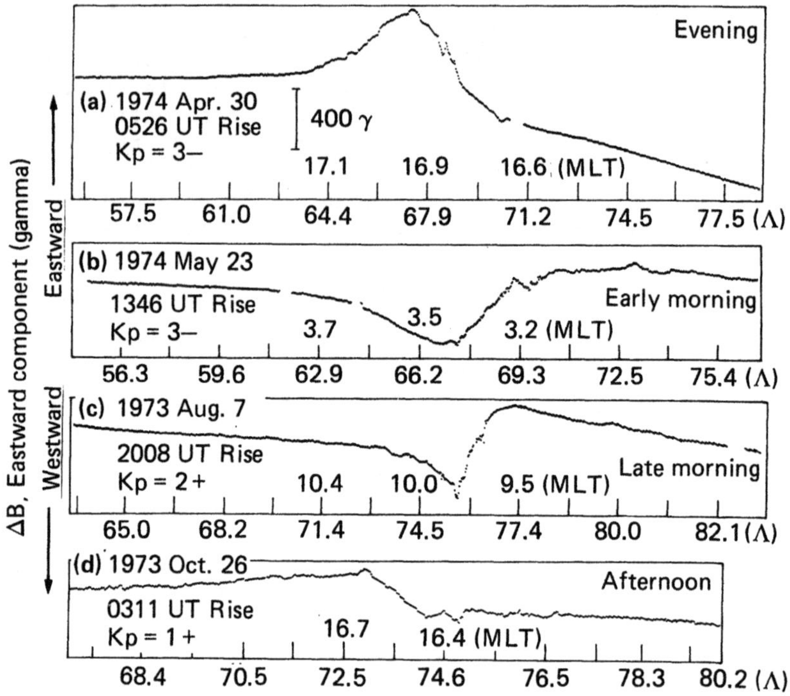

Fig. 1 Examples of magnetic perturbations in an approximately geomagnetic east-west direction observed by TRIAD in various local time sectors.

Figure 1c, and afternoon (1300–1600 MLT) in Figure 1d *(see Iijima and Potemra, 1976a,b or Sugiura and Potemra, 1976 for more examples)*.

The magnetic disturbance in the premidnight sector shown in Figure 1a is in the eastward direction, whereas the disturbance in the postmidnight sector shown in Figure 1b is in the westward direction. These disturbances are consistent with field-aligned currents directed away from the ionosphere in the poleward boundary of the magnetic disturbance region and into the ionosphere at the equatorward boundary in the premidnight sector and with currents directed into the ionosphere at the polar boundary and away from the ionosphere at the equator boundary in the postmidnight sector. This pattern was first determined by *Zmuda and Armstrong (1974b)*.

A striking feature which is apparent in the late morning sector shown in Figure 1c and in the afternoon sector shown in Figure 1d is an apparent difference in the level of ΔB before and after the pronounced disturbance. This apparent "level shift" as noted by *Sugiura and Potemra (1976)* is suggestive of an unbalanced or "net current" flow on one side of the field-aligned current region. This current imbalance will be discussed in more detail in the following sections.

The data displayed in Figure 1 for the different local time sectors include the following characteristics: (1) The disturbances in the late evening (1700-2200 MLT) and early morning (2300-0600) hours show the triangular variation noted in the earlier analyses of TRIAD data by *Zmuda and Armstrong (1974b)* in which the associated field-aligned currents consist of opposite and approximately equal current sheets. (2) The disturbances in the day side hours shown in Figure 1c and 1d are characteristically different from those on the night side shown in Figures 1a and 1b. These day side disturbances show a shift in the level of ΔB across the disturbance region suggestive of an unbalanced current system.

The magnitude and directions of equivalent field-aligned currents are determined from the ΔB variations with the relation $J = (1/\mu_0)\partial/\partial x\ \Delta By$, where J is the density of the field-aligned current, x is positive in the northward direction, and ΔBy is the observed disturbance in the east-west direction *(e.g., Zmuda and Armstrong, 1974b)*. Implicit in the derivation of this formula is the fact that the observed disturbances in the geomagnetic north-south direction are always smaller (less than 10%) than the east-west disturbances. In this case the formula is the same as that for the transverse magnetic disturbance due to current sheets of infinite longitudinal extent. At the altitude of TRIAD a change in ΔBy equal to 100 γ over 1° of latitude is equivalent to a field-aligned current density of 6.4×10^{-7} A/m^2 = 0.64 μA/m^2. The magnitude of the current intensity $\int J\,dx$ is given directly by the amplitude of the magnetic disturbance ΔBy by $\int J\,dx = (1/\mu_0)\Delta By$. A 100 γ disturbance corresponds to a current intensity of 0.08 A/m.

The "large-scale" features in the ΔB changes are described here which are defined by the following criteria: ΔB variations occurred over a latitude range greater than 0.5° and within a longitude span less than 1 hour in MLT: and only current densities greater than 0.25 μA/m^2 were considered (for example, at least a 40 γ change over a 1° change in latitude).

3. PRINCIPAL CHARACTERISTICS OF FIELD-ALIGNED CURRENTS

Figure 2 is a summary of the average spatial distribution of large-scale field-aligned currents in the northern high-latitude region determined from magnetometer data obtained on 493 TRIAD passes during weakly disturbed conditions (|AL| < 100 γ, the average |AL| being about 50 γ) *(from Iijima and Potemra, 1976b)*. The principal features determined by *Iijima and Potemra (1976a,b)* and *Sugiura and Potemra (1976)* include the following. Field-aligned currents are concentrated in two principal areas which encircle

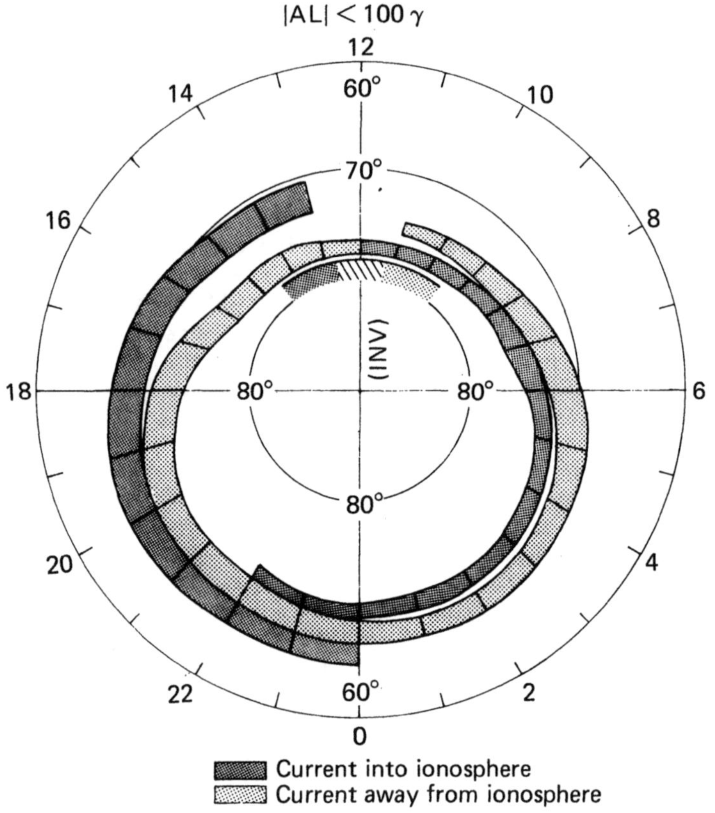

Fig. 2 The spatial distribution and flow directions of large-scale field-aligned currents determined from data obtained on 493 passes of TRIAD during weakly disturbed conditions (| AL | < 100 γ) from Iijima and Potemra (1976b). The hatched area shown between 1130 and 1230 MLT in the polar cusp region indicates that the current flow directions are often confused.

the geomagnetic pole, as does the so-called auroral oval: Region 1, located near the poleward boundary of the field-aligned current region; and Region 2, located near the equatorward boundary. The Region 1 field-aligned currents flow into the ionosphere in the morning sector and away from the ionosphere in the evening sector; whereas the Region 2 currents flow in the opposite direction at any given local time. A three-region flow pattern of field-aligned currents exists in the 2200 to 2400 MLT sector, and may be thought of as an "overlapping" of the two-region flow patterns usually observed in the surrounding MLT sectors. This same basic three-region flow pattern of field-aligned currents persists during disturbed substorm periods when the westward auroral electrojet is confined to the midnight sector *(Iijima and Potemra, 1977)*. However, during periods when the westward auroral electrojet has intruded deeply into the evening sector, the TRIAD magnetometer data exhibit very complicated and fine-structured variations indicating the presence of complex and multiple patterns of field-aligned currents which cannot be explained with the basic three-region patterns described earlier.

A region of field-aligned currents has been discovered in the dayside betweed 1000 and 1400 MLT and poleward of Region 1 between ~ 78° and 81° invariant latitude *(Iijima and Potemra, 1976b)* and is also shown in Figure 2. Because of their location and close correlation of their intensities with increases of the interplanetary magnetic field in the southward direction, these field-aligned currents are believed to be associated with the magnetospheric cusp. These "cusp" field-aligned currents flow away from the ionosphere in the pre-noon sector (1000 to 1130 MLT) and into the ionosphere

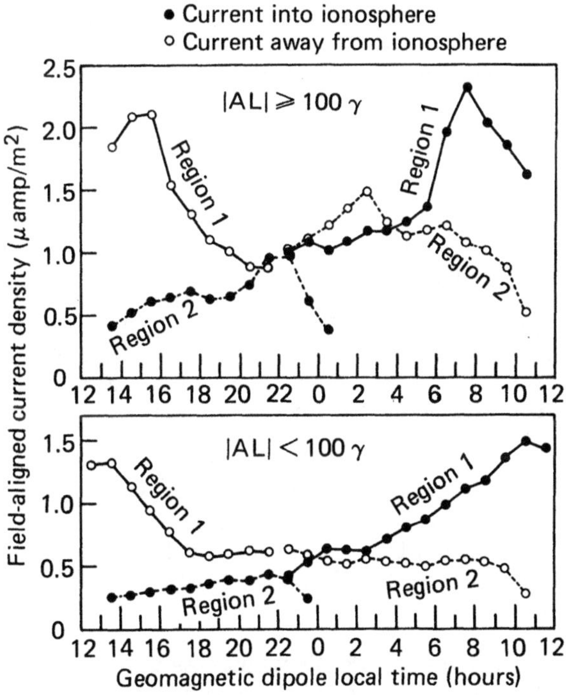

Fig. 3 Diurnal distribution of average field-aligned current densities during active periods (| AL | ≥ 100 γ) in the upper panel, and during weakly disturbed periods (| AL | < 100 γ) in the lower panel (from Iijima and Potemra, 1977).

in the post-noon sector (1230 to 1400 MLT). Both flow directions are observed in the one-hour interval centered about noon (and shown as the "hatched" area in Figure 2).

Figure 3 shows the average current densities in Region 1 and 2 plotted against magnetic local time *(from Iijima and Potemra, 1977)*. The average current densities during active periods ($|AL| \geq 100\ \gamma$) are plotted in the upper panel, and during relatively quiet periods ($|AL| < 100\ \gamma$) in the lower panel. In Region 1 during relatively quiet conditions, the largest current densities occur near 1000-1100 MLT and 1300-1400 MLT with average values of $\sim 1.4\ \mu amp/m^2$. These locations also correspond to the largest net field-aligned currents determined from a "step-like" level shift in the magnetic field component in the magnetic east-west direction *(e.g., Sugiura and Potemra, 1976)*.

During active periods, the locations of the largest Region 1 field-aligned currents shift toward the nightside and their average density increases to $\sim 2.2\ \mu amp/m^2$. The current density in Region 1 is statistically larger than the current density in Region 2 at all local times except in the pre-midnight to early morning sector. During active (substorm) periods, the densities of the Region 2 field-aligned currents show their largest enhancement in this local time sector, which is also the sector where the westward auroral electrojet is most active.

The average values of field-aligned current density and intensity in the various regions and sectors discussed above are listed in Table 1 *(from Iijima and Potemra, 1977)*. The total current, estimated by integrating the observed current intensity over the local time extent of the appropriate region, is also listed in Table 1. The total amount of current flowing into the ionosphere (including Regions 1 and 2) is always equal to the total current flowing away (within 5%) for every level of magnetic activity. The fact that current continuity is always preserved is important to the development of the models of the three-dimensional current system that couples the magnetosphere with the lower atmosphere and ionosphere. It is also significant that the total field-aligned current in Region 1 is significantly larger than the Region 2 total current during relatively quiet conditions, while during active periods, the total current flow in both regions becomes nearly equal.

Preliminary analysis of TRIAD magnetometer data acquired over the south polar regions and recorded at McMurdo, Antarctica, indicate that the same basic flow pattern shown in Figure 2 for the north pole also persists in the south pole *(Potemra et al., 1975)*. At any given local time and invariant latitude location, the field-aligned current flow is the same, as diagrammed in Figure 4 *(see also Potemra, 1977)*. This means that the current flow from both

TABLE 1

AVERAGE CHARACTERISTICS OF LARGE-SCALE FIELD-ALIGNED CURRENTS

Parameter	Afternoon to Midnight Region 1	Afternoon to Midnight Region 2	Midnight to Forenoon Region 1	Midnight to Forenoon Region 2
Current Density $\mu A/m^2$				
$\lvert AL \rvert < 100\ \gamma$	- 0.9	0.3	0.9	- 0.5
$\lvert AL \rvert \geq 100\ \gamma$	- 1.4	0.7	1.5	- 1.1
$2_- < K_p \leq 4_+$	- 1.1	0.6	1.3	- 0.9
Total Current 10^6 A				
$\lvert AL \rvert < 100\ \gamma$	- 1.3	0.8	1.5	- 1.1
$\lvert AL \rvert \geq 100\ \gamma$	- 2.8	2.6	3.2	- 3.3
$2_- \leq K_p \leq 4_+$	- 1.8	1.5	2.0	- 1.6

A positive value denotes current flow into the ionosphere;
A negative value denotes flow away from the ionosphere.
Values are averages over specified regions.

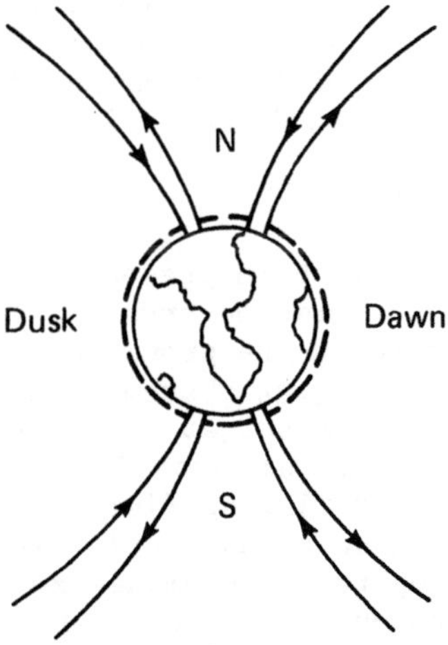

Fig. 4 A schematic diagram illustrating the general flow directions of large-scale field-aligned currents determined from TRIAD data recorded in the north and south high latitude regions.

hemispheres can be extended along geomagnetic field lines to the plasma sheet in the equatorial plane. This would support early suggestions *(e.g., Alfvén, 1939, 1940)* for the source of large-scale field-aligned currents being located in the plasma sheet.

The large-scale field-aligned currents comprise an important element in the three-dimensional system which couples energy from the magnetosphere to the lower atmosphere and ionosphere. The field-aligned currents are related to a variety of auroral and infrasonic waves. The importance of the field-aligned currents can be appreciated by comparing their magnitudes and the estimated power that they deposit in the auroral atmosphere with the values associated with auroral phenomena. For example, the total current available in the field-aligned currents (from Table 1) is greater than the eastward and westward auroral electrojet currents, estimated to range from 10^5 to 10^6 amp *(Nagata et al., 1966; Akasofu and Kamide, 1976)*. The total field-aligned current is comparable to that estimated in the ring current (2-4 x 10^6 amp) by *Akasofu (1969)* and *Sugiura and Poros (1973)* and it has been suggested that the source of the night side field-aligned currents is the plasma sheet region including the ring current *(Fairfield, 1973; Sugiura, 1975)*.

The total power dissipated by field-aligned currents can be estimated several ways. For example, if the voltage drop across the auroral oval is estimated to be 10 kV to 50 kV (corresponding to electric fields of 10 mV/m to 50 mV/m over a 100 km width; see for example the electric field observations of *Heppner, 1972* and calculations of *Boström, 1974)*, the total power deposited by currents ranging between 10^6 and 10^7 amps is 10^{10} to 5 x 10^{11} watts. This range is 0.03% to 2% of the estimated total power available in the solar wind *(Sharp and Johnson, 1968)*.

4. VISUAL AND RADAR AURORA

Several examples have been presented for the association of field-aligned currents determined from the TRIAD data with visual auroral features observed on the ground with all-sky cameras and with the DMSP satellite imaging system *(Armstrong et al., 1975; Kamide and Akasofu, 1976; Kroehl et al., 1976; Kamide et al.,1976)*. In particular, discrete auroral arcs have been observed near regions of intense, upward flowing field-aligned currents (consistent with the precipitation of energetic electrons producing the auroral arc). It is important to emphasize that not all field-aligned currents determined from the TRIAD data can be associated with visual auroral features. This is related to the fact that a particle with any energy can produce a field-aligned current, but only particles with energies greater than a certain threshold can produce visual auroral features. A consequence of this fact is that the visual emissions represent only a fraction of the total energy deposited in the auroral ionosphere by field-aligned currents.

Radar aurora observed with the 398 MHz phased-array located at Homer, Alaska have been compared with field-aligned currents determined by TRIAD on an event-by-event basis in the 1800 to 2100 MLT sector *(Tsunoda et al., 1976)*. The results of this study include the following findings: (1) the downward field-aligned currents in the evening sector are closely associated with diffuse radar aurora and the eastward electrojet; (2) the upward field-aligned currents in the same time sector are associated with the visual aurora; (3) part of the two oppositely directed field-aligned currents are connected in the ionosphere by a Pedersen current driven by a poleward-directed electric field applied across the field-aligned current region; (4) radar auroral echoes are not observed when the field-aligned current intensity is less than ~ 0.16 amp/m, which may represent an important threshold current level for the physical process producing the auroral echoes; and (5) during periods of disturbed geomagnetic conditions an imbalance in the intensities of the field-aligned currents is noticeable in the evening sector, with the outward field-aligned currents at the poleward boundary greater than the current at the equatorward boundary. This final point supports the conclusions of *Sugiura and Potemra (1976)* and *Iijima and Potemra (1976a)* concerning net field-aligned currents.

5. RELATION TO MAGNETOSPHERIC PROCESSES

The basic spatial distribution and flow direction pattern of field-aligned currents shown in Figure 2 (determined by *Zmuda and Armstrong, 1975; Iijima and Potemra, 1976a,b*) persists throughout a wide range of geomagnetic conditions (although their intensities are closely correlated with geomagnetic and interplanetary activity). Due to various uncertainties, it is difficult to accurately map the regions of field-aligned currents observed at relatively low altitudes to the distant magnetosphere. Figures 5 and 6 represent an attempt to map these field-aligned current regions to the equatorial plane using the magnetic field model of *Fairfield and Ness (1975)* for conditions corresponding to quiet and disturbed conditions. Also indicated on these figures is the boundary of the inner edge of the plasma sheet determined by *Vasyliunas (1968, 1972)* for quiet and disturbed periods. The boundary between the flow directions of the field-aligned currents on the dusk side statistically coincides with the inner edge of the plasma sheet shown in Figure 5. Furthermore, on the dusk side the field-aligned current flow *away* from the ionosphere maps to the region within the plasma sheet. The flow of electrons from the plasma sheet to the north and south auroral region can adequately account for the flow pattern of these currents. The

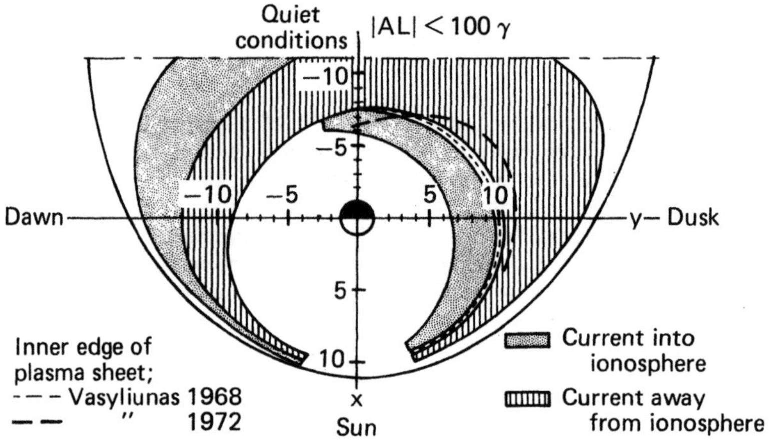

Fig. 5 The projection of the regions of large-scale field-aligned currents determined during relatively quiet conditions (| AL | < 100 γ) onto the equatorial plane using the geomagnetic field model of Fairfield and Mead (1975) for $K_p < 2$.

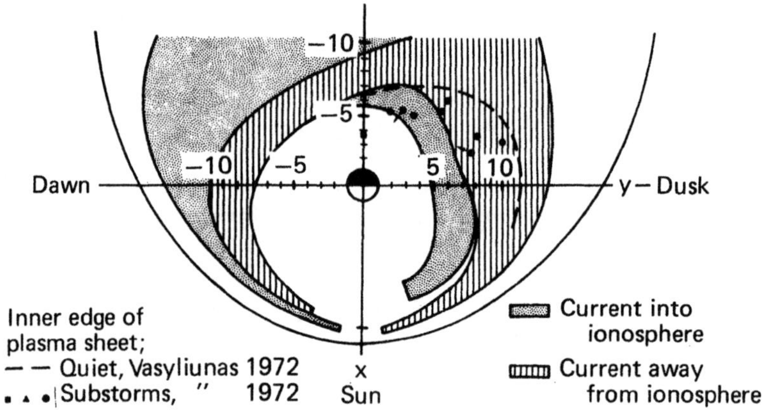

Fig. 6 The projection of the regions of large-scale field-aligned currents during disturbed conditions (| AL | ≥ 100 γ) onto the equatorial plane using the geomagnetic field model of Fairfield and Mead (1975) for $K_p \geq 2$.

field-aligned current flow *into* the auroral ionosphere on the dusk side maps to a region to the earthward side of the plasma sheet boundary (where no electrons are available for transport to the auroral region). This flow direction can be accounted for by the flow of positive particles from the equatorial plane into the auroral region, or the flow of electrons from the auroral ionosphere to the equatorial plane.

During disturbed conditions (particularly substorms), the equatorward shift of the field-aligned current regions observed from the TRIAD observations results in the somewhat distorted projection onto the equatorial plane as shown in Figure 6. During disturbed periods, the boundaries of the field-aligned currents in the equatorial plane move toward the earth, as does the inner edge of the plasma sheet.

The principal feature apparent from the statistical mapping discussed above is the association of the inner edge of the plasma sheet with regions of field-aligned currents. If this region of the plasma sheet is taken to be the *Alfvén layer* as defined by *Shield et al. (1969)*, the TRIAD observations support the mechanism originally proposed by *Alfvén (1939, 1940)* (and later expanded by *Shield et al., 1969*) to drive field-aligned currents. This mechanism is based on the fact that a certain amount of charge separation develops in the energetic plasma drifting toward the earth from the plasma sheet because the geocentric drift paths of ions and electrons are not coincident. The charge separation produces an electric field that is perpendicular to the geomagnetic field and is transmitted along field lines to the auroral ionosphere. Currents flow along the magnetic field lines in such a way to neutralize the charge separation. *Shield et al. (1969)* have computed the magnitude of the Birkeland current necessary to neutralize the *Alfvén layer* to be about 10^6 amps — comparable to the values determined from the TRIAD observations (see Table 1).

Whether this is the driving mechanism for large-scale field-aligned currents cannot be determined from a statistical survey of observations acquired by a single spacecraft and alternate source mechanisms have been proposed *(e.g., Stern, 1975)*. The statistical characteristics of the Region 1 currents, for example their relatively stable behavior, persistence during periods of geomagnetic quiet, dominance in intensity over the intensity of the Region 2 currents and their projection to the magnetospheric region (i.e., Figures 5 and 6) suggest that these currents are the *driven* or *primary* current system and are associated with the distant boundaries of the plasma sheet where the hot plasma and bulk plasma convection may contribute to the generation of large-scale field-aligned currents *(e.g., Sugiura, 1974; Stern, 1975)*. The Region 2 field-aligned currents which show greater variability, close association with electrojet activity, and map to the inner regions of the plasma sheet, appear to be controlled by more localized and variable mechanisms, such as the ionospheric conductivity, and may be the *response* to the Region 1 field-aligned currents.

The north and south polar regions are the focal points of a great variety of solar-terrestrial interactions. The permanent presence of large-scale field-aligned currents which flow into and away from the lower ionosphere has been confirmed by several rocket and satellite experiments. The total amplitude of these currents is measured in millions of amperes and they play an important role in a variety of auroral and magnetospheric phenomena. Many questions do, however, remain as to the exact source of these currents and even to the nature of the particles which comprise these currents. Simultaneous observations of charged particles, electric fields, and magnetic fields from a single satellite or multiple satellites are required to make substantial progress in understanding field-aligned currents.

ACKNOWLEDGEMENTS

Any contribution that the TRIAD magnetometer experiment may have made to an understanding of field-aligned currents is due to the efforts of a number of scientists. The late A. J. Zmuda and J. C. Armstrong played an important role in the initial development of the TRIAD magnetometer; S.-I. Akasofu assisted in the data acquisition from College, Alaska, A. J. Tucker assisted with the McMurdo data collection, and J. K. Walker assisted with the Resolute data collection. Participants in the data analysis presently include M. Sugiura, T. Iijima and N. A. Saflekos. Invaluable assistance in data processing was provided by S. Favin and the entire project was made possible by the Space Development Department of APL/JHU.

REFERENCES

Akasofu, S.-I. (1969) *Nature, 221,* 1020.
Akasofu, A.-I. and Kamide, Y. (1976) *Planet. Space Sci., 24,* 223.
Alfvén, H. (1939) *Kungl. Sv. Vet-Akademiens Handl., III,* 18(3).
Alfvén, H. (1940) *Kungl. Sv. Vet-Akademiens Handl., III,* 18(9).
Anderson, H. R. and Vondrak, R. R. (1975) *Rev. Geophys. Space Phys., 13,* 243.
Armstrong, J. C. (1974) p.155 in *Magnetospheric Physics* (ed.,B. M. McCormac) D. Reidel Pub. Co., Dordrecht, Netherlands.
Armstrong, J. C. and Zmuda, A. J. (1973) *J. Geophys. Res., 78,* 6802.
Armstrong, J. C., Akasofu, S.-I. and Rostoker, G. (1975) *J. Geophys. Res., 80,* 575.
Arnoldy, R. L. (1974) *Rev. Geophys. Space Phys., 12,* 217.
Birkeland, K. (1908), *The Norwegian Aurora Polaris Expedition 1902-1903, Vol. 1,* Section 1, Aschhoug, Oslo.
Boström, R. (1974) p.45 in *Magnetospheric Physics* (ed., B. M. McCormac) D. Reidel Pub. Co., Dordrecht, Netherlands.
Coleman, P. J. and McPherron, R. L. (1970) pp.171-194 in *Particles and Fields in the Magnetosphere* (ed., B. M. McCormac), D.

Reidel Pub. Co., Dordrecht, Netherlands.
Cummings, W. D. and Dessler, A. J. (1967), *J. Geophys. Res., 72,* 1007.
Fairfield, D. H. (1973) *J. Geophys. Res., 78,* 1553.
Fairfield, D. H. and Mead, G. D., (1975) *J. Geophys. Res., 80,* 535.
Feldstein, Y. I. (1966) *Planet. Space Sci., 14,* 121.
Fukushima, N. (1969) *Rept. Ionos. Space Res. Japan, 23,* 219.
Fukushima, N. (1976) *Rept. Ionos. Space Res. Japan, 30,* 35.
Heppner, J. P. (1972) *J. Geophys. Res., 77,* 4877.
Iijima, T. (1974) *Rept. Ionos. Space Res. Japan, 28,* 1973.
Iijima, T. and Potemra, T. A. (1976a) *J. Geophys. Res., 81,* 2165.
Iijima, T. and Potemra, T. A. (1976b) *J. Geophys. Res., 81,* 5971.
Iijima, T. and Potemra, T. A. (1977) submitted to *J. Geophys. Res.*
Kamide, Y. and Akasofu, S.-I. (1976) *J. Geophys. Res., 81,* 3999.
Kamide, Y., Kroehl, H. W.,Rostoker, G., Akasofu, A.-I. and Potemra, T. A. (1976) *EOS, 59,* 993.
Kroehl, H. W., Kamide, Y., Rostoker, G., Akasofu, A.-I. and Potemra, T. A. (1976) *EOS, 59,* 993.
Nagata, T., Kokubun, S. and Iijima, T. (1966) *Sci. Rept. Japan Antarctic Res. Exped., Ser. A, 3,* 1.
Potemra, T. A. (1977) *Smithsonian, 7,* (11), 64.
Potemra, T. A., Iijima, T. and Favin, S. (1975) *EOS, 56,* 617.
Sharp, R. D. and Johnson, R. G. (1968) *J. Geophys. Res., 73,* 969.
Shield, M. A., Freeman, J. W. and Dessler, A. J. (1969) *J. Geophys. Res., 74,* 247.
Stern, D. P. (1975) *NASA Goddard Space Flight Center Report, X-602-75-17.*
Sugiura, M. (1975) *J. Geophys. Res., 80,* 2057.
Sugiura, M. and Poros, D. (1973) *Planet. Space Sci., 21,* 1763.
Sugiura, M. and Potemra, T. A. (1976) *J. Geophys. Res., 81,* 2155.
Theile, B. and Praetorius, H. M. (1973) *Planet. Space Sci., 21,* 179.
Tsunoda, R. T., Presnell, R. I., and Potemra, T. A. (1976) *J. J. Geophys. Res. 81,* 3791.
Vasyliunas, V. M. (1968) *J. Geophys. Res., 73,* 5805.
Vasyliunas, V. M. (1972) p.192 in *Solar Terrestrial Physics, Part III,* (ed. Dyer) D. Reidel Pub. Co., Dordrecht, Netherlands.
Vestine, E. H. and Chapman, S.(1938) *Terr. Magn. Atmos. Elec., 43,* 351.
Zmuda, A. J., Martin, J. H. and Heuring, F. T. (1966) *J. Geophys. Res., 71,* 5033.
Zmuda, A. J., Heuring, F. T. and Martin, J. H., (1967) *J. Geophys. Res., 72,* 1115.
Zmuda, A. J. and Armstrong, J. C. (1974a), *J. Geophys. Res., 79,* 2501.
Zmuda, A. J. and Armstrong, J. C. (1974b), *J. Geophys. Res., 79,* 4611.

DISCUSSION

Reid: The picture you presented seems to be consistent with the classical theoretical picture of the field-aligned current systems associated with magnetospheric convection, as developed by Vasyliunas, Jaggi & Wolf, and others. The poleward current system is associated with the solar wind and should perhaps lie on open field lines. The equatorward system is associated with polarization of the partial ring current.

It would be interesting to know how far one has to travel outward along the field lines before the magnetic-field perturbation due to the field-aligned currents becomes comparable with the total field, assuming current conservation.

Shepherd: J.R. Burrows has measured field-aligned currents from the ISIS-II Spacecraft and D.M. Klumpar has searched for the corresponding current carriers in the particle fluxes measured on the same satellite. Very often the upward currents can be quantitatively identified with energetic electron fluxes, but at times it seems that the carriers must be below the 5 eV threshold of the detectors.

Potemra: I believe that the particle observations from the AE satellites and magnetic field observations from TRIAD support the ISIS-II measurements.

CHARACTERISTICS OF 1-500 eV ELECTRONS OBSERVED IN THE EARTH'S THERMOSPHERE FROM THE PHOTOELECTRON SPECTROMETER EXPERIMENT ON THE ATMOSPHERE EXPLORER SATELLITES*

W. K. Peterson and J. P. Doering

The Johns Hopkins University
Department of Chemistry
Baltimore, Maryland 21218 USA

T. A. Potemra and C. O. Bostrom
The Johns Hopkins University
Applied Physics Laboratory
Laurel, Maryland 20810

1. INTRODUCTION

Low energy electrons play a significant role in a variety of atmospheric, ionospheric, and magnetospheric processes. Those produced in the thermosphere by photoionization represent the primary source of ion and electron heating in this and higher-altitude regions. Photoelectrons produced in a sunlit hemisphere can, under certain circumstances, connect along geomagnetic field lines to a nonsunlit conjugate point in the opposite hemisphere, causing phenomena such as predawn heating. Secondary electrons produced by streams of high energy electrons which bombard the auroral atmosphere play an important role in the dynamics of the high latitude ionosphere. Low energy electrons are believed to be principal carriers of large-scale field-aligned, or Birkeland currents which are a basic element in the three-dimensional system coupling the magnetosphere with the lower atmosphere and ionosphere. The characteristics of low energy electrons are important to an understanding of the complex plasma processes that convey the solar wind plasma down through the magnetospheric cusps to the lower ionosphere. Low energy electrons precipitate over the earth's polar regions and can be used as an important tracer of the complicated

* This work was supported by the National Aeronautics and Space Administration under Contract NAS-5-11415.

B. Grandal and J. A. Holtet (eds.), Dynamical and Chemical Coupling, 353-364.

acceleration mechanisms which occur in the distant magnetospheric boundaries and tail region.

The Photoelectron Spectrometer Experiment on the Atmosphere Explorer series of spacecraft is providing high resolution energy spectra of electrons with energies between 1 and 500 eV at altitudes down to 140 km over the entire surface of the earth. These instruments onboard the three AE spacecraft have already been used to provide the following measurements: (1) daytime photoelectron spectra above 154 km with much higher energy resolution and over a much greater range of altitudes and solar zenith angles than have been obtained before *(Doering et al., 1976b)*; (2) high resolution measurements of secondary electrons during auroral precipitation events at various altitudes from 160 km to 732 km *(Peterson et al., 1977a)*; (3) the first direct confirmation of conjugate photoelectrons with a particle detector onboard a satellite at ionospheric altitudes *(Peterson et al., 1977b)*; (4) what is believed to be the first general classification of low-energy electrons in the high-latitude ionosphere *(Doering et al., 1976a)*; and (5) detailed properties of the cusp region at ionospheric altitudes *(Potemra et al., 1977)*. The purpose of this paper is to summarize some of these important results.

2. INSTRUMENTATION

Details of the Photoelectron Spectrometer (referred to as *PES* in the following discussion) on the Atmosphere Explorer satellites (referred to as *AE* below) have been described elsewhere in considerable detail *(e.g., Doering et al., 1973, 1975)*. The principal features of the spacecraft and detectors are summarized in the following.

The AE-C satellite was launched on December 13, 1973 into a 68° inclination orbit with initial perigee and apogee altitudes of 150 km and 4000 km. Perigee was maintained near 140 km with an onboard rocket motor. After about a year in orbit, AE-C was placed and maintained in a nearly circular orbit in the altitude range from 210 and 280 km by periodic operation of the rocket motor. The 68° orbit inclination allows a wide range of magnetic local times (MLT) to be sampled at a relatively constant invariant latitude. This complements the observations from the polar orbiting AE-D satellite which samples a wide range of invariant latitudes while at a relatively constant MLT.

The AE-D satellite was launched into a polar orbit on October 6, 1975 and AE-E was launched into a 22° inclination orbit on November 20, 1975. An onboard rocket motor was also used on these satellites to maintain a relatively low perigee.

The PES experiment consists of electrostatic electron energy analyzers with an energy bandpass $\Delta E/E = 2.5\%$ mounted on opposite sides of one end of the cylindrical spacecraft. The spacecraft operates in either a despun mode or a 4-rpm spinning mode. In the despun spacecraft mode of operation one sensor head looks upward and away from the earth while the other looks downward toward the earth. These detectors are mounted on each of the AE spacecraft and every detector has 3 selectable energy ranges (approximately 1 to 30 eV, 1-100 eV and 1-500 eV). Due to telemetry constraints either one 64-point linearly-spaced energy spectrum can be obtained every second, or four 16-point spectra every second.

3. PHOTOELECTRONS

The most important source of low energy electrons in the earth's upper atmosphere is the photoionization of the neutral atmosphere by the solar ultraviolet flux. Figure 1 is a day air-glow energy-time spectrogram obtained from AE-E on January 7, 1976. Data are shown in this figure for the portion of the orbit where the satellite moved from above a 1000 km altitude to perigee near 140 km and back up to above 600 km. Energy is plotted on the vertical scale from 0 to 35 eV and universal time (UT), altitude (in km), and the solar zenith angle (in degrees) are given on the horizontal scale. Variations in the differential electron flux (dJ/dE in elec/cm^{2}sec ster eV) are shown by the varying shades within this diagram. The most striking feature apparent in this diagram is the appearance of several spectral peaks (indicated by the white horizontal bands) between approximately 21 eV and 28 eV

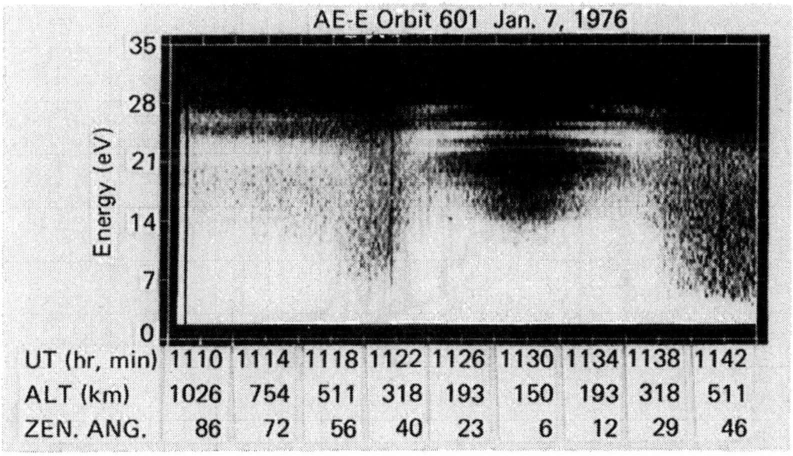

Fig. 1 Energy-time spectrogram obtained from AE-E orbit 601 on January 7, 1976 showing characteristic spectra of photo-electrons.

when the spacecraft was below ~ 250 km (from 1124 to 1136 UT on this diagram). These well resolved energy peaks are the signature of the photoelectron spectrum and correspond to production of O^+ and N_2^+ in various electronic states from photoionization of O and N_2 by solar He II (304 Å) radiation *(e.g., Doering et al., 1976a)*. The high resolution in energy of the PES instrument is emphasized by the width of the white bands in Figure 1, which for example is only ± 0.625 eV at 25 eV.

Figure 2 shows an example of the 15–30 eV region of the photoelectron spectrum in detail *(from Doering et al., 1976a)*. An energy scale has been added based on the best fit of the energies of the peaks in the spectrum to the laboratory calibration data. At the top of the figure are shown the energies of electrons expected from photoionization of atomic oxygen by 304 Å solar radiation. The intensities of the ^2P and ^2D lines relative to the ^4S line are shown as calculated by *Starace et al. (1974)*. The energies of electrons corresponding to production of N_2^+ in various final states by 304 Å radiation on N_2 are also shown. The intensity of the various vibrational bands of electronic states relative to the v' = 0 band of the $X^2\Sigma_g^+$ state are shown from the

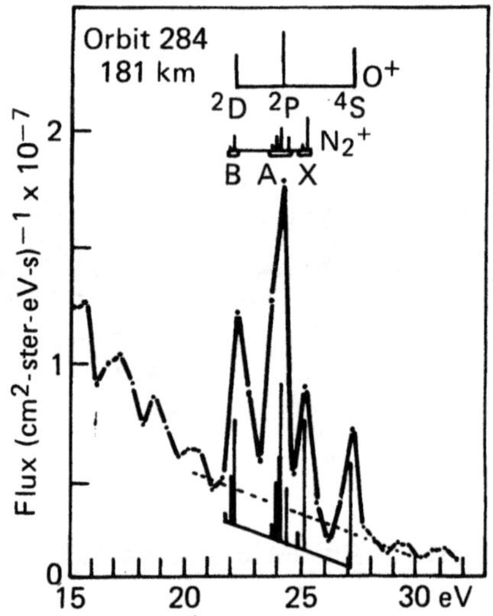

Fig. 2 Detailed spectrum of the 15–30 eV region determined from orbit 284 of AE-E (from Doering et al., 1976a). The solar zenith angle was 70° at the 181 km altitude of the observation.

experimental data of *Gardner and Samson (1973)*. Below the experimental spectrum (offset downwards for clarity) is shown the expected intensities added to a continuum background taken from the experimental data. In order to calculate the expected intensities, it has been assumed that the 27.2 eV line is solely due to O^+ ^4S and that the 25.2 eV line is solely due to N_2^+ $X^2\Sigma_g^+$ (v' = 0) and the intensities expected at the other energies have been calculated from the data above. The calculation also assumes that the electron energy loss function in the 22-28 eV region is a smooth and slowly varying function of the electron energy. The agreement between the observed 22.2 and 23.9 peak intensities and the sums of the expected O^+ and N_2^+ contributions at these energies is satisfactory — confirming the assignments.

4. CONJUGATE PHOTOELECTRONS

The idea that sizeable fluxes of photoelectrons might escape upwards from the sunlit ionosphere and flow along the geomagnetic field lines to the magnetic conjugate point was first advanced by *(Hanson, 1963)*. There is now sufficient evidence for the existence of such a flow of conjugate photoelectrons. The evidence has been obtained primarily from observations of the geophysical effects caused by precipitating fluxes of conjugate photoelectrons such as predawn enhancements in airglow and electron temperature *(e.g., Carlson, 1974 and Evans, 1968)*. Direct measurements of the photoelectron flux have been made at altitudes above 1400 km *(Rao and Maier, 1970 and Nagy et al., 1973)*, but these experiments had insufficient energy resolution to show whether the characteristic photoelectron energy spectrum (e.g., as shown in Figure 2) was present at these high altitudes.

Figure 3 shows the energy spectra acquired near an altitude of 263 km during orbit 5004 of AE-C by *Peterson et al. (1977b)*. The solar zenith angle was greater than 120° during the period when these data were acquired. This figure is offered as evidence for the precipitation of photoelectrons (generated at the magnetic conjugate sunlit point) into the dark ionosphere. Although the detailed spectral peaks in the 20-28 eV range (e.g., in Figure 2) have been "washed out", the distinct fall off above 60 eV identifies these as photoelectron spectra. That these are electrons produced in the conjugate sunlit hemisphere is proven by the relationship of the observed flux with the solar zenith angle at the magnetic conjugate point *(e.g., Peterson et al., 1977b)*.

An important implication of these observations is the fact that the statistical comparison of the photoelectron spectra measured in the dark ionosphere with those measured in the sunlit ionosphere show no acceleration (or deceleration) within ± 1 eV, of these low energy electrons as they travel the great distance

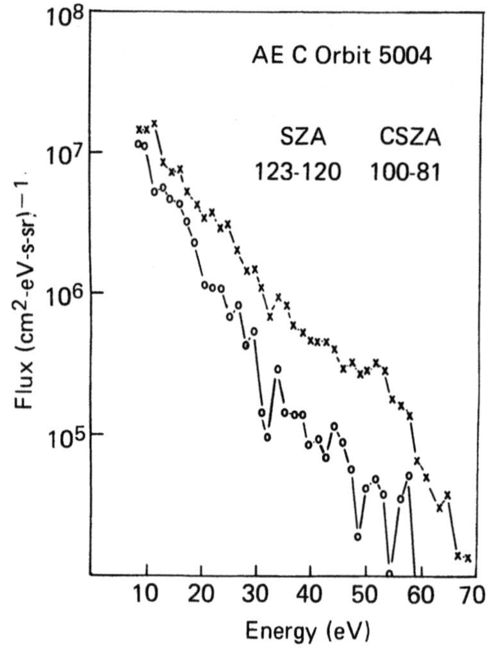

Fig. 3 Average energy spectra of electrons observed by AE-C at 263 km for a range of solar zenith angles (SZA) when the zenith angles at the magnetic conjugate point (CSZA) varied from 100° to 81°. Precipitating electrons are indicated by X's (pitch angles less than 60°) and back-scattered electrons are indicated by open circles (pitch angles greater than 120°). The dip latitude varied from 34° to 54° during the time these data were acquired.

along the magnetic field line. This effect has been observed in at least one case up to 67° invariant latitude (L = 6.5). This indicates that the net potential drop between conjugate points along a geomagnetic field lines is less than ± 1 eV. This is significant to the identification of various magnetospheric regions and emphasizes the important role that the low energy electrons have as tracer particles in magnetospheric studies.

5. LOW ENERGY ELECTRONS IN THE HIGH LATITUDE IONOSPHERE

Although the AE program was designed for aeronomy studies, the polar orbit of AE-D and the 68° inclined orbit of AE-C provide an opportunity to use the same set of detectors for studies of auroral, cusp, and polar cap phenomena. Numerous observations of low energy particles have been obtained in these regions in the past *(e.g., Frank, 1971; Frank and Ackerson, 1972; Winningham and Heikkila, 1972 and McDiarmid et al., 1975, 1976)*, but the PES observations are unique because they were obtained at lower electron energies, with higher energy resolution, and at lower altitudes than has been possible before. The low altitude observations are important because they permit *in situ* measurements of important magnetospheric boundaries in the high latitude ionosphere, eliminating the need to "map" down observations made at much higher altitudes. The lower energy measurements, combined with the high energy resolution, are essential to resolve present uncertainties about the cusp (or cleft) region which inherently involves low energy particles.

Properties of low energy electrons in the auroral, cusp and polar cap region acquired with the PES experiment have been discussed in detail by *Doering et al.,(1976b), Peterson et al., (1977a) and Potemra et al., (1977),* and some of the principal features will be summarized here.

Figure 4 shows a polar plot, in invariant latitude — MLT coordinates, with the tracks of AE-D orbits 210 and 232 superimposed upon the regions of large-scale field-aligned current regions determined by *Iijima and Potemra (1976)*. The large-scale field-aligned currents are concentrated in two major regions encircling the geomagnetic pole, and also in a region between 1000 MLT and 1400 MLT and 78° - 80° invariant which has been identified as the cusp *(Iijima and Potemra, 1976)*. AE-D orbit 210 occurred on October 24, 1975 when $K_p = 0^+$ and passed over the pre-noon cusp, the polar cap and nightside auroral region. Three areas of electron precipitation have been labelled "A, B and C" for reference in discussion of the data shown later in Figure 5. The spacecraft altitude ranged from about 1300 km at the beginning of the pass (near noon) down to 400 km (near midnight). AE-D orbit 232 occurred on October 26, 1975 when K_p = 3-, and passed over the postnoon cusp, the polar cap and nightside auroral region. The spacecraft altitude ranged from above 1300 km at the beginning of this pass down to 360 km (near midnight). In both orbits the data were obtained with the spacecraft in the "despun" mode, with the PES detector "looking" upward and away from the earth.

Figure 5 shows the energy-time spectrogram for AE-D orbit 210. Electron energy from 0 to 525 eV is indicated on the vertical axis and invariant latitude and magnetic local time are shown on the horizontal axis. Three areas of electron precipitation have been labelled on this figure and correspond to the areas shown on Figure 4. Region A extends from ~ 78° to 82° invariant latitude near 1100 MLT (the blank space near 79.5° is due to the loss of data). The largest flux in this region can be described with a maxwellian energy spectrum with most probable energy E_o = 50 eV and total flux = 7.5 x 10^8elec/cm^{2}sec ster. This region has been identified

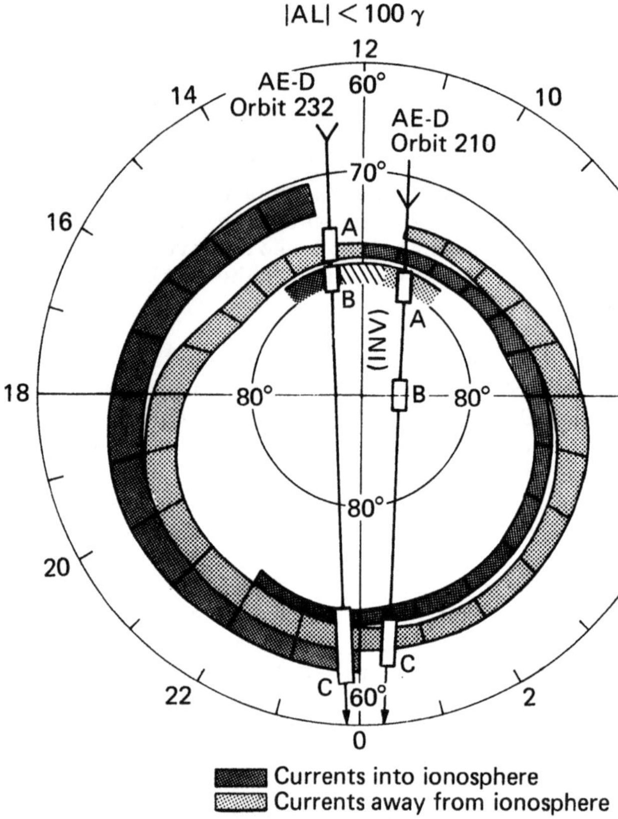

Fig. 4 Tracks of AE-D orbits 210 and 232 superimposed upon the statistical location of large-scale field-aligned currents from Iijima and Potemra (1976). Various areas of characteristic electron precipitation are labelled for discussion (see the text).

as the cusp and the precipitating electrons have been suggested to be the carriers of the upward-flowing field-aligned currents often observed in the pre-noon cusp *(Potemra et al., 1977)*.

Region B in Figure 5 extends up to 88° invariant latitude and is characterized by fluxes of electrons with energies below 200 eV. These fluxes may be similar to the low energy electron precipitation in the polar cap observed by *Winningham and Heikkila (1974)* primarily during world wide geomagnetic storms. The observations of Figure 5 were made during very quiet geomagnetic conditions (K_p = 0+) and may be related to the small-scale magnetic disturbances often observed in the dawn side of the polar cap region *(Saflekos and Potemra, 1976)*.

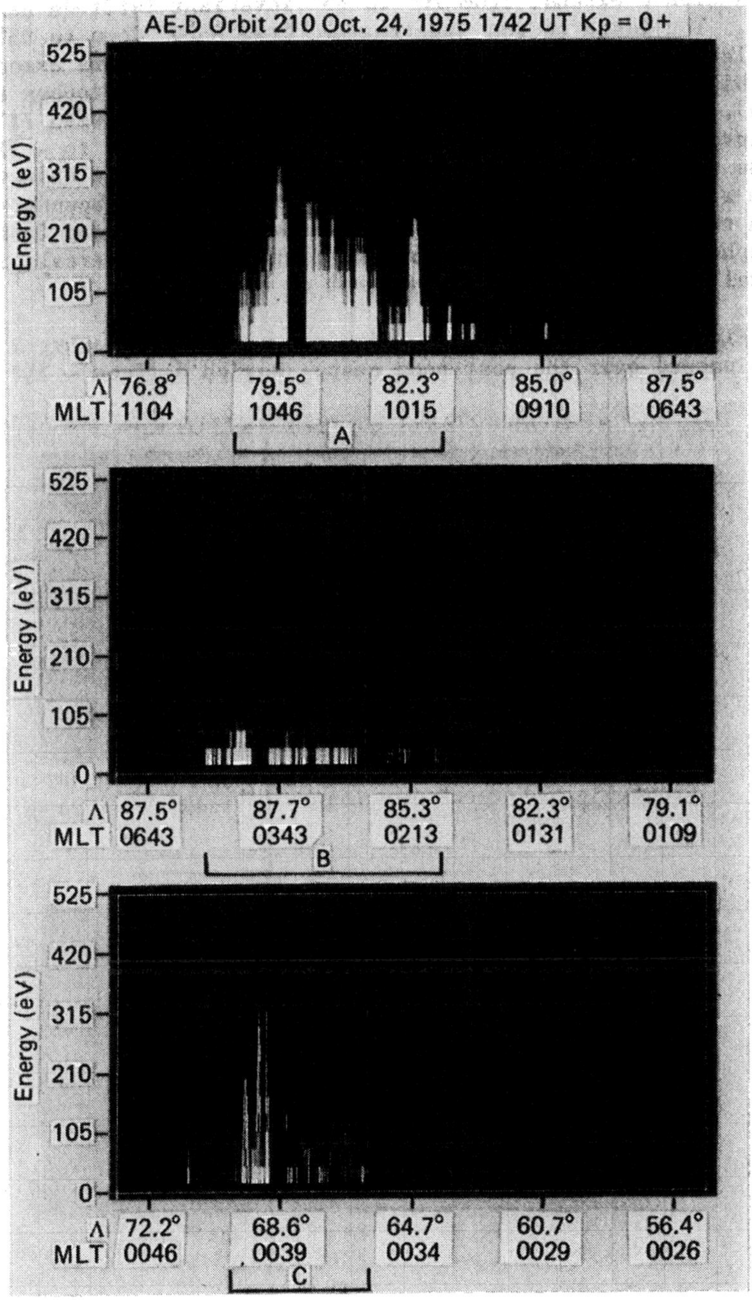

Fig. 5 Energy-time spectrogram acquired during AE-D orbit 210.

Region C extends from 70° to 65° invariant latitude near midnight. The image on the spectrogram from 68.6° down to 65° invariant latitude is characteristic of secondary electrons associated with higher energy (keV) auroral electrons (*e.g.*, *Peterson et al.*, *1977a*). Two inverted "V" structures can be seen between 71° and 69° invariant latitude, one of which extends only up to ~ 210 eV (close to the low energy limit of most other particle observations). These structures are statistically located near the boundary between the field-aligned currents directed into and away from the ionosphere and may be associated with shear flow reversals in the auroral plasma as suggested by *Burch et al.* *(1976)*.

Figure 6 shows the energy-time spectrogram for orbit 232 which passed over the post-noon cusp. Region A (from ~ 75° to

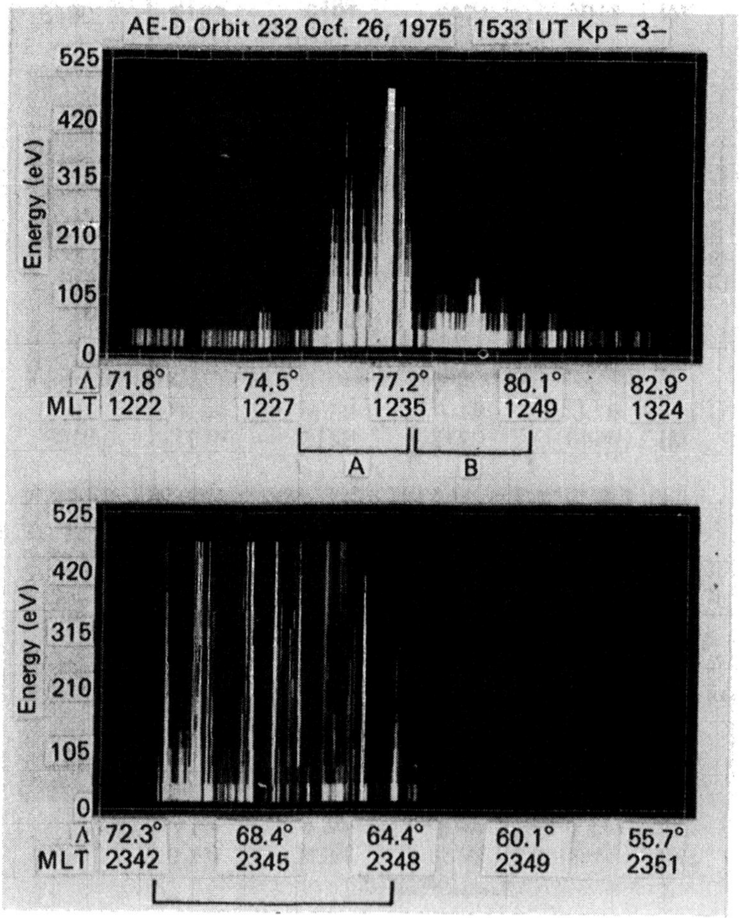

Fig. 6 Energy-time spectrogram acquired during AE-D orbit 232.

77.4° invariant latitude) statistically coincides with the "dayside auroral zone" where field-aligned currents (referred to as Region 1 by *Iijima and Potemra, 1976*) flow away from the ionosphere. A pronounced softening in the electron energy spectra occurs at 77.4° invariant latitude as shown in Figure 6. A region of "soft" electron precipitation is observed poleward of this boundary up to ~ 80° invariant latitude and is labelled "B" in this figure. Region B statistically coincides with the location of the cusp field-aligned currents (as shown in Figure 4). The electron spectra in Region B is close to maxwellian with E_0 = 12 eV, N = 82 cm^{-3}, and T = 140,000°K — similar to the values compiled for the magnetosheath/solar wind plasma *(e.g., Maynard and Johnstone, 1974)*.

Large fluxes of electrons with energies extending above the 525 eV limit of the PES detector are observed in Region C between ~ 72° and 64° invariant latitude near midnight and are probably associated with auroral particle precipitation.

The PES instruments on the AE spacecraft are providing measurements of electrons with lower energies, with higher energy resolution and at lower altitudes than have been possible in the high latitude ionosphere before. The high energy resolution of this instrument has already allowed the determination of the boundary between the dayside auroral zone and the post-noon cusp. These more detailed measurements should help identify the important magnetospheric/ionospheric regions and coupling mechanisms and aid in resolving uncertainties concerning the location of the cusp region, the nature of daytime auroral phenomena, and the source and charge carriers of field-aligned currents.

ACKNOWLEDGEMENTS

The authors gratefully acknowledge the assistance of S. Favin (APL) in data processing and thank N. A. Saflekos (APL) for stimulating and profitable discussions.

REFERENCES

Burch, J. L., Lennartsson, W., Hanson, W. B., Heelis, R. A., Hoffman, J. H. and Hoffman, R. A. (1976) *J. Geophys. Res., 81,* 3886.
Carlson, H. C. (1974) *Ann. Geophys., 30,* 59.
Doering, J. P., Bostrom, C. O. and Armstrong, J. C. (1973) *Radio Sci., 8,* 387.
Doering, J. P., Peterson, W. K., Bostrom, C. O. and Armstrong, J. C. (1975) *J. Geophys. Res., 80,* 3934.
Doering, J. P., Peterson, W. K., Bostrom, C. O. and Potemra, T. A. (1976a) *Geophys. Res. Lett., 3,* 129.
Doering, J. P., Potemra, T. A., Peterson, W. K. and Bostrom,

C. O. (1976b) *J. Geophys. Res., 81,* 5507.
Evans, J. V. (1968) *J. Geophys. Res., 73,* 3489.
Frank, L. A. (1971) *J. Geophys. Res., 76,* 5202.
Frank, L. A. and Ackerson, K. L. (1972) *J. Geophys. Res., 77,* 4116.
Gardner, J. L. and Samson, J. A. R. (1973) *J. Electron Spectroscopy and Rel. Phenomena, 2,* 259.
Hanson, W. B. (1963) *Space Res., 3,* 282.
Iijima, T. and Potemra, T. A. (1976) *J. Geophys. Res., 81,* 5971.
Maynard, N. C. and Johnstone, A. D. (1974) *J. Geophys. Res., 79,* 3111.
McDiarmid, I. B., Burrows, J. R. and Budzinski, E. E. (1975) *J. Geophys. Res., 80,* 73.
McDiarmid, I. B., Burrows, J. R. and Budzinski, E. E. (1976) *J. Geophys. Res., 81,* 221.
Nagy, A. F., Winningham, J. D. and Banks, P. M. (1973) *J. Atm. and Terr. Phys., 35,* 2289.
Peterson, W. K., Doering, J. P., Potemra, T. A., McEntire, R. W., Bostrom, C. O., Hoffman, R. A., Janetzke, R. W. and Burch, J. L. (1977a) *J. Geophys. Res., 82,* 43.
Peterson, W. K., Doering, J. P., Potemra, T. A., McEntire, R. W. and Bostrom, C. O. (1977b) *Geophys. Res. Lett., 4,* 1977.
Potemra, T. A., Peterson, W. K., Doering, J. P., Bostrom, C. O., McEntire, R. W. and Hoffman, R. A. (1977) *submitted to J. Geophys. Res.*
Rao, B. C. N. and Maier, E. J. R. (1970) *J. Geophys. Res., 75,* 816.
Saflekos, N. A. and Potemra, T. A. (1976) *EOS, 57,* 989.
Starace, A. F., Manson, S. T. and Kennedy, D. J. (1974) *Phys. Rev., A9,* 2453.
Winningham, J. D. and Heikkila, W. J. (1974) *J. Geophys. Res., 79,* 949.

DISCUSSION

Shepherd: Are the regions of soft and hard precipitation on the dayside always cleanly separated in latitude, or do they sometimes overlap ?

Potemra: From the limited amount of data we have examined, I believe that the soft and hard precipitation regions are well separated.

PART V: SPECIAL TOPICS

THE ADVANTAGES OF THE ANTARCTIC FOR IONOSPHERIC RESEARCH

W. R. Piggott

British Antarctic Survey, Cambridge, U.K.

The material of this lecture is based on several published reports (Piggott, 1977), and in particular on a review by Dudeney and Piggott (1977). Some of the main points and most useful figures are reproduced in this note.

The main advantages of the Antarctic zone for ionospheric research arises from the special relations between the magnetic field and the geographic co-ordinate systems which occur in the zone. The internal field of the earth may be approximated by an off-centre triangular shaped magnet with effectively a point pole in the South and a line pole (in total field, two displaced point poles) in the North. The Southern point pole is much further from the geographic pole than any part of the line pole in the North. Dynamic phenomena in the ionosphere, e.g. the effects of thermospheric winds, electric fields, or diffusion are mainly determined by a combination of the direction of the magnetic field in the ionosphere and the geographic co-ordinates. The former is best organised by using the magnetic dip co-ordinate system ($\tan(\text{lat}) = \frac{1}{2}\tan(\text{dip})$). In contrast magnetospheric phenomena are best organised using the invariant latitude, Λ, or corrected geomagnetic latitude systems, $\cos\Lambda = \sqrt{1/L}$ where L is the height of the field line in earth's radii above the centre of the earth at the equator. The contrast between the Northern and Southern hemispheres in the dip system can be seen in Figure 1. Figures 2 and 3 show the dip and invariant latitude systems in the Southern hemisphere. Not only do the magnetic systems show much greater displacement from the geographic system than in the North but they are almost equally displaced from each other. This gives the opportunity to test critically which factors are ordered in geographic,

B. Grandal and J. A. Holtet (eds.), Dynamical and Chemical Coupling, 367-372. *All Rights Reserved.*

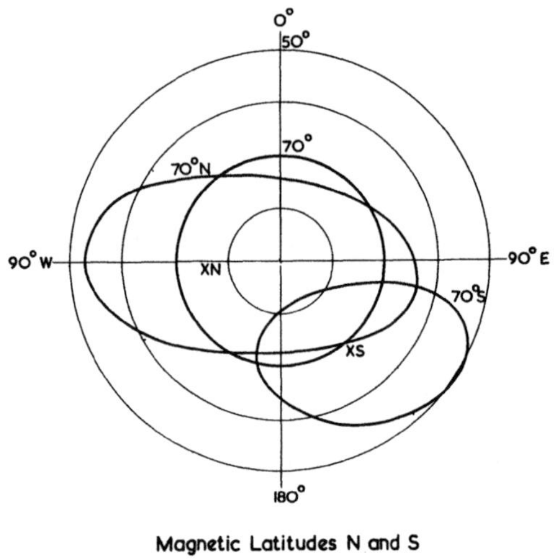

Fig. 1. Polar projection showing the relative positions of geographic and 70° dip latitudes in the Northern and Southern hemisphere. After Bellchambers and Piggott (1960).

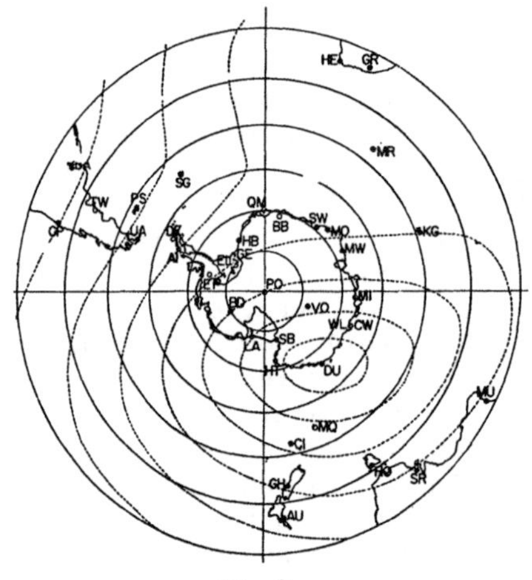

Fig. 2.

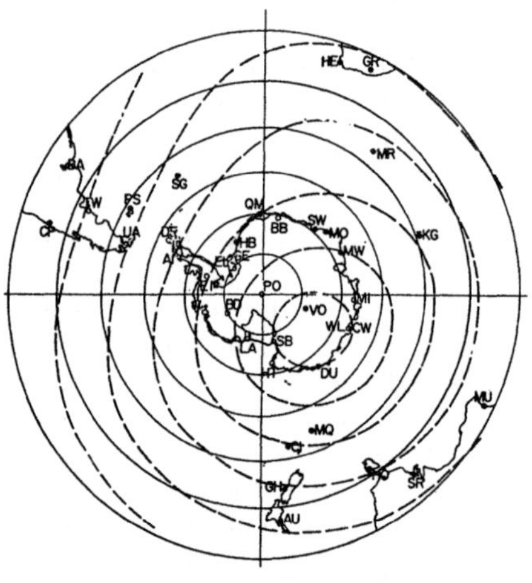

Fig. 3.

Fig. 2. Comparison of geographic co-ordinates and magnetic dip latitude in the Southern hemisphere. Geographic latitude is in 10° steps from 30° to 90°, dip latitude is in 10° steps from 20° to 80° (dotted curves). The positions of past ionosonde stations are indicated by open circles and active ones by filled circles.

Fig. 3. As Fig. 2 with the exception that the dashed curves are invariant latitude.

magnetic or invariant latitudes, which is important for distinguishing between different theories or to identify missing or ignored factors.

In geophysics, the analogue of the physicists' controlled experiment is to make measurements in the zones where the geophysical conditions are extreme and to wait for nature to do the critical experiment. The Southern hemisphere has great advantages from this point of view. For example, in Figure 4, the solid circles show the relation between geographical latitude and dip for all ionospheric stations which have been operated in the Northern hemisphere and the open circles show the corresponding family for the Southern hemisphere. The contrast in range of dip at constant latitude or range of latitude for constant dip is much greater in the Southern hemisphere, despite the fact that one extreme sector is mainly ocean with very few stations. The corresponding pattern in invariant latitude for the Northern hemisphere shows a closer correlation between geographic and invariant latitude than for the dip comparison but at least as great a range in the South as for dip, Figure 5. Even with the small number of stations available in Antarctica and the surrounding islands it is possible to make chains of stations at constant geographic latitude, constant dip latitude or constant invariant latitude which cover a wide range of latitudes in the remaining two systems.

A further effect of the displacement of the invariant axis from the geographic axis is that the angle between the solar wind and the magnetic axis makes much larger excursions, both diurnally and seasonally, in the South than in the North. This may be important for the interpretation of any solar-terrestrial interactions caused by this wind. Inevitably, phenomena associated with particle precipitation often sweep over an abnormally large range of geographic latitudes in the South as the earth rotates. However, the important noon sector of the auroral oval, in which cleft precipitation is present, describes a relatively small ellipse crossing the geographic pole.

The Weddell Sea-Antarctic Peninsula zone, in particular, shows some remarkable features which have not been fully exploited. For example, in this sector the auroral oval is usually well inside the Antarctic circle so that there is relatively little diurnal variation of solar zenith angle near the solstices to confuse its effects. Magnetic activity at constant invariant latitude is also abnormally small in this sector, e.g. K-indices average 1 to 2 units less than in the world as a whole. The contrast between quiet and disturbed conditions in the ionosphere is correspondingly large. The abnormally low dip latitudes give large dynamic effects in the ionosphere. For example, the diurnal variation of the height of the F-layer due to these features can be as large as

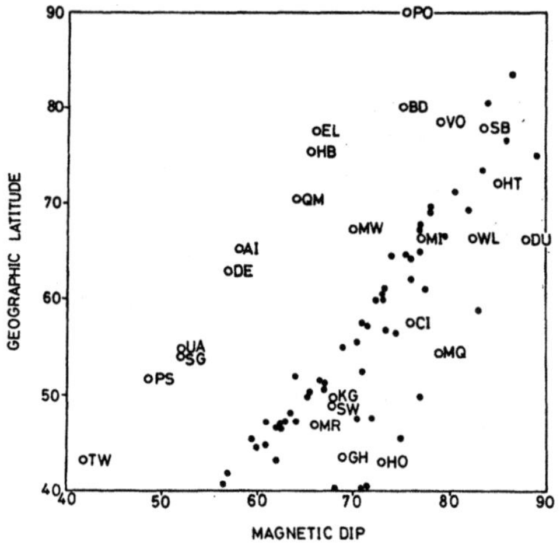

Fig. 4. Geographic latitudes of Northern (●) and Southern (o) hemisphere ionosonde stations as a function of magnetic dip. After Hill (1960).

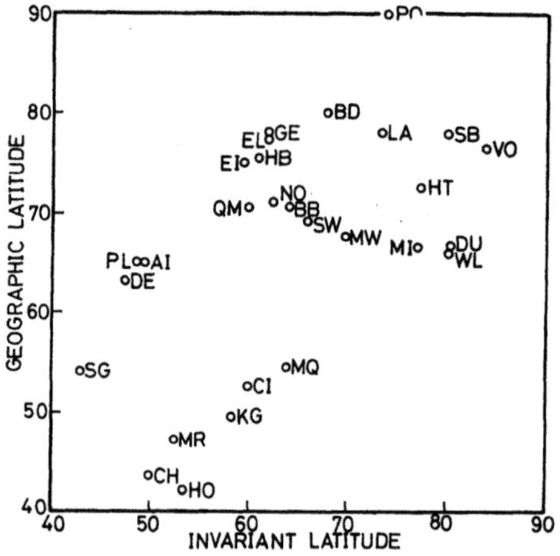

Fig. 5. Geographic latitudes of southern hemisphere stations as a function of invariant latitude. After Dudeney and Piggott (1977).

200 kms - so large that the effects on the direction of the thermospheric wind due to changes of ion drag with solar cycle can be readily measured.

The large displacement of the invariant latitude system relative to the geographic gives rise to a very skew incidence of solar and galactic cosmic rays. Thus it is easy to find stations at the same geographic latitude which are either strongly effected by cosmic rays or unaffected.

In the troposphere, temperatures at sea level are much lower at high latitudes in the South than in the North, so that the atmospheric circulation due to the difference in temperature between low and high latitudes is much more vigorous in the Southern hemisphere. Furthermore, the transport of energy is almost entirely over oceans. The gap between the Southern end of the Andes and the mountains of the Antarctic Peninsula is wider than the typical weather system so that, to a large extent, such systems can grow and decay without interference from topography. Furthermore, the surface albedo is remarkably uniform, either at a low value corresponding to sea or a very high value given by snow covered ice. In winter the latter area is more than twice the area of the Antarctic continent. At higher levels there are also considerable differences between the two hemispheres. For example, the frequent breakdowns of the polar stratospheric circulation seen in the North do not occur in the South where the heart of the polar vortex is usually maintained throughout the winter. This is centred, however, to the South of the Weddell Sea sector. Stratwarm events affect the outer parts of the circulation and are thus most evident at surprisingly low geographic and magnetic latitudes in this sector. The great displacement between the auroral zone and the zone where stratwarm effects in the D-region are seen greatly simplifies the separation of stratwarm changes from those due to storm after-effects. Of course the converse applies in the Australian sector which shows, in an exaggerated form, the complex effects seen in North America. The South Atlantic zone offers special advantages for testing theories of why the chemistry of the D-region is changed during winter anomaly events. It is interesting to note that in this sector the final stratwarm occurs much later than in the Northern hemisphere, in early summer.

The geographic latitudes of magnetic conjugate stations are particularly different in the geographic sectors containing and opposite to the magnetic pole. In the Weddell Sea-Peninsula area, one of the two extreme zones, the northern end is near the region of great thunderstorm activity near the east coast of the U.S.A. In the northern hummer there are therefore many lightning flashes generating whistlers which can be received at high geographic latitudes in the southern (winter) hemisphere. This gives exceptional opportunities to study movements of the plasmapause by whistler techniques. The fact that the Southern end can be in darkness through the day while the North is mainly sunlit, also offers special advantages for studying the interaction between magnetic pulsations and the ionosphere and, of course, the effects of the ionosphere on magnetic pulsations. The amplification of whistler mode signals from North to South near this sector enables whistler signals to be strong enough to be measured over very wide separations, 10 times greater than is possible in the Northern hemisphere. The effects of man-made radiation on the magnetosphere is, of course, much greater in the Southern hemisphere than in the North since most of the sources of radiation are in the Northern hemisphere.

The South Atlantic anomaly in the earth's magnetic field, (greatly reduced in this sector), causes trapped particles to mirror at abnormally low heights in the ionosphere. Effects which have been attributed to this have been reported from Sanae. There are important ionospheric bases close to both the dip pole (Terre Adelie) and the invariant pole (Vostok). The main limitation at present is the general lack of recognition of the advantages of research in Antarctica and the surrounding oceans due to the great distances from the major research groups of the Northern hemisphere. The development of satellite soundings of the atmosphere have greatly added to the potential but much of the new data deserve more intensive study. Thus the criticism, that the number of stations in Antarctica is too few to provide a firm basis for geophysical studies, is no longer true. Technological advances, particularly in the provision of unmanned equipment with very small power requirements enable additional measurements to be made relatively cheaply from unmanned stations. The deployment of stations of this type during the IMS is already considerable and increasing rapidly.

REFERENCES

Bellchambers, W.H., and Piggott, W.R., 1960, Proc. Roy. Soc., A, 256, pp. 200-218.

Dudeney, J.R., and Piggott, W.R., 1977, in Upper Atmospheric Research in the Antarctic, ed. L.H. Lanzerotti and C.G. Park, Antarctic Research Series, Publ. AGU.

Hill, G.E., 1960, J. Geophys. Res., 65, 2011-2023.

Piggott, W.R., 1977, Phil. Trans. R. Soc., Ser. B, 279, 275-285.

SPACELAB: A NEW TOOL FOR COOPERATIVE RESEARCH

E. R. Schmerling

ST/NASA Headquarters
Washington, DC 20546, USA

The Space Shuttle will enable large payloads to be placed in earth-orbit at lower costs than previously possible, due to the re-usability of the shuttle vehicle and the booster rockets. Many uses are envisaged for this vehicle, since the orbits achievable by the shuttle are suitable for a variety of tasks, including the first step in placing payloads into higher orbits, or for escaping from the earth entirely, and voyaging to more distant parts of the solar system. For work in earth orbit, the European Space Agency, ESA, in cooperation with NASA, has developed a flexible laboratory system called Spacelab, which will fit into the Space Shuttle, perform its assigned operations in earth orbit, for 7-28 days, and will then be returned and refitted for further flights. This flexible system will be made widely available after 1980 through both reimbursable and cooperative arrangements. Spacelab will offer some novel features for performing research into the behavior and properties of the neutral and ionized atmosphere, but will require a fresh approach in order to take full advantage of these new capabilities and to reduce the costs and lead-times now associated with space research, so that space research may be opened up to a much larger user community.

B. Grandal and J. A. Holtet (eds.), Dynamical and Chemical Coupling, 373-379.

ORBITS AVAILABLE

For launch from the Kennedy Space Center, the earliest Space Shuttle missions will be able to achieve orbits between 28° and 56° inclination. Up to 400 km altitude, the total payload weight will be in excess of 25,000 kg. Higher altitudes can be reached by adding propellant kits, which reduce the payload weight available. When the West Coast launch site (Vandenberg Air Force Center) becomes operational, probably by 1983, additional inclinations from 56 to 104° will become available, but there is a payload penalty for high inclination orbits. For example, about 15,000 kg total will be available for a polar orbit at 400 km circular altitude.

SPACELAB CHARACTERISTICS

Spacelab represents a set of modular laboratory fixtures which can be flown on the Space Shuttle. It consists of pressurized modules, in which people and equipment can operate under shirt-sleeve conditions of air pressure, composition and temperature, together with pallets for mounting instruments to be directly exposed to space. Several mixes of pressure modules and pallets are available, with the characteristics shown in Table 1.

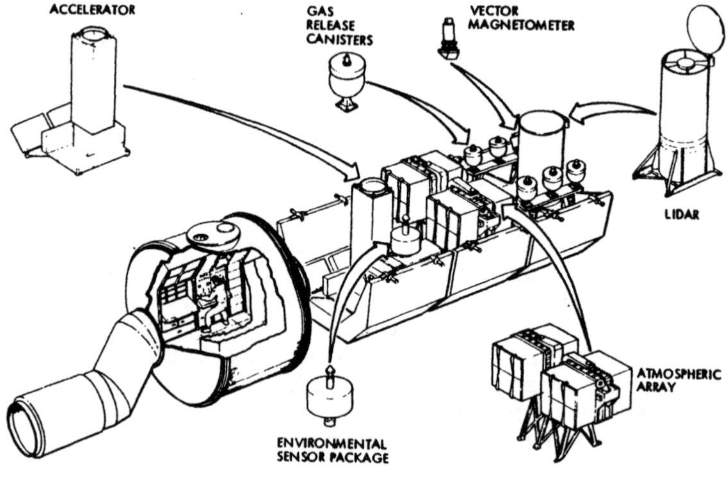

Possible AMPS Flight Payload Configuration

TABLE 1

Spacelab characteristics

Available to users \ Spacelab Configuration	 short module + 9 m pallet	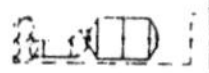 long module	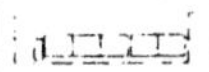 15 m pallet	 Independ. susp. pallet
o Payload weight (kg)	5500	5500	8000	9100
o Volume for experiment equipment				
Inside module (m^3)	8	22	–	–
On pallet (m^3)	100	–	160	100
o Pallet mounting area (m^2)	51	–	85	51
o Electrical power (28 V DC 115/200 V at 400 Hz AC)				
Average (kW)	3 – 4	3 – 4	4 – 5	4 – 5
Peak (kW)	8	8	9	9
Energy * (kWh)	300	300	500	500
o Exp. Support computer with central processing unit and data acquisition system	← 64 K core memory of 16-bit words → ← 320 000 operations per second. →			
o Data Handling				
Transmission through orbiter	← Up to 50 Mbps →			
Storage digital data	← Up to 30 Mbps →			
o Instrument pointing sub-system IPS	Mounted on pallet, will provide arc second pointing for payloads up to 5000 kg			

* Energy can be increased by the addition of payload-chargeable kits, each providing 840 kWh and weighing approx. 350 kg (at landing).

The main advantages of Spacelab consist of high payload weight, power, the return of the instrumentation, and the availability of man for real-time operation of the equipment. Limitations consist of the local contamination, mission durations of 1-4 weeks, limited altitudes and inclinations (for early missions). This makes it clear that Spacelab is not suited for many types of in-situ sensing, or for long-term monitoring. Should these be desired, they will be performed by conventional spacecraft which could be shuttle-launched, and may even be retrieved. Since so much exploratory and survey type work has already been performed from typical Spacelab orbits in the Atmospheric and Space Plasma Physics discipline, it is expected that the main thrusts of research in this area in the 1980's will be directed to specific physical problems which require a multi-instrument multi-parameter team approach. New features of especial interest are the ability to perform laboratory-type experiments, with Spacelab providing known inputs into the medium using electromagnetic pulses, particle beams and chemical stimuli for cause-and-effect studies, and the ability to carry heavy instrumentation for remote sensing of the atmosphere well below the vehicle to study atmospheric composition and dynamics. The Atmosphere, Magnetosphere and Plasmas-in-Space (AMPS) program typifies this approach, and groups together the varied specialized skills needed to emphasize the theme of this conference -- the coupling between the neutral and ionized portions of the earth's atmosphere.

It is clear that the costs of space instrumentation can be reduced by taking full advantage of Spacelab features, rather than following the automated spacecraft practice of the previous decade. The reduction in weight, volume and power constraints allows simpler construction procedures to be followed. Many component specifications can be reduced in stringency, because high radiation resistance, for example, will not be required. Since the instruments will be recovered, an evolutionary development can be programmed, which results in increasing capability from flight to flight, with each step in the development taking full account of the experience obtained in previous flights. Some allowance can even be made for occasional instrument failures: with critical resources, like power, fully used by less than a complete payload, it is possible to achieve 100% mission success with some instrument failures. Flexible flight plans and discipline trained specialists on board, in control, can assure best use of the functional instrumentation during a mission, and take full advantage of the critical abilities of man to detect and react to the unexpected. Last but not least, there is a place for electronic standardization, such as the nuclear-instrumentation module (NIM and NIM-CAMAC) concepts. Even when updated with modern electronics for operation in space, these can be fully compatible with older units used on the ground for testing, development and flight training.

PROPOSED SCIENCE PROGRAM

The NASA intends to issue Announcements of Opportunity for proposals, which will be issued by discipline, and not by flight. These will be reviewed periodically, and assigned to flights as appropriate. This procedure will allow facilities to be slowly built up so that proposals may then be submitted for the following types of research:

A. Using existing instruments, operating in the appropriate modes and orbits

B. Using existing instruments, supplemented by user supplied additions such as detector systems

C. Development of new instruments

It is hoped that the use of types A and B will greatly decrease the time needed to prepare for space flight after a reasonable equipment inventory has been built up. In International programs, it will be possible to propose in either the reimbursable or cooperative modes. In the former, payment will be pro-rated to the fraction of the payload capability used, and will be subject to only the conditions needed to assure the peaceful use of the facilities, and the safety of the flight. In the cooperative mode, each side pays only for its own costs, but has access to the results of the entire cooperative program. The NASA has a long history of cooperation in space, and a flexible posture in structuring these programs to meet the specific needs of each occasion. It is our belief that the best science program requires a pooling of the talents and resources of all nations working together for our mutual benefit.

PASSIVE OBSERVATIONS

New facilities being studied for remote sensing of the atmosphere will be able to measure major and minor constituents, and possibly also winds, as well as inputs from particle streams and ultraviolet radiation. To make a reasonably comprehensive package will require an assembly of instruments working closely together. But while a spacecraft provides a splendid platform for obtaining good geographical coverage, ground-based techniques will be able to add high time resolution to these data, and sounding rockets can provide valuable supplements by making insitu measurements of vertical profiles. It is therefore expected that, if closely coordinated with measurements made in space, there will be an important role for backscatter sounders, rockets, and lidars used both on the ground and in aircraft. It is hoped to use these as an integral part of a science program with their operation planned systematically from the beginning, rather than added haphazardly at a time too late to prepare for best coordination. All these elements can perform an essential function in helping us understand the complex forces which control our near-space environment.

CHEMICAL RELEASES AND ACTIVE EXPERIMENTS

The dynamics of both neutral motions and electric-field convected plasmas can be investigated by releasing chemical tracers. Large-scale dynamics can be studied by releasing tracers over a significant fraction of an orbit, and a large number of observers will then be needed to observe the motions of these tracers from other spacecraft, from aircraft, and from clear sites on the ground. These observers are an essential part of such a program, but their participation need not cost more than a small fraction of the money needed to effect the releases in space. For some purposes, these roles may be reversed, since Spacelab may provide a convenient platform to observe daytime releases made from sounding rockets, or for the large-area detectors needed to observe the entry of particles released in the solar wind or the magnetospheric tail in order to answer basic questions of solar wind entry into the magnetosphere or the energization of plasma from the magnetotail as it flows into the the auroral regions.

An exciting feature made possible by Spacelab is the ability to test the basic interactive processes which occur in the earth's environment by providing an active stimulus and observing its effects. Beams of electrons may be made to produce an artificial aurora, but with a known input. Electromagnetic energy or cold plasma injections may be used to precipitate trapped particles, and attempts may be made to test theories of the production of micropulsations by using the appropriate stimuli. In all these cases, important roles will be played by observers prepared to look for the predicted effects and to measure their magnitudes. An understanding of these interactive processes is not only important in understanding our own environment, but also in interpreting the signals observed from more distant parts of the Universe.

CONCLUSION

The introduction of Spacelab thus brings new capabilities to the tools available for space research. These capabilities are made up of both advantages and limitations, but the advantages are suitable for cooperative studies of the basic processes which control our environment, and for the team operation needed to answer basic physical questions by combining the many sub-discipline skills required for multi-parameter cause-and-effect studies. The advent of Spacelab thus represents a new opportunity for new types of research, and a challenge to use this new opportunity in innovative new ways not unduly constrained by the customs of the past.

SPECTRAL ANALYSIS OF IONOSPHERIC PHENOMENA THROUGH NARROW DATA WINDOWS

K. V. Paulson and M. G. Schmid

Department of Physics, University of Saskatchewan, Saskatoon, Sask., S7N OWO, Canada.

1. INTRODUCTION

Many ionospheric phenomena exhibit fluctuations in both time and space. In general, these fluctuations are best described in the frequency and wavenumber domains. Such descriptions require that the original data be transformed from one domain to another, the theory and techniques of such transformations being called spectral analysis. However, in the application of such techniques to ionospheric data, one is faced with the non-stationary and inhomogeneous character of these natural phenomena. For such cases, the resulting frequency and wavenumber spectra represent some type of average which may smear out valid structure and may lead to a serious misinterpretation of the phenomena under study.

On the other hand, what happens when limited amounts of data are employed so that the non-stationary and inhomogeneous characteristics are not evident? With conventional methods of spectral analysis (the fast fourier transform method, for example), one usually finds that resolvability or statistical reliability or both are affected to such a degree that the resulting spectra are again of doubtful physical significance.

2. THE CONCEPT OF THE DATA WINDOW

For the sake of simplicity, consider the data to be represented by a single trace, amplitude-versus-time display. This is the type of record produced by many analog chart recorders. A data window may now be defined which samples the trace at specified

B. Grandal and J. A. Holtet (eds.), Dynamical and Chemical Coupling, 381-386.

times to create a physical data sequence. (The word physical is used because the sampled amplitudes are assumed to be real numbers representative of some physical process.) Pictorially, one may superimpose a rectangle on the amplitude-versus-time display so that the base of the rectangle is parallel to the time axis and neither the base nor the side parallel to it intersects any part of the trace. The sides parallel to the amplitude axis will intersect the trace at two points and the corresponding times mark the beginning and the end of the physical data sequence. A number of vertical lines may then be drawn and their intersections with the trace define the intervening elements of the physical data sequence. It is customary that these lines are equidistant - the common time between them being called the sampling interval.

This simple idea of a "barred" data window can be very useful even when the data sequence is created directly with the aid of an analog-to-digital convertor. One cannot, of course, arbitrarily choose the sampling interval; its value must be chosen on the basis of the sampling theorem. However, the width of the data window may be a matter of individual choice - it may be wide or narrow. Roughly speaking, a narrow data window is one that contains only a few cycles of the lowest frequency of interest. Whenever an ionospheric process evolves in time, it will be advantageous to select a data window as narrow as possible prior to spectral analysis. Furthermore, a succession of narrow windows, with or without overlap, allows the construction of a time-varying spectrum. The proper physical interpretation of the latter can be very rewarding.

For some experiments, however, narrow data windows are not a matter of choice - they simply exist. This happens whenever the total experimental time is short relative to the time scale of the phenomenon. Examples will be given of narrow data windows of both the selected and the imposed variety.

3. MAXIMUM ENTROPY SPECTRAL ANALYSIS

As noted in the Introduction, conventional spectral analysis of data through a narrow data window is plagued with problems involving frequency resolution and statistical significance. One of the problems of these conventional techniques is that a spectral window must be selected and a judicious selection depends entirely on the experience of the individual performing the analysis. This introduces a certain amount of subjectivity into the analysis and in many cases this leads to an inaccurate spectrum.

There are today, however, several methods of spectral analysis called data adaptive (Lacoss 1971; Gersch 1970). With these methods, a spectral window is selected according to certain objective criteria. The best of these data adaptive methods is the one called maximum entropy spectral analysis (MESA), the conception and implementation of which is due to John Burg (1967, 1968, 1970, 1972). It is interesting to note that MESA was used in the commercial processing of exploration seismic data several years prior to Burg's first public disclosure.

When spectral analysis through narrow data windows is desired, MESA is of prime importance. When, for example, the fast-fourier-transform technique is employed, one must assume that the physical data sequence is periodic. In addition, spectral estimates are invariably made at equal-frequency intervals between zero and the Nyquist frequency. When it comes to tests of significance one finds that these are based, in general, on the assumption that the data sequence is a realization of a process with a normal distribution. MESA, on the other hand, avoids end-effect problems because no assumptions are made about elements beyond the limits of the data window. There is no restriction on the frequencies examined (provided they are below the Nyquist frequency) and the data-adaptive nature of the analysis ensures that an optimum spectral window is applied. For details of MESA, the reader is referred to the review by Bérubé (1975).

Computer programs were written in both FORTRAN and BASIC following the algorithm of Andersen (1974) and the order selection of Jones (1976). MESA was then applied to a variety of ionospheric data where narrow windows were either part of the experimental design or were judged appropriate because of the evolutive nature of the phenomenon. A few examples of these analyses are presented in the following sections.

4. GEOMAGNETIC MICROPULSATIONS

Calculation of the polarization state of a ULF electromagnetic wave is based on cross-spectral analysis (Paulson 1974) and an appropriate data window would necessarily have to provide information on all three orthogonal components of either the electric or the magnetic field. Observations of geomagnetic micropulsations show that the polarization state changes with time: the X-component may lead the Y-component at one time while at some later time the phase relationship is found to be reversed (Singh 1972). A similar evolutive behaviour is observed when micropulsation recordings are made at two different locations. When the X-components from the two stations are compared it is often found that a phase lag at one time becomes a phase lead at another (Kennedy 1974). Both of these situations typify nonlinear behaviour when viewed over a wide data window. Over a narrow data window, however, the relative phase relations are nearly constant and meaningful parameters based on spectral estimates may be evaluated.

No examples of cross-spectral analysis using the maximum entropy method are presented in this paper.

5. VLF TRANSMISSION THROUGH THE UPPER IONOSPHERE

Holtet (1974) has observed the effects of the ionospheric medium on VLF signals between the Omega transmitter at Aldra and a receiver on board the Alouette II satellite. Quasi-periodic fluctuations were observed at each of the four transmitted frequencies. Since the Omega transmissions have a well-defined pattern for the "on-periods" at the different frequencies, the data are recorded within windows which can only be classified as narrow.

A total of 36 of these recorded events were defined and subjected to MESA. The results are being studied to determine if the spectral variations depend on the frequency of transmission or if they are related to the time or latitude. An example of one of these fluctuation patterns is shown in Fig. 1 while its corresponding spectrum is depicted in Fig. 2. The reader can readily see that a crude "peak-to-peak" calculation yields a frequency approximately equal to that of the secondary peak in the

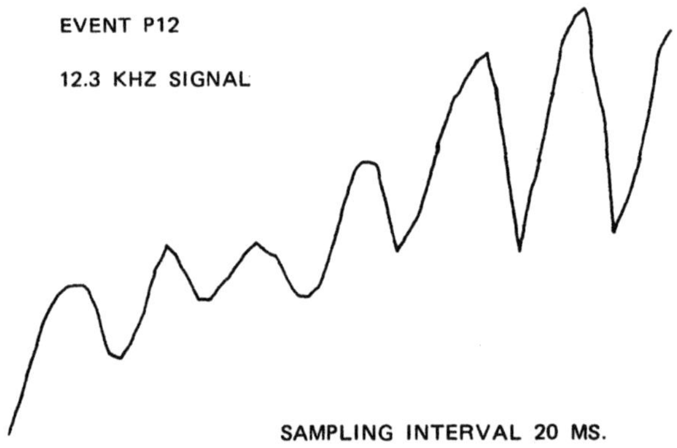

Fig. 1. Amplitude fluctuations of a VLF signal through a narrow data window. Adjacent samples are joined by straight lines to give a "trace-like" appearance. The sampling interval along the horizontal time axis is 20 ms.

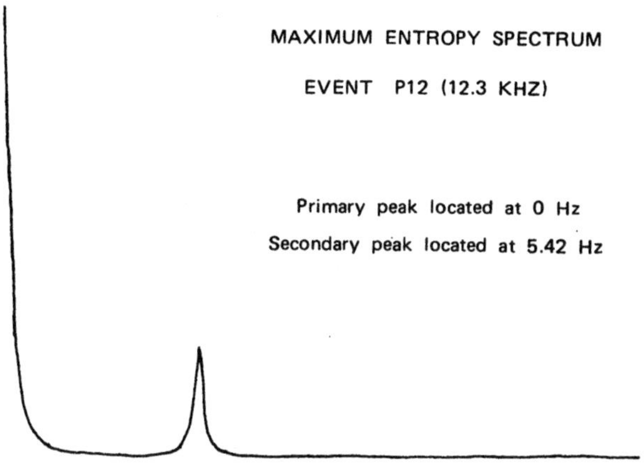

Fig. 2. The maximum entropy spectrum corresponding to the data shown in Fig. 1. The horizontal frequency axis extends from 0 to 18 Hz.

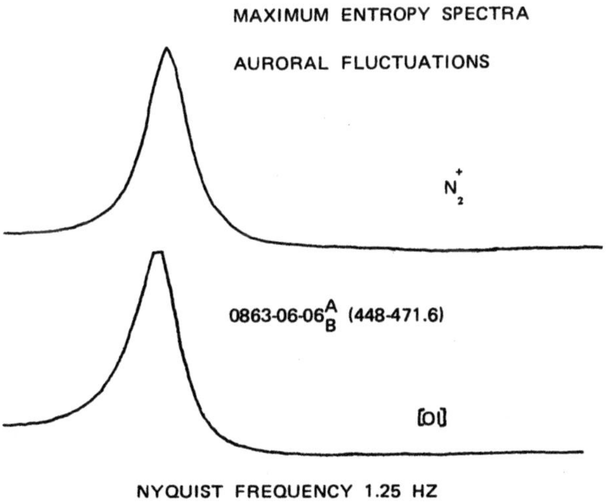

Fig. 3. The maximum entropy spectra corresponding to fluctuations in luminosity at 3914Å (N_2^+) and 5577Å ([OI]). The horizontal frequency axis extends from 0 to 1.25 Hz for both spectra.

spectrum. The upward trend is correctly reflected as a large low-frequency component in the spectrum.

6. MEAN LIFE OF THE $^1S_0$ LEVEL OF ATOMIC OXYGEN

Paulson and Shepherd (1965) developed a cross-spectral method for determining the lifetime of a metastable state of atomic oxygen from auroral measurements. Some of their lifetimes were evaluated using data windows of 160-sec duration with a 0.4-sec sampling interval. Two such consecutive, disjoint, data windows gave lifetimes of 0.73 and 0.65 sec. Since this suggests an evolutive behaviour, narrow data windows of 24-sec duration were applied to the same data and the corresponding MESA estimates were calculated. One of these spectral pairs is shown in Fig. 3 where a common spectral peak at 330 mHz is evident. The spectral pair for the preceding 24-sec window, however, has a major peak at 130 mHz and a secondary one at 360 mHz. The differences between these (and other) spectral pairs indicate that the statistics of the underlying process are changing with time.

It is possible to determine lifetimes from spectral density pairs at two different frequencies. This has been done using the foregoing MESA estimates but a high degree of scatter was observed. Work on this very interesting problem is continuing.

7. CONCLUSION

Spectral analysis of many ionospheric phenomena requires the application of narrow data windows and the maximum entropy method.

REFERENCES

Andersen, A. 1974, Geophysics, 39, 69.
Bérubé, L.P. 1975, Res. Memo. CL-75-10, Univ. of Saskatchewan
Burg, J.P. 1967, 37th Annual Meeting, S.E.G., Oklahoma City.
Burg, J.P. 1968, NATO Adv. Study Institute, Enschede.
Burg, J.P. 1970, 40th Annual Meeting, S.E.G., New Orleans.
Burg, J.P. 1972, Geophysics, 37, 375.
Gersch, W. 1970, Math. Biosci., 7, 205.
Holtet, J.A. 1974, p. 351 in ELF-VLF Radio Wave Propagation (Ed. J.A. Holtet), D. Reidel Publ. Co., Dordrecht.
Jones, R.H. 1976, Geophysics, 41, 771.
Kennedy, J.S. 1974, unpubl. M.Sc. thesis, Univ. of Saskatchewan.
Lacoss, R.T. 1971, Geophysics, 36, 661.
Paulson, K.V. and Shepherd, G.G. 1965, J. Atm. Terr. Phys., 27, 83.
Paulson, K.V. 1974, p. 399 in ELF-VLF Radio Wave Propagation (Ed. J.A. Holtet), D. Reidel Publ. Co., Dordrecht.
Singh, H. 1972, unpubl. M.Sc. thesis, Univ. of Saskatchewan.

SUBJECT INDEX

GPSR Compliance

The European Union's (EU) General Product Safety Regulation (GPSR) is a set of rules that requires consumer products to be safe and our obligations to ensure this.

If you have any concerns about our products, you can contact us on ProductSafety@springernature.com

In case Publisher is established outside the EU, the EU authorized representative is:

Springer Nature Customer Service Center GmbH
Europaplatz 3
69115 Heidelberg, Germany

Batch number: 10723233

Printed by Printforce, the Netherlands